Human Cell Culture Protocols

METHODS IN MOLECULAR MEDICINE™

John M. Walker, SERIES EDITOR

METHODS IN MOLECULAR MEDICINE™

Human Cell Culture Protocols

Second Edition

Edited by

Joanna Picot

Southampton, UK

Humana Press ✳ Totowa, New Jersey

© 2005 Humana Press Inc.
999 Riverview Drive, Suite 208
Totowa, New Jersey 07512

www.humanapress.com

This publication is printed on acid-free paper. ∞
ANSI Z39.48-1984 (American Standards Institute)

Permanence of Paper for Printed Library Materials.

Cover illustration: Figure 1 from Chapter 1, Establishment and Maintenance of Normal Human Keratinocyte Cultures, by Claire Linge.

Cover design by Patricia F. Cleary.

For additional copies, pricing for bulk purchases, and/or information about other Humana titles, contact Humana at the above address or at any of the following numbers: Tel.: 973-256-1699; Fax: 973-256-8341; E-mail: humana@humanapr.com; or visit our Website: www.humanapress.com

Printed in the United States of America. 10 9 8 7 6 5 4 3 2 1

E-ISBN 1-59259-861-7
Library of Congress Cataloging in Publication Data
Human cell culture protocols.— 2nd ed. / edited by Joanna Picot.
       p. ; cm. — (Methods in molecular medicine ; 107)
    Includes bibliographical references and index.
    ISBN 1-58829-222-3 (alk. paper)
    1. Human cell culture—Laboratory manuals.
    [DNLM: 1. Cells, Cultured—Laboratory Manuals. 2. Cytological Techniques—Laboratory Manuals. QS 525 H9175 2005] I. Picot, Joanna. II. Series.
    QH585.2.H85 2005
    616'.0277—dc22   2004007186

Preface

Since the publication of the first edition of *Human Cell Culture Protocols*, the field of human cell culture has continued to expand and increasing numbers of researchers find they need to dip their toes into the world of tissue culture, even if this is not the main focus of their work. Today, not only does a whole industry supply all the materials necessary for tissue culture, there is a growing number of companies specializing in the supply of a diverse range of human cells. For those without existing links to clinicians and hospital departments, the purchase of cells may be an attractive starting point, although this can be an expensive option. Alternatively, many researchers find an essential element in success is building close links with clinicians and hospital departments from which human tissue samples are obtained. In particular, close collaboration can enable cell culture to be initiated with the very minimum of delay, which is often the key to establishing a viable culture. Before any experimental work takes place, however, researchers must ensure that patients have provided informed consent and that local and/or national ethical and other guidelines for the procurement and use of human tissue are met. In addition, the tissues received are potentially biohazardous, possibly harboring infectious agents such as HIV, hepatitis, and tuberculosis, so appropriate safety measures must be in place.

A quick search of any of the literature databases reveals the breadth of uses that human cells are put to. Cell culture is the starting point for so many applications. Microarray technology continues to develop, helping to elucidate patterns of gene expression within cells. A wide range of techniques is available to help researchers identify and understand the complex web of protein–protein interactions within and between cells. Cell cultures are used to test approaches to gene therapy and to gain an understanding of the cell cycle, particularly in relation to the development of cancers. The construction of 3-D cell cultures and the field of tissue engineering are the subjects of many other texts and take us far beyond the scope of this volume. Advances in microscopy refine our ability to image live cells in culture. Ultimately, the pooling of many strands of knowledge over time allows the development of new therapeutic approaches for human disease.

The first edition of *Human Cell Culture Protocols* was published in 1996. Now in this second edition, the collection of chapters has been revised to bring the methods up to date. As in the first edition, it has not been possible to cover the vast array of distinct cell types in one volume. I have, however, kept to the ideals of the first edition in trying to ensure that protocols are provided for a selection of the major tissue groups. New to this edition are chapters on fibroblasts, Schwann cells, gastric and colonic epithelial cells, and parathyroid cells. This collection of protocols will provide researchers who are starting to use cell culture methods for the first time with the detailed knowledge and helpful pointers they need. It should enable them to achieve success quickly and with the minimum of difficulty. Even those familiar with cell culture techniques may find this book a useful resource.

Finally, I would like to thank Gareth E. Jones whose success in bringing together the first edition gave me a wonderful foundation. Grateful thanks to the many authors who agreed to update their chapters from the first edition, and to those authors who have contributed for the first time. Thanks also to Professor John Walker and the staff at Humana Press who were always extremely prompt in responding to any and every enquiry and who were also patient when my replies to them were less than punctual.

Joanna Picot

Contents

Contributors

FRANK ALBERT • *Charité, Department: Klinische Zell und Neurobiologie, Institute for Anatomie, Berlin, Germany*

MARTIN BAYLISS • *GlaxoSmithKline, Stevenage, Hertfordshire, UK*

MONICA BERRY • *Division of Ophthalmology, Bristol Eye Hospital, University of Bristol, UK*

ALISON M. J. BUCHAN • *Department of Physiology, University of British Columbia, Vancouver, Canada*

KATHERINE A. BUCKLEY • *Human Bone Cell Research Group, Department of Human Anatomy and Cell Biology, The University of Liverpool, UK*

KIMBERLIE A. BURNS • *Cystic Fibrosis/Pulmonary Research and Treatment Center, The University of North Carolina at Chapel Hill, NC, USA*

PIERRE CHAILLER • *CIHR Group on the Functional Development and Physiopathology of the Digestive Tract, Department of Anatomy and Cell Biology, Université de Sherbrooke, Québec, Canada*

BENJAMIN Y. Y. CHAN • *Department of Clinical Chemistry, Royal Liverpool University Hospital, UK*

JUDITH M CURRAN • *UK Centre for Tissue Engineering, Department of Clinical Engineering, University of Liverpool, UK*

SUSAN B. CURTIS • *Department of Physiology, University of British Columbia, Vancouver, Canada*

MONTE A. DELMONTE • *Sierra Sciences Inc., Reno, NV, USA*

JANE P. DILLON • *Human Bone Cell Research Group, Department of Human Anatomy & Cell Biology, University of Liverpool, UK*

CARSTEN EHRHARDT • *Department of Pharmaceutics and Pharmaceutical Technology, Trinity College, Dublin, Ireland*

EVA L. FELDMAN • *Sierra Sciences Inc., Reno, NV, USA*

DANIEL W. FRAGA • *Islet Transplant Laboratory, University of Tennessee, Memphis, TN, USA*

WILLIAM D. FRASER • *Department of Clinical Chemistry, Royal Liverpool University Hospital, UK*

M. LESLIE FULCHER • *Cystic Fibrosis/Pulmonary Research and Treatment Center, The University of North Carolina at Chapel Hill, NC, USA*

A. OSAMA GABER • *Islet Transplant Laboratory, University of Tennessee, Memphis, TN, USA*

SHERIF GABRIEL • *Cystic Fibrosis/Pulmonary Research and Treatment Center, The University of North Carolina at Chapel Hill, NC, USA*

JAMES A. GALLAGHER • *Human Bone Cell Research Group, Department of Human Anatomy and Cell Biology, The University of Liverpool, UK*

ALISON GARTLAND • *Human Bone Cell Research Group, Department of Human Anatomy & Cell Biology, University of Liverpool, UK*

MARY B. GOLDRING • *Beth Israel Deaconess Medical Center, Department of Medicine, Division of Rheumatology, New England Baptist Bone and Joint Institute, Harvard Institutes of Medicine, Boston, MA, USA*

DOUGLAS A. GREENE • *Sierra Sciences Inc., Reno, NV, USA*

EMANUELA GUSSONI • *Division of Genetics, Childrens Hospital, Boston, MA, USA*

HANS HAUNER • *Else-Kröner-Fresenius-Center for Nutritional Medicine of the Technical University of Munich, Germany*

GABRIELLE M. HAWKSWORTH • *Department of Medicine & Therapeutics, University of Aberdeen, Scotland, UK*

PER HELLMAN • *Department of Surgery, University Hospital, Uppsala, Sweden*

MEENHARD HERLYN • *The Wistar Institute, Philadelphia, PA, USA*

MEI-YU HSU • *The Wistar Institute, Philadelphia, PA, USA*

JOHN A. HUNT • *UK Centre for Tissue Engineering, Department of Clinical Engineering, University of Liverpool, UK*

YONA KEISARI • *Department of Human Microbiology, Sackler Faculty of Medicine, Tel Aviv University, Tel Aviv, Israel*

KWANG-JIN KIM • *Departments of Medicine, Physiology and Biophysics, Molecular Pharmacology and Toxicology, and Biomedical Engineering, and Will Rogers Institute Pulmonary Research Center, Schools of Pharmacy, Medicine, and Engineering, University of Southern California, Los Angeles, CA, USA*

MALAK KOTB • *Department of Surgery/Immunology, University of Tennessee, Memphis, TN, USA*

DENNIS D. LARKIN • *Sierra Sciences Inc., Reno, NV, USA*

CLAUS-MICHAEL LEHR • *Department of Biopharmaceutics and Pharmaceutical Technology, Saarland University, Saarbrücken, Germany*

LING LI • *The Wistar Institute, Philadelphia, PA, USA*

CLAIRE LINGE • *The RAFT Institute of Plastic Surgery, Mount Vernon Hospital, Northwood, Middlesex, UK*

MATTHEW J. LOZA • *Department of Microbiology and Immunology, Kimmel Cancer Center, Thomas Jefferson University, Philadelphia, PA, USA*

MARK P. MATTSON • *NIA Gerontology Research Center, Baltimore, MD, USA*

DANIEL MÉNARD • *CIHR Group on the Functional Development and Physiopathology of the Digestive Tract, Department of Anatomy and Cell Biology, Université de Sherbrooke, Québec, Canada*

VOLKER MESKE • *Department: Klinische Zell und Neurobiologie, Institute for Anatomie, Charité, Berlin, Germany*

HAMID A. MOHAMMADPOUR • *Sierra Sciences Inc., Reno, NV, USA*

THOMAS G. OHM • *Department: Klinische Zell und Neurobiologie, Institute for Anatomie, Charité, Berlin, Germany*

GRACE K. PAVLATH • *Department of Pharmacology, Emory University, Atlanta, GA, USA*

BICE PERUSSIA • *Department of Microbiology and Immunology, Kimmel Cancer Center, Thomas Jefferson University, Philadelphia, PA, USA*

JOANNA PICOT • *Southampton, UK*

MARCUS RADBURN-SMITH • *Division of Ophthalmology, Bristol Eye Hospital, University of Bristol, UK*

SCOTT H. RANDELL • *Cystic Fibrosis/Pulmonary Research and Treatment Center, The University of North Carolina at Chapel Hill, NC, USA*

THOMAS SKURK • *Else-Kröner-Fresenius-Center for Nutritional Medicine of the Technical University of Munich, Germany*

GRAHAM SOMERS • *GlaxoSmithKline, Stevenage, Hertfordshire, UK*

MARTIN J. STEVENS • *Sierra Sciences Inc., Reno, NV, USA*

KEITH N. STEWART • *Department of Medicine & Therapeutics, University of Aberdeen, Scotland*

VICTOR J. TURNBULL • *Van Cleef Centre for Neuroscience, Alfred Hospital, Melbourne, Australia*

VANESSA VAN HARMELEN • *Else-Kröner-Fresenius-Center for Nutritional Medicine of the Technical University of Munich, Germany*

HEATHER M. WILSON • *Department of Medicine & Therapeutics, University of Aberdeen, Scotland*

JAMES R. YANKASKAS • *Cystic Fibrosis/Pulmonary Research and Treatment Center, The University of North Carolina at Chapel Hill, NC, USA*

1

Establishment and Maintenance of Normal Human Keratinocyte Cultures

Claire Linge

1. Introduction

Keratinocytes are the major cell type of the epidermis, which is the stratified squamous epithelium forming the outermost layer and thus providing the barrier function of the skin. The keratinocytes lie on a highly specialized extracellular matrix structure known as the basement membrane and are organized into multiple layers of cells. These layers are formed into distinct regions or strata that differ both morphologically and biochemically. From the basement membrane outward, they are: the stratum basale, the stratum spinosum, the stratum granulosum, and finally at the skin's surface, the stratum corneum. Cellular proliferation is restricted to the s. basale and results in the production of replacement progenitor cells, which remain within this layer, and also cells that are committed to undergo the process of terminal differentiation. The latter cells leave the s. basale and progressively migrate through each layer of the epidermis, simultaneously maturing along the differentiation pathway as they go. Finally, they reach the outer surface of the epidermis in the form of fully functional anuclear cells known as corneocytes. The function of these mature cells is the protection of the underlying viable tissues from the external milieu.

Initial attempts to grow keratinocytes in vitro were limited to the use of organ/explant culture (1). Using these techniques, whole pieces of skin could be kept alive in the short term, and growth was generally confined to the tissue fragment or onto the plastic immediately adjacent to the explant. The major drawback with these cultures is that they had an extremely limited lifespan and inevitably a limited application, because mixed cultures of keratinocytes and fibroblasts were obtained. Indeed, the presence of fibroblasts can often

From: *Methods in Molecular Medicine, vol. 107: Human Cell Culture Protocols, Second Edition*
Edited by: J. Picot © Humana Press Inc., Totowa, NJ

present a major problem where keratinocyte culture is concerned. Even small amounts of contamination with fibroblasts can lead to them overgrowing the keratinocyte cultures. This is generally because of the relatively higher proliferation rates of fibroblasts compared to keratinocytes, even when under culture conditions optimized for keratinocyte proliferation.

The greatest advance in the development of a long-term keratinocyte culture method came in 1975, when Rheinwald and Green reported the serial cultivation of pure cultures of keratinocytes from a single-cell suspension of epidermal cells *(2)*. This was achieved by growing the cells in serum-containing medium on a feeder cell layer of an established cell line known as 3T3 (murine embryo fibroblasts). The 3T3 cells were pretreated in such a way (such as irradiation) as to render them nonproliferating, but still viable (in the short term) and metabolically active. These mesenchymally derived cells provided an actively secreting cell layer, which dramatically improved the proliferation and lifespan of the keratinocyte cultures while simultaneously reducing the attachment and growth of any contaminating fibroblasts. The longevity of these keratinocyte cultures was further improved by the addition of a variety of mitogens determined as being important for the healthy maintenance of keratinocyte cultures, the most vital of these being epidermal growth factor (EGF) *(3)*. A list of these cytokines and the relevant references are given in **Subheading 2.**

Since the introduction by Rheinwald and Green of a method of long-term culture of keratinocytes, alternative culture methods have been developed, each being designed with specific experimental requirements in mind. The degree to which the pattern of keratinocyte differentiation in vivo is reproduced in vitro depends on the method of culture. Keratinocyte cultures can vary from undifferentiated monolayers *(4)* under low calcium conditions (<0.06 mM) to fully stratified multilayers achieved when grown in skin-equivalent cultures *(5–17)*. In general, the more closely the culture conditions mimic the tissue environment (with regard to acidic pH, calcium levels, extracellular matrix components and architecture, the presence of the correct subtype of mesenchymal cells, and being at the air/liquid interface of cultures), the more complete the expression of epidermal differentiation characteristics.

For some uses, the original Rheinwald and Green method, although reliably producing healthy cultures of human keratinocytes, is still not ideal, mainly because of its use of animal serum and animal-derived cell feeder layers to promote exponentially growing cultures. This is of specific concern when cultured keratinocytes are to be used clinically (zoonosis being a particular worry) or when more thoroughly defined culture conditions are required for experimentation. Alternative methods of keratinocyte culture that are either more defined or that avoid the use of serum and feeder layers have therefore been sought. Most of these, particularly the commercially available formulations, are based on modifications of the original serum-free keratinocyte culture medium (MCDB 151–153) developed by Peehl and Ham *(18–21)* and do not require the use of feeder layers to support serial propagation of keratinocytes. However, the majority of these media are not ideal for the long-term culture of keratinocytes, and indeed reduce their culture lifespan by approximately half compared to standard Rheinwald and Green conditions. In addition, although the latest of these commercially available serum-free media claim to have now optimized the replicative lifespan of keratinocytes in culture, their formulation still requires the addition of relatively undefined components, which are animal derived, such as bovine pituitary extract, and as such remain far from ideal.

The methods detailed in this chapter are adapted from the original "Rheinwald and Green" *(2)* method of cultivating human keratinocytes, which is still the most reliable method of establishing relatively pure cultures of keratinocytes that can be maintained long term. This method supplies a stock of healthy, proliferating cells that can be used either as they are or in any of the alternative methods mentioned above for experimentation. The required methodology includes the following.

1. Routine maintenance of the 3T3 cell line (required to produce the feeder layer important for healthy keratinocyte growth).
2. Production of feeder layers from 3T3 cultures.
3. Initiation of primary human keratinocyte cultures.
4. Routine maintenance of keratinocyte cultures once established.

2. Materials

1. 3T3 cells [either the original Swiss embryo *(22)* or NIH *(23)* variety]: available from a number of sources such as the European Collection of Animal Cell Cultures (ECACC) or the American type tissue collection (ATTC).
2. 3T3 culture medium: Dulbecco's modified Eagle's medium (DMEM) supplemented with 10% (v/v) fetal calf serum (FCS), 4 mM L-glutamine, 100 U/mL penicillin, 100 µg/mL streptomycin (*see* **Notes 1–4**).
3. Freshly isolated normal human skin (*see* **Note 5**) from consenting patients or volunteers (as authorized by the relevant regulatory authorities).
4. Sterile saline-soaked gauze or skin transport medium: DMEM supplemented with 10% FCS, 100 U/mL penicillin, 100 µg/mL streptomycin, 2.5 µg/mL fungizone, 50 µg/mL gentamycin.
5. Keratinocyte growth medium (KGM): 3 vol Ham's F12 medium, 1 vol DMEM, 10% FCS, 4 mM L-glutamine, 100 U/mL penicillin, 100 µg/mL streptomycin, 0.4 µg/mL hydrocortisone *(24)*, 10^{-10} M cholera enterotoxin *(25)*, 5 µg/mL transferrin *(26)*, 2×10^{-11} M liothyronine (also known as Triiodo-L-thyronine) *(26)*, 1.8×10^{-4} M adenine *(27)*, 5 µg/mL insulin *(26)*, and 10 ng/mL EGF *(24)* (*see* **Note 6**).
6. Phosphate-buffered saline (PBS): PBS referred to in this text lacks calcium and magnesium ions and is made up of the following: 1% (w/v) NaCl, 0.025% KCl,

0.144% Na_2HPO_4, and 0.025% KH_2PO_4. This solution is adjusted to pH 7.2, autoclaved at 121°C (15 psi) for 15 min, and stored at room temperature.

7. Trypsinization solution: 1 vol trypsin stock is added to 4 vol ethylene diamine tetraacetic acid (EDTA) stock and used immediately. Trypsin stock is made up of 0.25% (w/v) trypsin (Difco, Detroit, MI, 1:250) in Tris-saline pH 7.7 (0.8% NaCl, 0.0038% KCl, 0.01% Na_2HPO_4, 0.1% dextrose, 0.3% Trizma base). Stocks are filter sterilized and stored aliquoted at –20°C. EDTA stock is made up of 0.02% (w/v) EDTA in Ca- and Mg-free PBS, autoclaved at 121°C (15 psi) for 15 min, and stored at 4°C.

8. The following sterile equipment is required: forceps, scalpel, iris scissors, hypodermic needles, medical gauze, tissue-culture flasks, and Petri dishes.

9. Mitomycin-C stock solution (*see* **Note 7**): dissolve in sterile H_2O to a concentration of 400 µg/mL. Store at 4°C in the absence of light. Solution is stable for 3–4 mo. Or use a gamma-radiation source (*see* **Note 8**).

10. Dispase medium: 3T3 medium containing 2 mg/mL Dispase (approx 8 U/mg) and filter sterilized. Use immediately.

3. Methods

3.1. Routine Maintenance of 3T3 Cell Line

These adherent cells are grown in 3T3 media at 37°C to near confluence (*see* **Note 9**) and passaged as follows:

1. Remove media from flask, and wash cells with an equivalent volume of PBS.
2. Add trypsinization mixture to the flask at approx 1.5 mL/25 cm^2 surface area, and incubate at 37°C for approx 5 min, or until all cells have rounded up.
3. Add 4 vol medium to deactivate the trypsin and EDTA, and disperse the cells with repeated pipetting.
4. Estimate the cell number using a hemocytometer, and pellet the cells at approx 200*g* for 5 min.
5. Resuspend the cells in fresh media, and seed into flasks or Petri dishes at approx 3×10^3 cells/cm^2 of surface area. Density of cells at seeding can be varied depending on when confluence is required.
6. Cells should reach confluence in approx 3–5 d.

3.2. Production of Feeder Layers

1. Select flasks of exponentially growing 3T3, which have no more than approx 50% of the flask's surface area covered by cells, refresh the media, and incubate for a further 24 h (to ensure a rapid rate of proliferation).
2. Add approx 1–10 µg of mitomycin-C/mL of medium (*see* **Note 7**), and incubate for a further 12 h.
3. Wash the flask three times with fresh medium, incubating the cells with the final wash for approx 10–20 min at 37°C.

4. Harvest the cell immediately by trypsinization in the usual manner (detailed in **Subheading 3.1.**) and seed in fresh flasks at approx 2.5×10^4 cells/cm^2 in KGM (1 mL media/5 cm^2 of flask surface area).
5. Incubate at 37°C for approx 12 h to allow the cells to adhere and spread before seeding with keratinocytes.

Alternatively (*see* **Note 8**):

1. Select flasks of exponentially growing 3T3, which have no more than approx 50% of the flask's surface area covered by cells, refresh the media and incubate for a further 24 h (to ensure a rapid rate of proliferation).
2. Harvest the cell immediately by trypsinization in the usual manner (detailed in **Subheading 3.1.**) and resuspend in fresh 3T3 medium at approx $2–4 \times 10^6$ cells/mL.
3. Subject the cells to approx 6000 rads of γ-radiation from a ^{60}Co source or similar (*see* **Note 8**).
4. The cell suspension can either be used immediately or stored refrigerated for up to 48 h before use. Obviously the viability of the cells decreases with increased storage time.
5. Seed in fresh flasks at approx 2.5×10^4 cells/cm^2 in KGM (1 mL media/5 cm^2 of flask surface area).
6. Incubate at 37°C for approx 12 h to allow the cells to adhere and spread before seeding with keratinocytes.

3.3. Initiation of Keratinocyte Cultures

1. Immediately on removal from patient (*see* **Note 5**), place skin sample as sterilely as possible into transport media (important if the sample is thought likely to carry a considerable microbial load) or wrap it in sterile saline-soaked gauze, and stored at 4°C until use (up to 24 h is acceptable).
2. Place skin into a shallow sterile container (a 10-cm-diameter Petri dish is perfect for small skin samples), and using fine forceps and iris scissors, trim away as much hypodermis as possible (adipose and loose connective tissue), until only the epidermis and the relatively dense and white dermis remain (*see* **Notes 10** and **11**).
3. Flatten the skin (epidermis down) onto the surface of the Petri dish and using a sterile scalpel, cut the skin into long 2–3 mm thin strips.
4. Place the strips into a centrifuge tube (50 mL capacity) containing at least a covering volume of dispase medium (*see* **Note 12**), and incubate either for 2–4 h at 37°C or overnight at 4°C.
5. After incubation, remove the strips of skin from the dispase medium, dabbing excess medium off on the inside of the lid of a 10-cm Petri dish, and place the relatively media-free strips into the Petri dish. Peel the epidermis away from the dermis using two sterile hypodermic needles. The epidermis is the semi-opaque thin layer, whereas the connective tissue of the dermis will have absorbed fluid

and will appear as a thick swollen slightly gelatinous layer. These two layers should come apart easily. If sections remain attached, then either the strips were too thick or further incubation in fresh dispase is required. (Note that this shouldn't be a problem with overnight incubations, although lower viability of the resulting cell suspension is often a problem.)

6. Quickly place the epidermal strips only into 5 mL trypsin stock solution, and shake rapidly for approx 1 min. Add 15 mL of DMEM/10% FCS to inactivate the trypsin and remove the undissociated portions of the epidermal strips either manually (if dealing with a small skin sample) or via passing through sterile medical gauze into a centrifuge tube.

7. Pellet the single-cell suspension by centrifugation at approx $200g$ for 5 min. Resuspend in KGM and count using a hemocytometer. Seed at approx $2–5 \times 10^4$ viable cells/cm^2 onto preplated feeder layers.

3.4. Routine Culture of Keratinocyte Strains

1. Change the medium twice per week.
2. With time, the feeder cells will begin to die and detach from the flask. Replace these with fresh feeder cells as necessary (*see* **Note 13**).
3. The cultures should look rather messy at first, with the keratinocytes only beginning to form obvious colonies after 3–7 d. The keratinocyte cells are only easily distinguishable from the surrounding feeders once they have begun to form colonies, where the cells are closely adherent to each other forming a distinctive "crazy paving" pattern (*see* **Fig. 1**).
4. These cultures should reach confluence within 10–14 d. It is extremely important to passage the keratinocytes before they reach absolute confluence and ideally when they only cover 70–80% of the surface area of the flask (*see* **Note 14**).
5. To passage the keratinocytes, proceed as described for passaging of 3T3 cells in **Subheading 3.1., steps 1–4**, with the exceptions that keratinocytes require additional washing (repeat **step 1**), may often take longer to trypsinize (10–15 min), and, even once rounded up, will require vigorous agitation of the flask to detach the cells from the surface.
6. Once counted, seed the keratinocytes onto fresh feeder layers at a density of approx $5–50 \times 10^3$ viable cells/cm^2. The density seeded depends on when confluence is required. Healthy passaged keratinocyte cultures should reach confluence within 7–10 d (*see* **Note 15**).

4. Notes

1. The shelf life of culture media is dictated by the stability of its essential components. In the case of complex media containing cocktails of added growth factors such as KGM, it should be used as fresh as possible but preferably within 7–10 d. The shelf life of less sophisticated culture media (such as 3T3 medium) is simply limited by the instability of L-glutamine (an essential and common media component) at 4°C in aqueous neutral conditions. However, fresh L-glutamine can be added to extend the shelf life to several months.

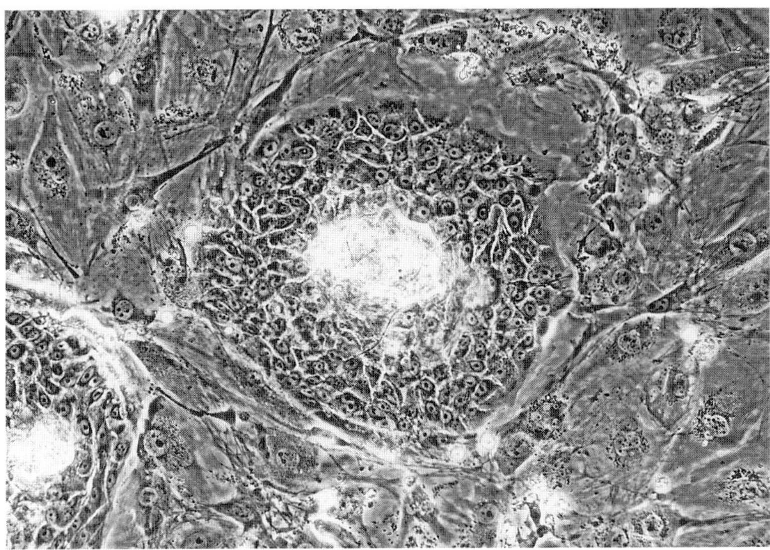

Fig. 1. A healthy keratinocyte colony (center) surrounded by dying feeder cells in a 7-d-old primary culture. Note the symmetrical appearance of the colony, the smooth rounded edges of which are typical of rapidly growing keratinocyte colonies, and the "crazy paving" pattern of the closely adherent keratinocytes. The phase-bright debris located at the center of the colony is commonly seen in primary cultures particularly, and is thought to be because of a cellular aggregation of terminally differentiated cells attaching to the proliferating cells that have adhered to the plastic. Magnification ×200.

2. Culture media, supplements, and other culture solutions are available from most tissue culture retailers unless otherwise specified.
3. Note that each batch of FCS has different properties with regard to promoting the healthy growth and proliferation of cells, and different batches can prove optimal for different types of cells. FCS must be batch tested to obtain optimal serum for the health and growth of normal human keratinocytes. It is possible that the optimum FCS for the 3T3 cell line is different from that identified for keratinocytes, however, this is less important as this cell line will grow sufficiently in suboptimal FCS.
4. Culture media (both as supplied and once supplemented) are stored at 4°C. Media supplements (concentrated) are stored as recommended by suppliers: antibiotics and L-glutamine are stored at <–20°C, the more temperature-sensitive proteins such as growth factors are generally made up to stock concentrations, aliquoted, and stored at <–80°C, FCS (large stock bottles) are stored at <–80°C, whereas working stock aliquots of 50 mL can be stored at <–20°C for up to 1 mo.
5. The type of skin used depends on your experimental needs and ethical permission, but keratinocytes can be successfully cultured from most body sites. A good source of skin is from the redundant skin left over after routine operative proce-

dures such as abdominoplasties, breast reductions or (for younger skin) circumcisions, hyperspadias repair, prominent ear correction. Generally, keratinocyte cultures from younger patients (<20 yr old) have a greater growth potential. It is advised that culture initiation from patients older than 65 yr be avoided.

6. Most stock supplements are made up in PBS containing a carrier protein, such as 0.1% bovine serum albumin (BSA) with the following exeptions due to relative insolubility: highly concentrated stocks of liothyronine dissolve in 1 part HCl and 2 parts ethanol, this can then be further diluted in aqueous solution, adenine dissolves in NaOH, pH 9.0, insulin dissolves in 0.05 M HCl, and hydrocortisone dissolves in EtOH.

7. The exact concentration of mitomycin-C required to produce viable, yet nonproliferating feeder cells should be titrated, because it can vary with batch.

8. The use of irradiation to produce feeder cells is preferable as it is both rapid and simple, but its use is limited by availability of a sufficient source. Note that great care should be taken on use of radiation sources such as ^{60}Co and that radiochemical safety procedures should be strictly adhered to at all times. It is also necessary to intermittently titrate the exact dosage required to produce a healthy and viable feeder layer and yet one which is incapable of proliferating.

9. The 3T3 cell line is an undemanding cell line to maintain in culture, and a competent tissue culturist should encounter few problems. The only thing that must be strictly adhered to during cultivation is the cells must not be allowed to reach confluence. Cells that have been allowed to become over confluent begin to form foci (piling up and escaping density inhibition) and must not be used either for the continuation of stocks or for the production of feeder layers, because the cells appear to transform further and can become resistant to irradiation or mitomycin-C treatment, maintaining their proliferative ability and thereby overrunning keratinocyte cultures.

10. Skin samples are often contaminated with bacteria or yeast. Submerging relatively intact skin samples briefly in alcohol before processing should reduce most contamination problems. However, pockets of bacteria and the like are often found trapped in sweat or sebaceous pores (particularly common in foreskins). Fortunately, once the skin is stretched upside down across the Petri dish, the presence of these blocked pores becomes obvious (depending on the density of the hypodermis). The affected portion of skin sample should be carefully dissected out and discarded, taking particular care not to rupture the blocked pore.

11. The density of hypodermis varies with body site. For foreskins, the dermis is relatively shallow and the hypodermis is particularly loose and thus easy to dissect, whereas skin taken from the back has an extremely dense hypodermis that is hard to distinguish from the thick dermis, making it difficult to dissect and remove in entirety. In the latter case, simply remove as much extraneous connective tissue as possible without cutting into the epidermis itself (often easier with iris scissors from underneath).

12. Alternatively, if the sample being dealt with is either small in size (approx <10 cm^2) and/or is extremely thin (in the form of leftover split thickness skin

graft or thin skin from certain body sites) once any extraneous subcutaneous (sc) tissue has been removed, the sample can simply be finely minced using the iris scissors (pieces ideally no bigger than approx 1 mm^2) and incubated directly in a 1:4 (v/v trypsin:EDTA) trypsinization mixture for approx 1–3 h (until small squares of semi-opaque dermis are seen sticking to the sides of the tube after vigorous shaking). The trypsin is then inactivated and the cells retrieved as detailed in the rest of the protocol.

13. An adequate feeder layer density is extremely important to maintain, both for the continued growth of the keratinocytes and the inhibition of fibroblasts growth. The ideal density of cells within the feeder layer should cover approx 70–80% of the surface area (i.e., slightly higher density than that seen surrounding the keratinocytes colonies in **Fig. 1**) and should not be allowed to go below 50% even when the keratinocyte colonies are established and large.

14. In order to maintain healthy cultures of rapidly growing keratinocytes long term, it is imperative that keratinocyte cultures are passaged well before full confluence is reached. If this is not done, the proliferative ability of keratinocytes begins to reduce dramatically with passage. This phenomena is presumably due to the cellular stratification that takes place in this media, resulting in the production of multiple layers of differentiated keratinocytes overlying the viable, proliferating layer of cells. The formation of this relatively impermeable barrier between the medium and the proliferating cells would considerably reduce their access to nutrients, feasibly reducing culture viability.

15. Keratinocyte stocks can be successfully stored in liquid nitrogen. Only cultures of rapidly growing keratinocytes should be chosen for the production of cryogenic stocks and ideally these should be approx 50% confluent only. Trypsinize, count, and pellet the cells as usual. Resuspend the cells at approx $1–5 \times 10^6$ cells/mL in a rich freezing mixture of 90% FCS and 10% dimethyl sulfoxide. Aliqout into cryotubes immediately and insulate the tubes by wrapping them individually in several layers of tissue or placing them into polystyrene containers and freeze overnight at –80°C, before placing into liquid nitrogen.

References

1. Cruickshank, C., Cooper, J., and Hooper, C. (1960) The cultivations of cells from adult epidermis. *J. Invest. Dermatol.* **34,** 339–342.
2. Rheinwald, J. G. and Green, H. (1975) Serial cultivation of strains of human epidermal keratinocytes: The formation of keratinizing colonies from single cells. *Cell* **6(3),** 331–343.
3. Rheinwald, J. G. and Green, H. (1977) Epidermal growth factor and the multiplication of cultured human epidermal keratinocytes. *Nature* **265,** 421–424.
4. Boyce, S. T. and Ham, R. G. (1983) Calcium-regulated differentiation of normal human epidermal keratinocytes in chemically defined clonal culture and serum-free serial culture. *J. Invest. Dermatol.* **81,** 33s–40s.
5. Bell, E., Sher, S., Hull, B., et al. (1983) The reconstitution of living skin. *J. Invest. Dermatol.* **81(Suppl. 1),** 2s–10s.

6. Prunieras, M., Regnier, M., and Woodley, D. (1983) Methods for cultivation of keratinocytes with an air-liquid interface. *J. Invest. Dermatol.* **81(Suppl. 1),** 28s–33s.

7. Boyce, S. T., Christianson, D. J., and Hansbrough, J. F. (1988) Structure of a collagen-GAG dermal skin substitute optimized for cultured human epidermal keratinocytes. *J. Biomed. Mater. Res.* **22(10),** 939–957.

8. Boyce, S. T. and Hansbrough, J. F. (1988) Biologic attachment, growth, and differentiation of cultured human epidermal keratinocytes on a graftable collagen and chondroitin-6-sulfate substrate. *Surgery* **103(4),** 421–431.

9. Yannas, I. V., Lee, E., Orgill, D. P., Skrabut, E. M., and Murphy, G. F. (1989) Synthesis and characterization of a model extracellular matrix that induces partial regeneration of adult mammalian skin. *Proc. Natl. Acad. Sci. USA* **86(3),** 933–937.

10. Bouvard, V., Germain, L., Rompre, P., Roy, B., and Auger, F. A. (1992) Influence of dermal equivalent maturation on the development of a cultured skin equivalent. *Biochem. Cell Biol.* **70(1),** 34–42.

11. Fransson, J. and Hammar, H. (1992) Epidermal growth in the skin equivalent. *Arch. Dermatol. Res.* **284(6),** 343–348.

12. Medalie, D. A., Eming, S. A., Collins, M. E., Tompkins, R. G., Yarmush, M. L., and Morgan, J. R. (1997) Differences in dermal analogs influence subsequent pigmentation, epidermal differentiation, basement membrane, and rete ridge formation of transplanted composite skin grafts. *Transplantation* **64(3),** 454–465.

13. Sorensen, J. C. (1998) Living skin equivalents and their application in wound healing. *Clin. Podiatr. Med. Surg.* **15(1),** 129–137.

14. Pouliot, R., Germain, L., Auger, F. A., Tremblay, N., and Juhasz, J. (1999) Physical characterization of the stratum corneum of an in vitro human skin equivalent produced by tissue engineering and its comparison with normal human skin by ATR-FTIR spectroscopy and thermal analysis (DSC). *Biochim. Biophys. Acta* **1439(3),** 341–352.

15. Boelsma, E., Gibbs, S., Faller, C., and Ponec, M. (2000) Characterization and comparison of reconstructed skin models: morphological and immunohistochemical evaluation. *Acta Derm. Venereol.* **80(2),** 82–88.

16. Lee, D. Y., Ahn, H. T., and Cho, K. H. (2000) A new skin equivalent model: dermal substrate that combines deepidermized dermis with fibroblast-populated collagen matrix. *J. Dermatol. Sci.* **23(2),** 132–137.

17. Yannas, I. V. (2000) Synthesis of organs: in vitro or in vivo? *Proc. Natl. Acad. Sci. USA* **97(17),** 9354–9356.

18. Peehl, D. M. and Ham, R. G. (1980) Growth and differentiation of human keratinocytes without a feeder layer or conditioned medium. *In Vitro* **16(6),** 516–525.

19. Tsao, M. C., Walthall, B. J., and Ham, R. G. (1982) Clonal growth of normal human epidermal keratinocytes in a defined medium. *J. Cell. Physiol.* **110,** 219–229.

20. Shipley, G. D. and Pittelkow, M. R. (1987) Control of growth and differentiation in vitro of human keratinocytes cultured in serum-free medium. *Arch. Dermatol.* **123(11),** 1541a–1544a.

21. Shipley, G. D., Keeble, W. W., Hendrickson, J. E., Coffey, R. J., Jr., and Pittelkow, M. R. (1989) Growth of normal human keratinocytes and fibroblasts in serum-free medium is stimulated by acidic and basic fibroblast growth factor. *J. Cell. Physiol.* **138(3),** 511–518.
22. Todaro, G. and Green, H. (1963) Quantitative studies of the growth of mouse embryo cells in culture and their development into established lines. *J. Cell Biol.* **17,** 299–313.
23. Jainchill, J. L., Aaronson, S. A., and Todaro, G. J. (1969) Murine sarcoma and leukemia viruses: assay using clonal lines of contact-inhibited mouse cells. *J. Virol.* **4(5),** 549–553.
24. Rheinwald, J. G. (1980) Serial cultivation of normal human epidermal keratinocytes. *Methods Cell Biol.* **21A,** 229–254.
25. Green, H. (1978) Cyclic AMP in relation to proliferation of the epidermal cell: a new view. *Cell* **15(3),** 801–811.
26. Watt, F. M. and Green, H. (1981) Involucrin synthesis is correlated with cell size in human epidermal cultures. *J. Cell Biol.* **90,** 738–742.
27. Wu, Y. J., Parker, L. M., Binder, N. E., et al. (1982) The mesothelial keratins: a new family of cytoskeletal proteins identified in cultured mesothelial cells and nonkeratinizing epithelia. *Cell* **31(3 Pt. 2),** 693–703.

2

Cultivation of Normal Human Epidermal Melanocytes in the Absence of Phorbol Esters

Mei-Yu Hsu, Ling Li, and Meenhard Herlyn

1. Introduction

An important approach in studies of normal, diseased, and malignant cells is their growth in culture. The isolation and subsequent culture of human epidermal melanocytes has been attempted since 1957 *(1–5)*, but only since 1982 have pure normal human melanocyte cultures been reproducibly established to yield cells in sufficient quantity for biological, biochemical, and molecular analyses *(6)*. Selective growth of melanocytes, which comprise only 3–7% of epidermal cells in normal human skin, was initially achieved by suppressing the growth of keratinocytes and fibroblasts in epidermal cell suspensions with the tumor promoter 12-O-tetradecanoyl phorbol-13-acetate (TPA) and the intracellular cyclic adenosine 3′, 5′ monophosphate (cAMP) enhancer cholera toxin, respectively, which both also act as melanocyte growth promoters. However, phorbol ester is metabolically stable and has prolonged effects on multiple cellular responses *(6)*. Recent progress in basic cell-culture technology, along with an improved understanding of culture requirements, has led to an effective and standardized isolation method, and special TPA-free culture media for selective growth and long-term maintenance of human melanocytes. The detailed description of this new method is aimed at encouraging its use in basic and applied biological research.

2. Materials

1. Normal skin-transporting medium: The medium for collecting normal skin is composed of Hanks balanced salt solution (HBSS) without Ca^{2+} and Mg^{2+} (HBSS; Gibco-BRL Grand Island, NY, cat. no. 21250-089) supplemented with penicillin (100 U/mL; USB, Cleveland, OH, cat. no. 199B5), streptomycin (100 µg/mL;

From: *Methods in Molecular Medicine, vol. 107: Human Cell Culture Protocols, Second Edition*
Edited by: J. Picot © Humana Press Inc., Totowa, NJ

USB, cat. no. 21B65), gentamicin (100 µg/mL; BioWittaker, Walkersville, MD, cat. no. 17-518Z), and fungizone (0.25 µg/mL; JRH Biosciences, Lenexa, KS, cat. no. 59-604-076). After sterilization through a 0.2-µm filter, the skin-transporting medium is transferred into sterile containers in 20-mL aliquots and stored at 4°C for up to 1 mo.

2. Epidermal isolation solution: Dissolve 0.48 g of dispase (grade II, 0.5 U/mg; Boehringer Mannheim, Indianapolis, IN, cat. no. 165859) in 100 mL of phosphate-buffered saline (PBS) without Ca^{2+} and Mg^{2+} (Cellgro® by Mediatech, Herndon, VA, cat. no. MT21-031-CM) containing 0.1% bovine serum albumin (BSA) (fraction V; Sigma, St. Louis, MO, cat. no. A9418) to yield a final dispase activity of 2.4 U/mL. Sterilize the enzyme solution through a 0.2-µm filter, aliquot into 5-mL tube, and store at –20°C for up to 3 mo.

3. Cell-dispersal solution: The cell-dispersal solution contains 0.25% trypsin and 0.1% ethylenediaminetetraacetic acid (EDTA) and is purchased from Cellgro by Mediatech, cat. no. 25-053-CI. Store at 4°C for up to 1 mo.

4. TPA-free melanocyte growth medium (TPA-free MGM): The following stock solutions and reagents are required:

   a. MCDB153 (Sigma, cat. no. M7403): Dissolve MCDB153 powder in ~approx 700 mL ddH$_2$O, add 1.18 g sodium bicarbonate (Sigma, cat. no. S5761), adjust pH to 7.4 ± 0.02, bring the total volume to 1 L with ddH$_2$O, sterilize through a 0.2-µm filter, and store light-protected at 4°C for up to 3 wk. Use 87 mL per 100 mL complete MGM.

   b. Heat-inactivated fetal bovine serum (FBS; Cansera, Etobicoke, ON, Canada, cat. no. CS-C08-100): Heat FBS in manufacturer's bottle at 56°C water bath for 20 min and store at 4°C for up to 3 wk. Use 2 mL per 100 mL complete MGM.

   c. Chelated FBS: Add 15 g Chelax 100 (Sigma, cat. no. C7901) to 500 mL FBS, stir to mix at 4°C for 1.5 h, filter-sterilize with a 0.2-µm filter, prepare 10-mL single-use aliquots, and store at –20°C for up to 3 mo.

   d. L-Glutamine, 200 m*M* stock (Cellgro by Mediatech, cat. no. MT25-005-C1): Prepare 1-mL single-use aliquots and store at –20°C for up to 6 mo.

   e. Cholera toxin (Sigma, cat. no. C3021), 40 n*M* (3.33 µg/mL) stock: Dissolve 500 µg of cholera toxin in 150 mL ddH$_2$O, sterilize through a 0.2-µm low protein-binding filter (Millipore, Marlborough, MA, cat. no. SLGV025LS), divide into 250-µL aliquots, and store at 4°C for up to 1 yr. Use 50 µL/100 mL MGM to give a final cholera toxin concentration of 20 p*M*.

   f. Recombinant human basic fibroblast growth factor (rh-bFGF; Research Diagnostics, Flanders, NJ, cat. no. RDI-118bx), 0.57 µg/mL stock: Dissolve 4 µg of rh-bFGF in 7 mL of 0.1% BSA; Sigma, cat. no. A9647) in Ca^{2+}- and Mg^{2+}-free PBS. Pre-wet a 0.2-µm low protein-binding filter with Ca^{2+}- and Mg^{2+}-free PBS containing 0.1% BSA before filter-sterilizing rh-bFGF stock solution to avoid loss of the recombinant protein due to nonspecific binding to the filter. Prepare 500-µL aliquots and store at –20°C for up to 3 mo. Add 200 µL of rh-bFGF stock per 100 mL MGM to yield a final concentration of rh-bFGF of 1.14 ng/mL.

g. Recombinant human, rat, pig, or rabbit endothelin-3 (rET-3; American Peptide Company, Sunnyvale, CA, cat. no. 88-5-10), 100 μM (264 μg/mL) stock: Dissolve 500 μg of rET-3 in 1.89 mL of 0.1% BSA in Ca^{2+}- and Mg^{2+}-free PBS. Filter-sterilize rET-3 stock solution by passage through a 0.2-μm low protein-binding filter pre-wet with PBS containing 0.1% BSA. Make 200-μL single-use aliquots and store at –20°C for up to 3 mo.

h. Recombinant human stem cell factor (rhSCF; R&D systems, Minneapolis, MN, cat. no. 255-SC-050), 10 μg/mL stock: Add 50 μg rhSCF to 5 mL of 0.1% BSA in Ca^{2+}- and Mg^{2+}-free PBS, filter-sterilize through a 0.2-μm low protein-binding membrane pre-wet with PBS containing 0.1% BSA, prepare 100-μL single-use aliquots, and store at –20°C for up to 3 mo.

i. Heparin (Sigma, cat. no. H3149), 1 μg/mL stock: Prepare heparin storage stock at 1 mg/mL by dissolving 1 mg of heparin sodium salt in 1 mL of Ca^{2+}- and Mg^{2+}-free PBS, filter-sterilizing through a 0.2-μm filter, dividing into 10-μL aliquots, and storing at 4°C for up to 6 mo. Before making up the complete medium, 1 μg/mL heparin stock solution is prepared fresh by diluting 1 μL of 1 mg/mL storage stock with 1 mL of PBS. Use 100 μL of 1 μg/mL heparin stock for 100 mL complete medium to yield a final heparin concentration of 1 ng/mL.

TPA-free MGM is prepared as follows: Mix 87 mL of MCDB153 with 2 mL heat-inactivated FBS, 10 mL chelated FBS, 1 mL L-glutamine (200 mM stock), 50 μL cholera toxin (40 nM stock), 200 μL bFGF (0.57 μg/mL stock), 200 μL ET-3 (100 μM stock), 100 μL SCF (10 μg/mL stock), and 100 μL heparin (1 μg/mL stock) to give final concentrations of 12% FBS, 2 mM L-glutamine, 20 pM cholera toxin, 1.14 ng/mL bFGF, 100 nM ET-3, 10 ng/mL SCF, and 1 ng/mL heparin. Store TPA-free MGM at 4°C for up to 8 d.

5. Trypsin–versene solution: Make a 5X stock by mixing 0.5 mL of 2.5% trypsin solution (BioWittaker, cat. no. 17-160E) with 100 mL of versene composed of 0.1% EDTA (Fisher, Pittsburgh, PA, cat. no. 02793-500) in Ca^{2+}- and Mg^{2+}-free PBS (pH 7.4). To prepare trypsin–versene solution, dilute 5X stock with Ca^{2+}- and Mg^{2+}-free HBSS to give a final concentration of 0.0025% trypsin and 0.02% EDTA.

6. Cell-preservative medium: Prepare 5% (v/v) dimethyl sulfoxide (DMSO; Sigma, cat. no. D2650) in 95% heat-inactivated FCS as needed.

3. Methods

3.1. Day 1

1. Prepare the following in a laminar flow hood: one pair each of sterile forceps, curved scissors, and surgical scalpel blade; 5 mL of epidermal isolation solution (*see* **Subheading 2.2.**) in a sterile centrifuge tube; 10 mL of Ca^{2+}- and Mg^{2+}-free HBSS in a sterile nontissue-culture Petri dish; and 10 mL of 70% ethanol in a separate sterile Petri dish.

2. Soak the skin specimens in 70% ethanol for 1 min. Transfer skin to the Petri dish containing HBSS to rinse off ethanol (*see* **Notes 1** and **2**).

3. Cut skin-ring open, and trim off fat and sc tissue with scissors (*see* **Note 3**).
4. Cut skin into pieces (approx 5 × 5 mm²) using the surgical scalpel blade with one-motion cuts (*see* **Note 4**).
5. Transfer the skin pieces into the tube containing epidermal isolation solution. Cap, invert, and incubate the tube in the refrigerator at 4°C for 18–24 h (*see* **Note 5**).

3.2. Day 2

1. Remove the tube containing the sample from the refrigerator and incubate at 37°C for 5 min.
2. Prepare the following in a laminar flow hood: two pairs of sterile forceps and a surgical scalpel blade; two empty sterile nontissue-culture Petri dishes; 5 mL of cell-dispersal solution; and 10 mL of Ca^{2+}- and Mg^{2+}-free HBSS in a 15-mL centrifuge tube.
3. Pour tissue in epidermal isolation solution into one of the empty Petri dishes. Separate the epidermis (thin, brownish, translucent layer) from the dermis (thick, white, opaque layer) using the forceps. Hold the dermal part of the skin piece with one pair of forceps, and the epidermal side another. Gently tease them apart. Discard the dermis immediately (*see* **Note 5**). Transfer the harvested epidermal sheets to an empty Petri dish, add a drop of Ca^{2+}- and Mg^{2+}-free HBSS to prevent tissue from drying. Repeat the above described procedure for each piece of tissue and then mince them into smaller pieces (approx 2 × 2 mm²) with a surgical scalpel blade (*see* **Note 5**).
4. Transfer the collected epidermal sheets from the Petri dish to the centrifuge tube containing 5 mL of cell-dispersal solution. Incubate the tube at 37°C for 5 min. Vortex the tube vigorously or use repetitive pipet motions to release single cells from epidermal sheets. Wash the resulting single-cell suspension once with 10 mL of Ca^{2+}- and Mg^{2+}-free HBSS. Centrifuge for 5 min at 800*g* at room temperature. Carefully aspirate the supernatant, which may contain remaining stratum corneum. Resuspend the pellet with 5 mL TPA-free MGM (*see* **Note 6**).
5. Plate the resulting epidermal cell suspension in a T25 cell-culture vessel. Incubate at 37°C in 5% CO_2/5% air for 48–72 h without disturbance.

3.3. Subsequent Maintenance, Subcultivation, Cryopreservation, and Thawing

1. Wash culture with MGM on d 4 to remove nonadherent cells, which may include but are not limited to keratinocytes and fragments of stratum corneum. Medium change should be performed twice a week thereafter. Seventy percent confluent primary melanocyte cultures can be obtained in approx 1 wk (*see* **Note 7**).
2. Subcultivation: Primary cultures established from foreskins usually reach 70% confluence within 7–9 d after plating. At this point, cultures are treated with trypsin–versene solution (*see* **Subheading 2., step 5**) at room temperature for 2–3 min, harvested with Leibovittz's L-15 (Gibco-BRL, cat. no. 41300-070) containing 10% heat-inactivated FBS, centrifuged at 2000 rpm for 3 min, resuspended in

TPA-free MGM, reinoculated at approx 10^4 cells/cm², and serially passaged. Medium is changed twice each week.

3. Cryopreservation: Melanocyte suspensions harvested by trypsin–versene and Leibovitz's L-15 containing 10% heat-inactivated FBS are centrifuged at 800*g* for 5 min and resuspended in cell-preservative medium (*see* **Subheading 2., step 6**) containing 5% DMSO as a cryopreservative. Cells are normally suspended at a density of 10^6/mL and transferred to cryotubes. The tubes are then placed in a plastic sandwich box (Nalgene™ Cryo 1°C Freezing container; Nalge, Rochester, NY, cat. no. 5100-0001), which is immediately transferred to a –70°C freezer. The insulation of the freezing container ensures gradual cooling of the cryotubes and results in more than 80% viability of cells upon thawing. After overnight storage in the –70°C freezer, the cryotubes are placed in permanent storage in liquid nitrogen.

4. Thawing: The melanocyte suspension is thawed by incubating the cryotube in a 37°C water bath. When the cell-preservative medium is almost, but not totally, defrosted, the outside of the tube is wiped with 70% alcohol. The cell suspension is then withdrawn, quickly diluted in TPA-free MGM at room temperature, centrifuged, and resuspended in fresh TPA-free MGM. Cell viability is determined by Trypan blue exclusion. The resulting melanocytes are then seeded at a density of 10^4 cells/cm².

3.4. Results

3.4.1. Minimal Growth Requirements

Earlier studies of normal melanocytes *(6–8)* were done using media containing bovine pituitary extracts, which provides a host of poorly characterized growth-promoting activities. Deprivation of serum and brain tissue extracts from media has led to the delineation of four groups of chemically defined melanocyte mitogens.

1. Peptide growth factors, including bFGF *(9–12)*, which is the main growth-promoting polypeptide in bovine hypothalamus and pituitary extracts, insulin/insulin-like growth factor-1 (IGF-1; *13*), epidermal growth factor (EGF; *14,15*), transforming growth factor-α (TGF-α; *16*), endothelins (ET; *17–21*), hepatocyte growth factor/scatter factor (HGF/SF; *22–24*), and stem cell factor (SCF; *10,25–28*).
2. Calcium, because reduction of Ca^{2+} concentrations in TPA-containing MGM from an optimal 2.0 to 0.03 m*M* reduces cell growth by approx 50% *(20)*, and cation-binding proteins, such as tyrosinase at 10^{-11} *M*, and ceruloplasmin at 0.6 U/mL *(29)*.
3. Enhancers of intracellular levels of cAMP, including α-melanocyte stimulating hormone (α-MSH) at 10 ng/mL *(30)*; forskolin at 10^{-9} *M* *(29)*, follicle stimulating hormone (FSH) at 10^{-7} *M* *(31)*; and cholera toxin at 10^{-12} *M* *(6,29,32,33)*.
4. Activators of protein kinase C (PKC), such as TPA *(34)*, which is lipophilic and cannot be removed by simple washing, and 20-oxo-phorbol-12,13-dibutyrate

(PDBu; *34*), which is a similar derivative, but more hydrophilic. Recent data suggest that the tigliane class phorbol compounds, such as 12 deoxyphorbol, 13 isobutyrate (DPIB), and 12 deoxyphorbol, 13 phenylacetate (DPPA), which possess diminished tumor-promoting activity, are also able to activate PKC as well as stimulate melanocyte proliferation *(35)*.

3.4.2. Morphology

Human epidermal melanocytes grown in TPA-free MGM normally exhibit a dendritic morphology with varying degrees of pigmentation (*see* **Fig. 1**). By contrast, melanocytes maintained in the conventional TPA medium *(36)* are bi- or tri-polar.

3.4.3. Expression of Antigens

Extensive studies have been done to characterize the antigenic phenotype of malignant melanoma cells *(37)*. On the other hand, very few attempts have been made to produce monoclonal antibodies (mAbs) to normal melanocytes *(15,38)*. Cultured melanocytes share with melanoma cells the expression of a variety of cell-surface antigens (melanoma-associated antigens), including p97 melanotransferrin, integrin β_3 subunit of the vitronectin receptor, gangliosides GD_3 and 9-O-acetyl GD_3, chondroitin sulfate proteoglycan *(15)*, and MelCAM/ MUC18/CD146 *(39,40)*. However, these antigens are not expressed by normal melanocytes *in situ* *(41,42)*. **Table 1** summarizes the expression of antigens on melanocytes *in situ* and in culture. The observed divergent antigenic phenotype in culture and *in situ* suggested a role for the epidermal microenvironmental signals in controlling the melanocytic phenotype. Indeed, accumulating evidence indicates that undifferentiated keratinocytes can control proliferation, morphology, pigmentation, and antigen expression of melanocytes in coculture *(43–47)*. Using coculture and three-dimensional (3D) skin reconstruct models, we have begun to characterize the molecular bases of the crosstalk between keratinocytes and melanocytes *(48–50)*.

3.4.4. Growth Characteristics

Melanocytes from neonatal foreskin can be established with a success rate of 80% and have a maximum lifespan of 60 doublings, with a doubling time of

---

Fig. 1. *(see facing page)* Morphology of normal human epidermal melanocytes maintained in TPA-free medium supplemented with bFGF, ET-3, SCF, cholera toxin, and serum. A, post-plating d 1: Admixed in the background cellular and tissue debris, some cells (including melanocytes and occasional keratinocytes) though still rounded attach to the substratum. B, post-plating d 3: Attached melanocytes spread out on the substratum giving a bi- or tri-polar morphology. The occasional surviving keratinocytes

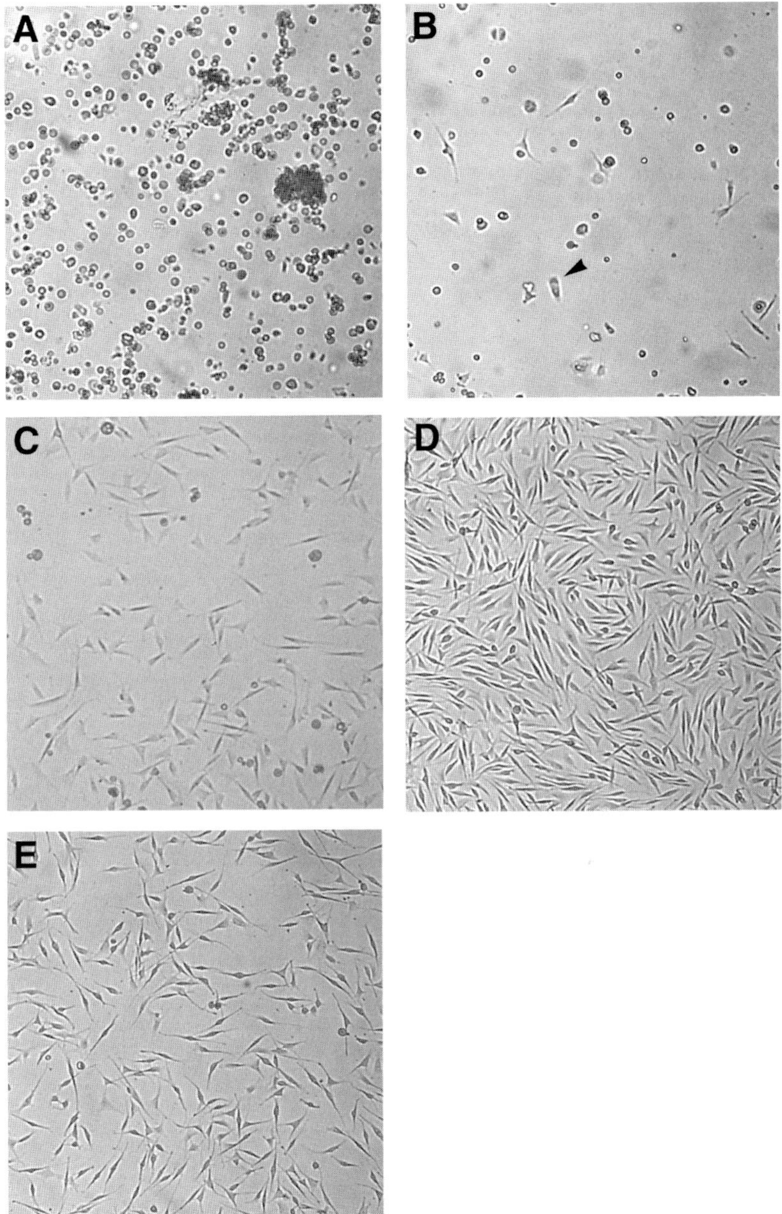

Fig. 1. *(continued)* are identified by their cobblestone/polygonal morphology (arrowhead). C, post-plating d 7: High serum concentration in the medium eliminates contaminated keratinocytes by induction of terminal differentiation. Pure melanocyte culture is usually established at this point. Melanocytes now appear more flattened and multidendritic. D, postplating d 9: Greater than 70% confluence is usually achieved by d 9, at which point the cells are ready to be subcultured. E, passage 1 melanocyte culture (1 d after splitting): Cells display the characteristic dendritic morphology.

Table 1
Expression of Antigens on Melanocytes *in situ* **and in Culture**[a]

Antigens	*in situ*	in culture
CD26	++++	++++
gp145	++	++
c-kit	++++	++++
TRP-1 (gp 75)	++++	++++
TRP-2 (DCT)	++++	++++
MITF	++++	++++
E-cadherin	++++	++++
α-catenin	++++	++++
β-catenin	++++	++++
integrin β3 subunit	–	++++
tenascin	–	++++
fibronectin	–	++++
chondroitin sulfate proteoglycan	±	++++
p97 melanotransferrin	–	++++
NGF-receptor (p75)	±	++++
9-O-acetyl GD3	±	+++
GD3	±	++++
HLA-DR	–	–
MelCAM/MUC18/CD146	–	++++

[a]–, lack of expression; ±, 0–20%; +, 20–40%, ++, 40–60%; +++, 60–80%; ++++, 80–100%.

1.5–4 d. Heavily pigmented cells isolated from black individuals have a shorter doubling time and tend to senesce after 20–30 doublings. By contrast, epidermal melanocytes from adult skin only grow in about 10% of cases and for no more than 10 doublings with a doubling time of 7–14 d. The cells do not grow beyond 70% confluence and exhibit signs of growth arrest by contact inhibition. Normal melanocytes do not proliferate anchorage independently in soft agar and are nontumorigenic in athymic nude mice **(12,14)**.

4. Notes

1. Tissue source and collection: The source of tissue for melanocyte cultures are human neonatal foreskins obtained from routine circumcision and normal adult skin acquired from reduction mammoplasty. At the time of excision, the skin is placed into a sterile container with 20 mL of normal skin-transporting medium (*see* **Subheading 2.1.**) supplied in advance and kept near the surgical area at 4°C. Specimens are delivered immediately to the tissue-culture laboratory or stored at

4°C. Neonatal foreskins can be kept for up to 48 h, and normal adult skin, for up to 24 h. However, the fresher the specimens, the higher the yield of live cells on isolation.

2. Sterilization of skin specimens: Reduce contamination by a short treatment (1 min) of intact skin with 70% ethanol in a laminar flow hood. After sterilization, rinse samples thoroughly with Ca^{2+}- and Mg^{2+}-free HBSS.

3. Preliminary tissue preparation: Place tissue on a 100-mm nontissue-culture Petri dish, and remove most of the sc fat and membranous material with curved scissors.

4. Adjustment of tissue size for enzymatic digest: To improve reagent penetration, cut the skin samples into small pieces (approx 5×5 mm^2) rinsed in Ca^{2+}- and Mg^{2+}-free HBSS.

5. Dispase treatment: Because melanocytes are located just above the basement membrane in the epidermis, successful isolation requires effective separation of epidermis from dermis. Pieces of skin are incubated in epidermal isolation solution for up to 24 h at 4°C to allow detachment of epidermis from dermis. As originally described *(51)*, dispase splits epidermis from dermis along the basement membrane. Each piece of skin is secured with two pairs of forceps; one holds the epidermis and the other the dermis. The epidermal sheet is then peeled apart from the dermis, transferred to a Petri dish, and minced with a scalpel blade to smaller fragments to expedite the subsequent cell dispersal. To prevent the epidermal sheets from drying, a drop of Ca^{2+}- and Mg^{2+}-free HBSS can be added to the Petri dish. To avoid potential sources of fibroblast contamination, dermal pieces should be discarded immediately once they are separated from the epidermis, and the forceps used to hold the dermis should never come in contact with the epidermal sheets and vise versa. Contaminated dermal fragments are easily recognized by their white opaque color in contrast to the yellowish-brown semi-transluscent epidermis.

6. Cell dispersal techniques: A single-cell suspension is generated from clumps of epidermal tissue by enzymatic treatment with cell-dispersal solution containing trypsin at 37°C for 5 min followed by mechanical dissociation. After washing the cells once with Ca^{2+}- and Mg^{2+}-free HBSS to remove the enzyme, cells are then pelleted by centrifugation, resuspended and seeded in a T25 culture vessel. Extra caution should be taken to remove the supernatant when washing the cells, as the cells tend not to form a solid pellet because of the presence of remaining fragments of cornified materials. It is suggested that manual pipeting be used in place of suctioning.

7. Selective growth: Most methods for growing pure cultures of melanocytes from epidermal cell suspensions depend on optimal conditions that enable melanocytes, but not keratinocytes, to attach to a substrate and proliferate. These conditions include high oxygen tension *(52)*, high seeding density *(53)*, high Ca^{2+} concentration *(54–56)*, and the presence of sodium citrate *(57)*, 5-fluorouracil *(58)*, and phorbol esters *(6)*. The presence of phorbol esters not only suppresses the growth

of keratinocytes, but also promotes melanocyte growth. However, phorbol esters have been shown to reduce the numbers of melanosomes in human melanocytes in culture and to delay the onset of melanization *(6)*. Thus, although these reagents support long-term culture of human melanocytes, they may have limited use in studies of melanocyte differentiation.

In our original report dated back in 1987, when melanocytes were established in medium without TPA, they grew at doubling times of 4–7 d for the first 2–3 passages and senesced by passage 5. Initially, the cells assumed a spindle morphology, which changed by passage 3–5 to a flat, polygonal morphology *(59)*. The flat, polygonal cells were unpigmented and proliferated slowly. Concomitant with the morphological and proliferative changes, there was a decrease in expression of the nerve growth factor (NGF) receptor and an increase in expression of gp145 (*see* **Table 2**). Recently, with the advance in melanocyte biology, we have devised a growth medium for human melanocytes (TPA-free MGM) based on the use of more physiologic mitogens that substituted for routinely used artificial and undefined agents. Important features of this method include the following. First, long-term culture of melanocytes in the absence of phorbol esters is achieved. Second, contamination by dermal fibroblasts, a common problem in establishing melanocyte culture, is dramatically reduced (from 15 to 20% to less than 5%) by minimal tissue manipulation during the isolation process. Third, melanocytes maintained in TPA-free MGM exhibit a more physiologic morphology (dendritic vs bi- or tri-polar) and a shorter population doubling time (1.5–4 vs 2–6 d) comparing to their counter parts grown in the conventional TPA medium (*see* **Table 2**). Abdel-Malek and co-workers also reported successful long-term proliferation sustained by TPA-free medium supplemented with 0.6 ng/mL bFGF, 10 n*M* ET-1, and 10 n*M* α-MSH *(60)*.

There are other alternative media for melanocyte culture. TIP medium, a TPA-containing medium, consists of 85 n*M* TPA, 0.1 m*M* isobutylmethyl xanthine (IBMX), and 10–20 μg protein/mL placental extract in Ham's F-10 medium supplemented with 10% newborn calf serum *(61)*. TPA-free medium *(62)*, composed of Ca^{2+}-free M199 medium supplemented with 5–10% chelated FCS, 10 μg/mL insulin, 10 g/mL EGF, 10^{-9} *M* triiodothyronine, 10 μg/mL transferring, 1.4×10^{-6} *M* hydrocortisone, 10^{-9} *M* cholera toxin, and 10 ng/mL bFGF *(62)*, can also support short-term culture of melanocytes.

Acknowledgment

We thank Dr. P. Donatien and Dr. F.M. Meier for their contribution in the development of TPA-free MGM. Dr. Dong Fang is acknowledged for helping with the antigenic phenotyping of cultured melanocytes. This work was sup-

Table 2
Phenotype of Neonatal Foreskin Melanocytes in Culture[a]

	Growth Passage			Morphology[b] Passage			Pigmentation Passage			Antigen expression					
										NGFR Passage			gp145 Passage		
Culture conditions	1	5	8	1	5	8	1	5	8	1	5	8	1	5	8
TPA-MGM without TPA[c]	+++	+	–	S	F	F	+	–	–	++++	–	–	–	+++	nt
TPA-MGM[d]	+++	+++	+++	B	B	B	+++	+++	+++	++	+++	+++	–	–	+
TPA-free MGM[e]	++++	++++	++++	D	D	D	++	++	++	nt	nt	nt	nt	nt	nt

[a]+ to ++++, Degree of growth, pigmentation, or antigen expression; –, no growth or >14 d doubling time, no pigmentation, and no expression of antigen; nt, not tested.

[b]S, spindle; F, flat, polygonal; B, bi- or tri-polar; D, dendritic.

[c]TPA-MGM without TPA *(36)* consists of four parts of MCDB153 supplemented with 2 mM CaCl$_2$, one part of Leibovitz's L-15, 2% heat-inactivated FCS, 5 µg/mL of insulin, 5 ng/mL EGF, and 40 µg protein/mL bovine pituitary extract.

[d]TPA-MGM *(36)* is composed of four parts of MCDB153 supplemented with 2 mM CaCl$_2$, one part of Leibovitz's L-15, 2% heat-inactivated FCS, 5 µg/mL of insulin, 5 ng/mL EGF, 40 µg protein/mL bovine pituitary extract, and 10 ng/mL TPA.

[e]TPA-free MGM, as described in **Subheading 2., step 4**.

ported by grants from the National Institutes of Health numbers: CA25874, CA47159, CA76674, CA80999, and CA10815 to M. Herlyn.

References

1. Hu, F., Staricco, R.J., Pinkus, H., and Fosnaugh, R. (1957) Human melanocytes in tissue culture. *J. Invest. Dermatol.* **28,** 15–32.

2. Karasek, M. and Charlton, M. E. (1980) Isolation and growth of normal human skin melanocytes. *Clin. Res.* **28,** 570A.

3. Kitano, Y. (1976) Stimulation by melanocyte stimulating hormone and dibutyryl adenosine 3′, 5′-cyclic monophosphate of DNA synthesis in human melanocytes in vitro. *Arch. Derm. Res.* **257,** 47–52.

4. Mayer, T. C. (1982) The control of embryonic pigment cell proliferation in culture by cyclic AMP. *Dev. Biol.* **94,** 509–614.

5. Wilkins, L. M. and Szabo, G. C. (1981) Use of mycostatin-supplemented media to establish pure epidermal melanocyte culture (abstract). *J. Invest. Dermatol.* **76,** 332.

6. Eisinger, M. and Marko, O. (1982) Selective proliferation of normal human melanocytes in vitro in the presence of phorbol ester and cholera toxin. *Proc. Natl. Acad. Sci. USA* **79,** 2018–2022.

7. Herlyn, M., Herlyn, D., Elder, D. E., et al. (1983) Phenotypic characteristics of cells derived from precursors of human melanoma. *Cancer Res.* **43,** 5502–5508.

8. Herlyn, M., Thurin, J., Balaban, G., et al. (1985) Characteristics of cultured human melanocytes isolated from different stages of tumor progression. *Cancer Res.* **45,** 5670–5676.

9. Halaban, R., Kwon, B. S., Ghosh, S., Delli Bovi, P., and Baird, A. (1988) bFGF as an autocrine growth factor for human melanomas. *Oncogene Res.* **3,** 177–186.

10. Hachiya, A., Kobayashi, A., Onuchi, A., Takema, Y., and Imokawa, G. (2001) The paracrine role of stem cell factor/c-kit signaling in the activation of human melanocytes in ultraviolet-B-induced pigmentation. *J. Invest. Dermatol.* **116,** 578–586.

11. Hedley, S. J., Gawkrodger, D. J., Weetman, A. P., and MacNeil, S. (1998) alpha-MSH and melanogenesis in normal human adult melanocytes. *Pigment Cell Res.* **11,** 45–56.

12. Nesbit, M., Nesbit, H. K., Bennett, J., et al. (1999) Basic fibroblast growth factor induces a transformed phenotype in normal human melanocytes. *Oncogene* **18,** 6469–6476.

13. Edmondson, S. R., Russo, V. C., McFarlane, A. C., Wraight, C. J., and Werther, G. A. (1999) Interactions between growth hormone, insulin-like growth factor 1, and basic fibroblast growth factor in melanocyte growth. *J. Clin. Endocrinol. Metab.* **84,** 1638–1644.

14. Herlyn, M., Rodeck, U., Mancianti, M. L., et al. (1987) Expression of melanoma-associated antigens in rapidly dividing human melanocytes in culture. *Cancer Res.* **47,** 3057–3061.

15. Herlyn, M., Clark, W. H., Rodeck, U., Mancianti, M. L., Jambrosic, J., and Koprowski, H. (1987) Biology of tumor progression in human melanocytes. *Lab Invest.* **56,** 461–474.

16. Pittelkow, M. R. and Shipley, G. D. (1989) Serum-free culture of normal human melanocytes: growth kinetics and growth factor requirements. *J. Cell Physiol.* **140,** 565–576.

17. Imokawa, G., Yada, Y., and Miyagishi, M. (1992) Endothelins secreted from human keratinocytes are intrinsic mitogens for human melanocytes. *J. Biol. Chem.* **267,** 24,675–24,680.

18. Hirobe, T. (2001) Endothelins are involved in regulating the proliferation and differentiation of mouse epidermal melanocytes in serum-free primary culture. *J. Invest. Dermatol. Symp. Proc.* **6,** 25–31.

19. Tada, A., Suzuki, I., Im, S., et al. (1998) Endothelin-1 is a paracrine growth factor that modulates melanogenesis of human melanocytes and participates in their responses to ultraviolet radiation. *Cell Growth Differ.* **9,** 575–584.

20. Lahav, R., Ziller, C., Dupin, E., and Le Douarin, N. M. (1996) Endothelin-3 promotes neural crest cell proliferation and mediates a vast increase in melanocyte number in culture. *Proc. Natl. Acad. Sci. USA* **93,** 3892–3897.

21. Reid, K., Turley, A. M., Maxwell, G. D., et al. (1996) Multiple roles for endothelin in melanocyte development: regulation of progenitor number and stimulation of differentiation. *Development* **122,** 3911–3919.

22. Halaban, R., Rubin, J. S., Funasaka, Y., et al. (1992) Met and hepatocyte growth factor/scatter factor signal transduction in normal melanocytes and melanoma cells. *Oncogene* **7,** 2195–2206.

23. Matsumoto, K., Tajima, H., and Nakamura, T. (1991) Hepatocyte growth factor is a potent stimulator of human melanocyte DNA synthesis and growth. *Biochem. Biophys. Res. Commun.* **176,** 45–51.

24. Imokawa, G., Yada, Y., Morisaki, N., and Kimura, M. (1998) Biological characterization of human fibroblast-derived mitogenic factors for human melanocytes. *Biochem. J.* **330,** 1235–1239.

25. Grichnik, J. M., Burch, J. A., Burchette, J., and Shea, C. R. (1998) The SCF/KIT pathway plays a critical role in the control of normal melanocyte homeostasis. *J. Invest. Dermatol.* **111,** 233–238.

26. Halaban, R. (2000) The regulation of normal melanocyte proliferation. *Pigment Cell Res.* **13,** 474.

27. Imokawa, G., Kobayasi, T., and Miyagishi, M. (2000) Intracellular signaling mechanisms leading to synergistic effects of endothelin-1 and stem cell factor on proliferation of cultured human melanocytes. Cross-talk via trans-activation of the tyrosine kinase c-kit receptor. *J. Biol. Chem.* **275,** 33,321–33,328.

28. Kapur, R., Everett, E. T., Uffman, J., et al. (1997) Overexpression of human stem cell factor impairs melanocyte, mast cell and thymocyte development: A role for receptor tyrosine kinase-mediated mitogen activated protein kinase activation in cell differentiation. *Blood* **90,** 3018–3026.

29. Herlyn, M., Mancianti, M. L., Jambrosic, J., Bolen, J. B., and Koprowski, H. (1988) Regulatory factors that determine growth and phenotype of normal human melanocytes. *Exp. Cell Res.* **179,** 322–331.
30. Abdel-Malek, Z. A. (1988) Endocrine factors as effectors of integumetal pigmentation. *Dermatol. Clin.* **6,** 175–184.
31. Adashi, E. Y., Resnick, C. E., Svoboda, M. E., and Van Wyk, J. J. (1986) Follicle-stimulating hormone enhances somatomedin C binding to cultured rat granulosa cells. *J. Biol. Chem.* **261,** 3923–3926.
32. Gilchrest, B. A., Vrabel, M. A., Flynn, E., and Szabo, G. (1984) Selective cultivation of human melanocytes from newborn and adult epidermis. *J. Invest. Dermatol.* **83,** 370–376.
33. Medawar, P. B. (1941) Sheets of pure epidermal epithelium from human skin. *Nature* **148,** 783.
34. Niedel, J. E. and Blackshear, P. J. (1986) Protein kinase C, in *Phosphoinositides and Receptor Mechanisms* (Putney, J. W., Jr., ed.), Liss, New York, pp. 47–88.
35. Cela, A., Leong, I., and Krueger, J. (1991) Tigliane-type phorbols stimulate human melanocyte proliferation: potentially safer agents for melanocyte culture. *J. Invest. Dermatol.* **96,** 987–990.
36. Hsu, M.-Y. and Herlyn, M. (1996) Cultivation of normal human epidermal melanocytes, in *Human Cell Culture Protocols* (Jones, G., ed.), Humana, Totowa, NJ, pp. 9–20.
37. Herlyn, M. and Koprowski, H. (1988) Melanoma antigens: immunological and biological characterization and clinical significance. *Ann. Rev. Immunol.* **6,** 283–308.
38. Houghton, A. N., Eisinger, M., Albino, A. P., Cairncross, J. G., and Old, L. J. (192) Surface antigens of melanocytes and melanomas: markers of melanocyte differentiation and melanoma subsets. *J. Exp. Med.* **156,** 1755–1766.
39. Shih, I.-M., Elder, D. E., Hsu, M.-Y., and Herlyn, M. (1994) Regulation of Mel-CAM/MUC18 expression on melanocytes of different stages of tumor progression by normal keratinocytes. *J. Am. Pathol.* **145,** 837–845.
40. Shih, I.-M., Nesbit, M., Herlyn, M., and Kurman, R. J. (1998) A new Mel-CAM (CD146)-specific monoclonal antibody, MN-4, on paraffin-embedded tissue. *Mod. Pathol.* **11,** 11,098–11,106.
41. Elder, D. E., Rodeck, U., Thurin, J., et al. (1989) Antigenic profile of tumor progression stages in human melanocytes, nevi, and melanomas. *Cancer Res.* **49,** 5091–5096.
42. Hsu, M.-Y., Wheelock, M. J., Johnson, K. R., and Herlyn, M. (1996) Shifts in cadherin profiles between human normal melanocytes and melanomas. *J. Invest. Dermatol. Symp. Proc.* **1,** 188–194.
43. Valyi-Nagy, I. and Herlyn, M. (1991) Regulation of growth and phenotype of normal human melanocytes in culture, in *Melanoma 5, Series on Cancer Treatment and Research* (Nathanson, L., ed.), Kluwer Academic, Boston, MA, pp. 85–101.
44. Scott, G. A. and Haake, A. R. (1991) Keratinocytes regulate melanocyte number in human fetal and neonatal skin equivalents. *J. Invest. Dermatol.* **97,** 776–781.

45. DeLuca, M., D'Anna, F., Bondanza, S., Franzi, A. T., and Cancedda, R. (1988) Human epithelial cells induce human melanocyte growth in vitro but only skin keratinocytes regulate its proper differentiation in the absence of dermis. *J. Cell Biol.* **107,** 1919–1926.
46. Valyi-Nagy, I., Hirka, G., Jensen, P.J., Shih, I.-M., Juhasz, I., and Herlyn, M. (1993) Undifferentiated keratinocytes control growth, morphology, and antigen expression of normal melanocytes through cell-cell contact. *Lab. Invest.* **69,** 152–159.
47. Herlyn, M. and Shih, I.-M. (1994) Interactions of melanocytes and melanoma cells with the microenvironment. *Pigment Cell Res.* **7,** 81–88.
48. Hsu, M.-Y., Meier, F., Nesbit, M., et al. (2000) E-cadherin expression in melanoma cells restores keratinocyte-mediated growth control and down-regulates expression of invasion-related adhesion receptors. *Am. J. Pathol.* **156,** 1515–1525.
49. Hsu, M.-Y., Andl, T., Li, G., Meinkoth, J. L., and Herlyn, M. (2000) Cadherin repertoire determines partner-specific gap junctional communication during melanoma progression. *J. Cell Sci.* **113,** 1535–1542.
50. Hsu, M.-Y., Meier, F., and Herlyn, M. (2002) Melanoma development and progression: A conspiracy between tumor and host. *Differentiation* **70,** 522–536.
51. Kitano, Y. and Okada, N. (1983) Separation of the epidermal sheet by dispase. *Br. J. Dermatol.* **108,** 555–560.
52. Riley, P. A. (1975) Growth inhibition in normal mammalian melanocytes in vitro. *Br. J. Dermatol.* **92,** 291–304.
53. Mansur, J. D., Fukuyama, K., Gellin, G. A., and Epstein, W. L. (1978) Effects of 4-tertiary butyl catechol on tissue cultured melanocytes. *J. Invest. Dermatol.* **70,** 275–279.
54. Stanley, R. and Yuspa, S. H. (1983) Specific epidermal protein markers are modulated during calcium-induced terminal differentiation. *J. Cell Biol.* **96,** 1809–1814.
55. Price, F. M., Taylor, W. G., Camalier, R. F., and Sanford, K. K. (1983) Approaches to enhance proliferation of human epidermal keratinocytes in mass culture. *J. Natl. Cancer Inst.* **70,** 853–861.
56. Hennings, H. and Holbroot, K. A. (1983) Calcium regulation of cell-cell contact and differentiation of epidermal cells in culture. An ultrastructural study. *Exp. Cell Res.* **143,** 127–142.
57. Prunieras, M., Moreno, G., Dosso, Y., and Vinzens, G. (1976) Studies on guinea pig skin cell cultures: V. Co-cultures of pigmented melanocytes and albino keratinocytes, a model for the study of pigment transfer. *Acta Dermatovenereol.* **56,** 1–9.
58. Tsuji, T. and Karasek, M. (1983) A procedure for the isolation of primary cultures of melanocytes from newborn and adult human skin. *J. Invest. Dermatol.* **81,** 179–180.
59. Herlyn, M., Clark, W. H., Rodeck, U., Mancianti, M. L., Jambrosic, J., and Koprowski, H. (1987) Biology of tumor progression in human melanocytes. *Lab. Invest.* **56,** 461–474.

60. Swope, V. B., Medrano, E. E., Smalara, D., and Abdel-Malek, Z.A. (1995) Long-term proliferation of human melanocytes is supported by the physiologic mitogens alpha-melanotropin, endothelin-1, and basic fibroblast growth factor. *Exp. Cell Res.* **217,** 453–459.
61. Halaban, R., Langdon, R., Birchall, N., et al. (1988) Basic fibroblast growth factor from keratinocytes is a natural mitogen for melanocytes. *J. Cell Biol.* **107,** 1611–1619.
62. Tang, A., Eller, M. S., Hara, M., Yaar, M., Hirohashi, S., and Gilchrest, B. A. (1994) E-cadherin is the major mediator of human melanocyte adhesion to keratinocytes in vitro. *J. Cell Sci.* **107,** 893–992.

3

Isolation and Culture of Human Osteoblasts

Alison Gartland, Katherine A. Buckley, Jane P. Dillon, Judith M. Curran, John A. Hunt, and James A. Gallagher

1. Introduction

Bone is a complex tissue that contains at least four different cell types of the osteoblastic lineage. (1) Active osteoblast—a plump, polarized, cuboidal cell rich in organelles involved in the synthesis and secretion of matrix proteins. (2) Osteocyte—an osteoblast with low metabolic activity that has been engulfed in matrix during bone formation and entombed in lacunae. (3) Bone-lining cell—osteoblasts that have avoided entombment in lacunae and lose their prominent synthetic function; these cells cover most of the bone surfaces in mature bone. (4) Preosteoblast—a fibroblastic proliferative cell with osteogenic capacity. In addition, bone contains cells of a distinct lineage, the osteoclast (reviewed in Ch. 4).

The complex structure of bone tissue, the heterogeneity of cell types, as well as the crosslinked extracellular matrix and the mineral phase, all combine to make bone a difficult tissue from which to extract cells and to study at the cellular and molecular level *(1)*. Consequently, early attempts to culture osteoblasts relied on enzymic digestion of poorly mineralized, highly cellular fetal or neonatal tissue from experimental animals, and avoided mature, mineralized human bone. Although these studies undoubtedly furthered our knowledge of bone cell biology, they had obvious drawbacks due to the differences in cell physiology between the species, and also between adults and neonates within a species. In order to understand fully the pathological mechanisms that underlie bone diseases, including age-related bone loss, the ability to culture human bone cells is essential.

From: *Methods in Molecular Medicine, vol. 107: Human Cell Culture Protocols, Second Edition*
Edited by: J. Picot © Humana Press Inc., Totowa, NJ

Early attempts to isolate cells from adult human bone were reported by Bard et al., using demineralization and collagenase digestion *(2)*, and by Mills et al. using the alternative approach of explant culture *(3)*. The first successful attempts to isolate large number of cells that expressed an osteoblastic phenotype from human bone were undertaken in the early 1980s in Graham Russell's laboratory at the University of Sheffield. The defining characteristics of these studies were the use of explant culture, which avoided the need for digestion of the tissue, and the availability of an appropriate phenotypic maker. Successful culture of any cell type can only be achieved if there is a specific marker of the phenotype that can be used to confirm the identity of the cells in vitro. In the case of bone, the marker was the then recently discovered, bone gla protein as measured by a radioimmunoassay developed by Jim Poser *(4,5)*. Twenty years later, bone gla protein, now known as osteocalcin, undoubtedly remains the most specific marker of the osteoblastic phenotype.

Although formation of bone is the most conspicuous function of the osteoblasts, research over the past decade has revealed that cells of the osteoblastic lineage play a major role in regulating bone resorption. Researchers have increasingly used tissue culture techniques to investigate the roles of osteoblasts in bone formation and bone resorption and the culture of human bone cells is now the most predominant technique used in the investigation of bone biology.

Researchers culture osteoblasts for many reasons including:

1. To investigate the basic biochemistry and physiology of bone formation.
2. To investigate the molecular and cellular basis of human bone disease.
3. To investigate the roles of cells of the osteoblastic lineage in regulating bone resorption.
4. To screen for potential therapeutic agents.
5. To test and develop new biomaterials.
6. To use cell therapy and tissue engineering for bone transplantation.

The technique for culturing primary human osteoblasts have been reviewed widely over the past few years *(6–8)*; however, in this rapidly moving field there have been many new developments. The aim of this chapter is threefold: to describe the basic techniques of bone cell culture (both primary and clonal); to update researchers on new developments in the methodology used to characterize these cells; and to highlight the investigative value of bone cell cultures, including new techniques in which progenitor cells, particularly mesenchymal stem cells, are used to assess the effects of environmental cues on osteoblast differentiation. For those readers who would like a more detailed review of the history of the development of human bone cells, they should consult one of these previous reviews *(6–8)*.

2. Materials

2.1. Tissue-Culture Media and Supplements

1. Phosphate-buffered saline (PBS) without calcium and magnesium, pH 7.4 (Invitrogen).
2. Dulbecco's modification of minimum essential medium (DMEM) (Invitrogen) supplemented to a final concentration of 10% with fetal calf serum (FCS), 2 mM L-glutamine, 50 U/mL penicillin, 50 µg/mL streptomycin.
3. In cultures in which matrix synthesis or mineralization are investigated, 50 µg/mL freshly prepared L-ascorbic acid (*see* **Note 1**).
4. Serum-free DMEM (SF-DMEM).
5. FCS (*see* **Note 2**).
6. Tissue-culture flasks (typically 75 cm^2) or Petri dishes (typically 900- or 100-mm diameter) (*see* **Note 3**).

2.2. Preparation of Explants

1. Bone rongeurs from any surgical instrument supplier.
2. Solid stainless-steel scalpels with integral handles (BDH Merck).

2.3. Passaging and Secondary Culture

1. Trypsin-EDTA solution: 0.05% trypsin and 0.02% EDTA in Ca^{2+}- and Mg^{2+}-free Hank's BSS (HBSS), pH 7.4 (Invitrogen).
2. 0.4% Trypan blue in 0.85% NaCl (Sigma Aldrich).
3. 70 µm "Cell Strainer" (Becton Dickinson).
4. Neubauer Hemocytometer (BDH Merck).
5. Collagenase (Sigma type VII from Clostridium histolyticum).
6. DNAse I (Sigma Aldrich).

2.4. Phenotypic Characterization

1. 1,25(OH)$_2$D$_3$ (Leo Pharmaceuticals or Sigma Aldrich).
2. Menadione (vitamin K$_3$) (Sigma Aldrich).
3. Alkaline phophatase assay kit (Sigma Aldrich).
4. Staining Kit 86-R for alkaline phosphatase (Sigma Aldrich).
5. Osteocalcin radioimmunassay (IDS Ltd., Boldon, U.K.) (*see* **Note 4**).
6. Sybr-Green PCR mastermix (Bio-Rad), 96-well plates (Bio-Rad), gene-specific PCR primers for a panel of osteoblastic markers (Vh-Bio).

2.5. In Vitro Mineralization

1. L-Ascorbic acid (*see* **Note 1**).
2. Dexamethasone (Sigma Aldrich).
3. Hematoxylin (BDH Merck).
4. Alizarin Red (Sigma Aldrich).
5. DPX (BDH Merck).

3. Methods

3.1. Bone Explant Culture System

*3.1.1. Establishment of Primary Explant Cultures
of Human Osteoblasic Cells*

In this chapter, we use the term human bone-derived cells (HBDCs) to describe cells expressing osteoblastic characteristics derived from human bone. Since the description of the original methods, many investigators have turned to the use of HBDCs (extensively reviewed in *8*). Whereas some have developed techniques for the isolation and culture of HBDCs that differ significantly from the original method (*see* **Note 5**), most have continued to use the explant technique with only a few minor modifications. A scheme outlining the basic explant culture technique is shown in **Fig. 1**.

1. Transfer tissue, removed at surgery or biopsy, into a sterile container with PBS or serum-free medium (SFM) for transport to the laboratory with minimal delay, preferably on the same day (*see* **Note 6**). An excellent source is the upper femur of patients undergoing total hip replacement surgery for osteoarthritis. Cancellous bone is removed from this site prior to the insertion of the femoral prosthesis and would otherwise be discarded. The tissue obtained is remote from the hip joint itself, and thus from the site of pathology, and is free of contaminating soft tissue (*see* **Note 7**).

2. Remove extraneous soft connective tissue from the outer surfaces of the bone by scraping with a sterile scalpel blade. Rinse the tissue in sterile PBS and transfer to a sterile Petri dish containing a small volume of PBS (5–20 mL, depending on the size of the specimen). If the bone sample is a femoral head, remove cancellous bone directly from the open end using sterile bone rongeurs or a solid stainless-steel blade with integral handle. Disposable scalpel blades may shatter during this process. With some bone samples (e.g., rib), it may be necessary to gain access to the cancellous bone by breaking through the cortex with the aid of the sterile surgical bone rongeurs.

3. Transfer the cancellous bone fragments to a clean Petri dish containing 2–3 mL of PBS and dice into pieces 3–5 mm in diameter. This can be achieved in two stages using a scalpel blade first, and then fine scissors. Decant the PBS and transfer the bone chips to a sterile 30-mL "Universal container" with 15–20 mL of PBS. Vortex the tube vigorously three times for 10 s and then leave to stand for 30 s to allow the bone fragments to settle. Carefully decant off the supernatant containing hematopoietic tissue and dislodged cells, add an additional 15–20 mL of PBS, and vortex the bone fragments as before. Repeat this process a minimum of three

---

Fig. 1. *(see facing page)* Technique used to isolate cells expressing osteoblastic characteristics (HBDCs) from explanted cancellous bone.

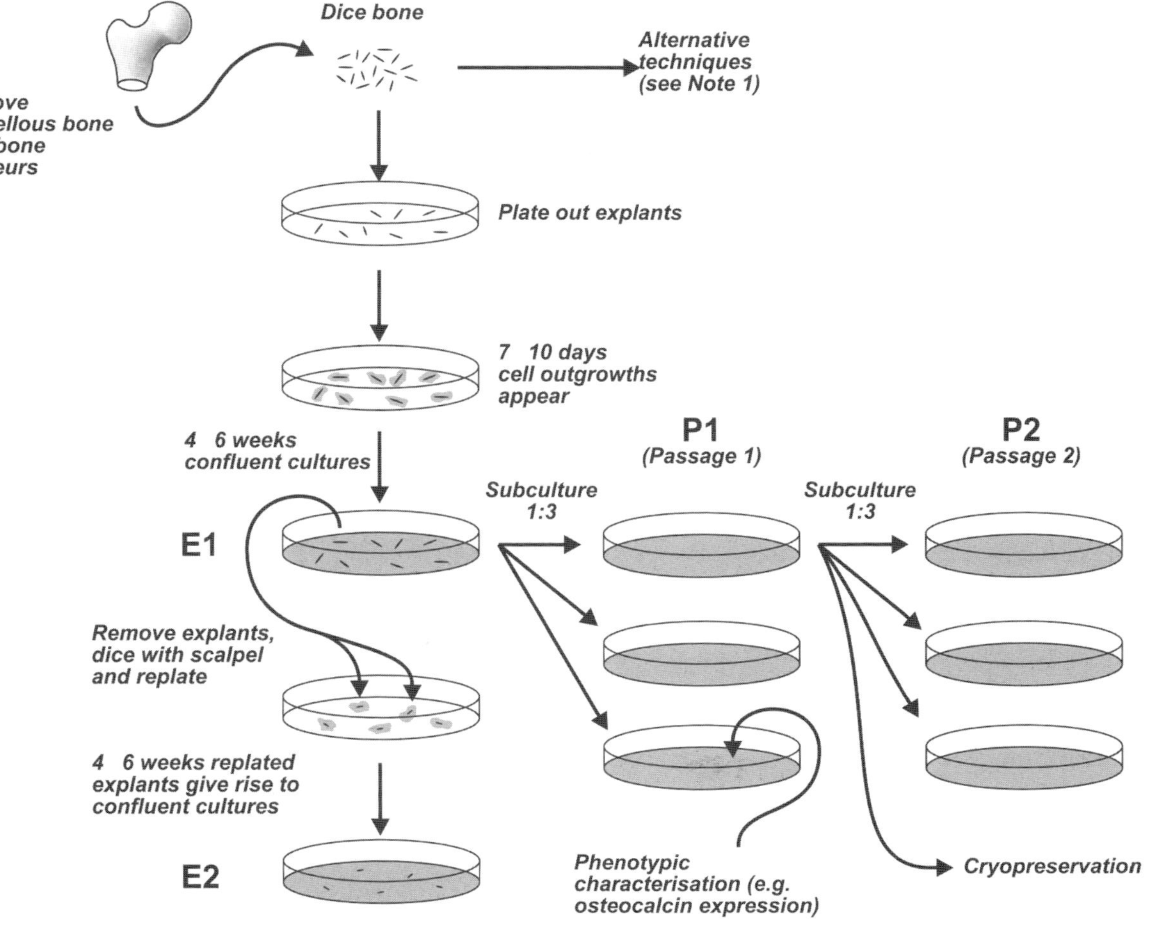

Remove cancellous bone with bone rongeurs
Dice bone
Alternative techniques (see Note 1)
Plate out explants
7–10 days cell outgrowths appear
4–6 weeks confluent cultures
E1
Subculture 1:3
P1 (Passage 1)
Subculture 1:3
P2 (Passage 2)
Remove explants, dice with scalpel and replate
4–6 weeks replated explants give rise to confluent cultures
E2
Phenotypic characterisation (e.g. osteocalcin expression)
Cryopreservation

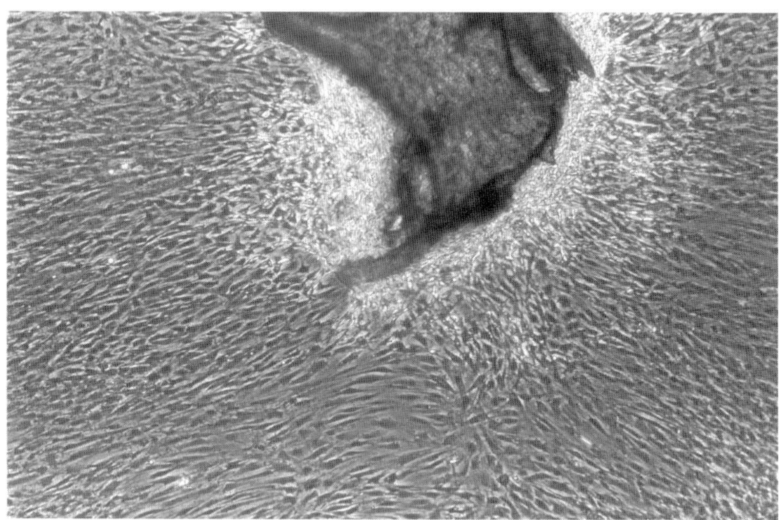

Fig. 2. Migration of cells expressing osteoblastic characteristics (HBDCs) from explanted cancellous bone.

   times or until no remaining hematopoietic marrow is visible and the bone fragments have assumed a white, ivory-like appearance.
4. Culture the washed bone fragments as explants at a density of 0.2–0.6 g
   of tissue/100-mm-diameter Petri dish or 75-cm^2 flask (*see* **Note 3**) in 10 mL of
   medium at 37°C in humidified atmosphere of 95% air, 7% CO_2.
5. Leave the cultures undisturbed for 7 d after which time replace the medium with
   an equal volume of fresh medium taking care not to dislodge the explants. Feed
   again at 14 d and twice weekly thereafter.

With the exception of small numbers of isolated cells, which probably
become detached from the bone surface during the dissection, the first evidence
of cellular proliferation is observed on the surface of the explants, and this normally occurs within 5–7 d of plating. After 7–10 d, cells can be observed
migrating from the explants onto the surface of the culture dish (*see* **Fig. 2**). If
care is taken not to dislodge the explants when feeding, and they are left undisturbed between media changes, they rapidly become anchored to the substratum
by the cellular outgrowths. Typical morphology of the cells is shown in **Fig. 3**,
but cell shape varies between donors, from fibroblastic to cobblestone-like (*see*
Note 8). Cultures generally attain confluence 4–6 wk postplating, and typically
achieve a saturation density of 29,000 ± 9000 cells/cm^2 (mean ± SD, n = 11
donors).

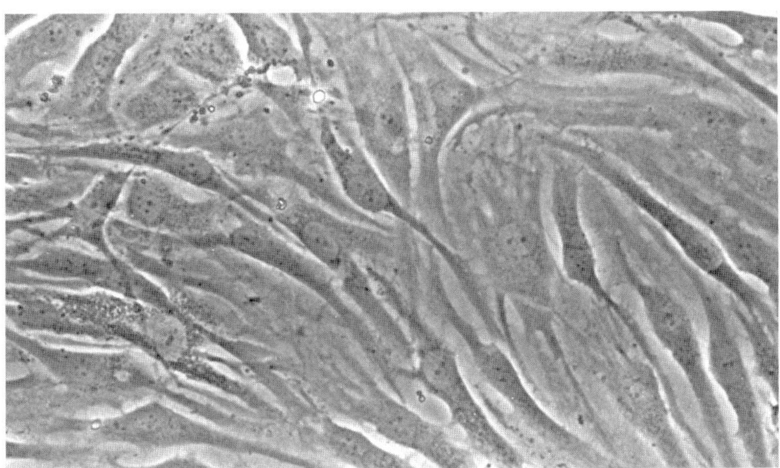

Fig. 3. Typical morphology of cells expressing osteoblastic characteristics (HBDCs) from explanted cancellous bone.

3.1.2. Passaging and Secondary Culture

1. Remove and discard the spent medium.
2. Gently wash the cell layers three times with 10 mL of PBS without Ca^{2+} and Mg^{2+}.
3. To each flask add 5 mL of freshly thawed trypsin-EDTA solution at room temperature (20°C) and incubate for 5 min at room temperature with gentle rocking of the flask every 30 s to ensure that the entire surface area of the flask and explants is exposed to the trypsin-EDTA solution.
4. Remove the flasks from the incubator and examine under the microscope. Look for the presence of rounded, highly refractile cell bodies floating in the trypsin-EDTA solution. If none, or only a few, are visible, tap the base of the flask sharply on the bench top in an effort to dislodge the cells. If this is without effect, incubate the cells for a further 5 min at 37°C.
5. When the bulk of the cells have become detached from the culture substratum, transfer the cells to a "Universal container" with 10 mL of DMEM with 10% FCS to inhibit tryptic activity. Wash the flask two to three times with 10 mL of SFM and pool the washings with the original cell isolate. To recover the cells centrifuge at 250*g* for 5 min at 15°C.
6. Remove and discard the supernatant, invert the tube, and allow it to drain briefly. Resuspend the cells in 2 mL of SFM. If the cells are clumping, *see* **Note 9**. If required, the cell suspension can be filtered through a 70-µm "Cell Strainer" (Becton Dickinson) to remove any bone spicules or remaining cell aggregates. For convenience and ease of handling, the filters have been designed to fit into the

neck of a 50-mL polypropylene tube. Wash the filter with 2–3 mL of SFM and add the filtrate to the cells.

7. Take 20 μL of the mixed cell suspension and dilute to 80 μL with SFM. Add 5 μL of trypan blue solution, mix, and leave for 1 min before counting viable (round and retractile) and nonviable (blue) cells in a Neubauer Hemocytometer. Using this procedure, typically $1–1.5 \times 10^6$ cells are harvested per 75-cm^2 flask of which ≥75% are viable.

8. Plate the harvested cells at a cell density suitable for the intended analysis. We routinely subculture at $5 \times 10^3–10^4$ cells/cm^2 and achieve plating efficiencies measured after 24 h of ≥70% (*see* **Note 10**).

If dishes have reached confluence but the cells are not required immediately, the cells can be stored by cryopreservation (*see* **Subheading 3.4.**).

3.1.3. Phenotypic Stability in Culture

As a matter of routine, we perform our studies on cells at first passage. Other investigators have studied the effects of repeated subculture on the phenotypic stability of HBDCs and found that they lose their osteoblast-like characteristics. In practical terms, this presents real difficulties, because it is often desirable to obtain large numbers of HBDCs from a single donor.

As an alternative to repeated subculture, trabecular explants can be replated at the end of primary culture into a new flask (*see* **Fig. 1**). Using this technique, it is possible to obtain additional cell populations that continue to express osteoblast-like characteristics, including the ability to mineralize their extracellular matrix, and maintain their cytokine expression profile *(7)*. Presumably, these cultures are seeded by cells that are situated close to the bone surfaces, and that retain the capacity for extensive proliferation and differentiation. The continued survival of these cells may be related to the gradual release over time in culture of the cytokines and growth factors that are known to be present in the extracellular bone matrix, many of which are known to be produced by mature cells of the osteoblast lineage *(9–12)*.

3.1.4. Modifications to Basic Culture System: Importance of Ascorbate

L-ascorbic acid (vitamin C) functions as a cofactor in the hydroxylation of lysine and proline residues in collagen and is essential for its normal synthesis and secretion. In addition, it increases procollagen mRNA gene transcription and mRNA stability *(13)*. Addition of L-ascorbic acid (50 μg/mL–25 μM) (*see* **Note 1**) to HBDCs in secondary culture (E1P1) increases proliferation, produces a sustained increase in the steady-state levels of α1 (I)-procollagen mRNA, and dramatically increases the secretion of type I collagen. There is also an increase in noncollagenous protein synthesis, most importantly bone sialoprotein and osteocalcin, and increased deposition of matrix.

3.1.5. Passaging Cells Cultured in the Continuous Presence of Ascorbate

Because of their synthesis and secretion of an extensive collagen-rich extracellular matrix, HBDCs cultured in the continuous presence of ascorbate cannot be subcultured using trypsin-EDTA alone. They can, however, be subcultured if first treated with purified collagenase. The basic procedure is as follows:

1. Rinse the cell layers twice with SFM (10 mL/75-cm^2 flask).
2. Incubate the cells for 2 h at 37°C in 10 mL of SFM containing 25 U/mL purified collagenase (Sigma type VII) and 2 mM additional calcium (1:500 dilution of a filter-sterilized stock solution of 1 M CaCl$_2$). Gently agitate the flask for 10–15 s every 30 min.
3. Terminate the collagenase digestion by discarding the medium (check that there is no evidence of cell detachment at this stage). Gently rinse the cell layer twice with 10 mL of Ca^{2+}- and Mg^{2+}-free PBS.
4. Proceed from **Subheading 3.1.2., step 3**.

Typically this procedure yields ~approx 3.5–4 × 10^6 cells/75-cm^2 flask after 28 d in primary culture. Cell viability is generally ≥90%.

3.2. Osteoblastic Cell Lines

Osteosarcomas are malignant tumors of bone derived from cells of the osteoblast lineage. An osteosarcoma is a relatively uncommon tumor, although it is the most frequently occurring primary malignant tumor of bone, predominantly occurring in children and adolescents. It is also reported to occasionally occur in elderly patients with long-standing Paget's disease of bone. Histologically, the tumor is heterogeneous, the cells being associated with immature bone matrix or osteoid. The tumor cells are generally poorly differentiated and pleomorphic with high mitotic activity. The tumors are highly vascularized and the cells have a tendency to metastasize. As osteosarcoma cells express osteoblastic genes, synthesize bone matrix proteins, and, in many cases, respond to calcium-regulating hormones, including parathyroid hormone, they are extremely useful as an osteoblast cell model. Their widespread use has meant that researchers do not always take into account that some of the characteristics of these cells are related to their oncogenic potential rather than osteoblastic characteristics. Nevertheless, many significant advances in our knowledge of bone biology have been made via experiments with these cells.

The most commonly used human clonal osteoblast cell lines are SaOs2, MG63, Te85 (also referred to as HOS), and U2 OS (HTB96). These cell lines were all derived from osteosarcomas; however, they differ in their responsiveness to certain hormones, i.e., estrogen and progesterone. This is most likely because of their stage of differentiation along the osteoblastic phenotype. Therefore, the choice of which cell line to use is solely dependent on the hypothesis being tested and preference of the investigator. An advantage of transformed human osteoblasts cell lines is that they are easily manipulated and, as such, can be utilized to make stable reporter cell lines, which are powerful research tools *(14)*. Alternatives to the "classic" osteosarcoma-derived cell lines are the cell line hFOB 1.19 *(15)* and, more recently, the osteoprecursor cell line OCP1 *(16)*, both of which were derived from human fetal bone tissue. (*See* **Note 11**.)

3.2.1. Routine Culture of the Human Osteoblast-Like Cell Line SaOS-2

SaOS-2 cells are grown in basic medium, DMEM (Invitrogen) supplemented to a final concentration of 10% with FCS, 2 m*M* L-glutamine, 50 U/mL penicillin, 50 µg/mL streptomycin. A confluent 100-mm Petri dish of SaOS-2 cells split with a ratio of 1:6 will result in the newly passaged cells reaching confluence a week after plating. If left beyond confluence, SaOS-2 cells will form mineralized nodules, a process that can be accelerated with the addition of osteogenic factors *(17)*.

3.3. Mesenchymal Stem Cells

Mesenchymal stem cells (MSC) are, by definition, pluripotent *(18–21)*, and as our knowledge of MSC increases, so does the potential sources and associated plasticity of the cells *(22)*. One of the most common sources of MSC used is bone marrow, allowing a potentially homogeneous population of MSC to be isolated, expanded, and differentiated into a predetermined phenotype. Manipulating the phenotype of MSC has become a focus of bone/osteoblast biology both in vitro and in vivo—if a source of homogeneous osteoblasts were available, the risk of rejection during prosthetic implantation would be greatly reduced on an immunological basis. A standard method for inducing osteogenic differentiation of MSC was introduced by Jaiswal et al. in 1997 *(23)* and is now widely reported throughout an array of studies *(24–32)*. This method takes into account the effects of cell concentration, culture vessel and supplemented medium. The optimum medium for inducing MSC (*see* **Note 12**) to the osteogenic phenotype in vitro, regardless of passage number and/or cryopreservation is: DMEM containing 10% FCS base medium, 100 n*M* dexamethasone, 0.05 m*M* ascorbic acid-2-phosphate, 10 m*M* β-glycerophosphate.

Culturing MSC with this supplemented medium resulted in optimal osteogenic differentiation as determined by morphology, reactivity with antiosteogenic antibodies, osteocalcin mRNA expression, and the formation of a mineralized extracellular matrix containing hydroxyapatite. This technique is now widely used, in parallel with the same source of cells cultured under the same conditions, to determine the osteogenic potential of the cells. End-point analysis, i.e., von Kossa (*see* **Fig. 5**) or alizarin red staining is commonly used to determine the true osteogenic potential of the cells.

3.4 Cryopreservation of Cells

If required, HBDCs and clonal osteoblasts can be stored frozen for extended periods in liquid nitrogen or in ultralow temperature (–135°C) cell freezer banks. For this purpose, subconfluent cells are removed from the culture vessel using trypsin-EDTA and following centrifugation, are resuspended in a solution of (v/v) 90% serum and 10% DMSO. For best results, the cells should be frozen gradually at a rate of –1°C/min in a –70°C freezer using one of the many devices available that allow the rate of cooling to be controlled precisely (e.g., Mr. Frosty, Nalgene).

The cells will be adherent to the tissue culture plastic, therefore, they must undergo trypsinization to obtain a cell pellet (**Subheading 3.1.2., steps 1–5**).

1. Resuspend cell pellet ($1–2 \times 10^6$ cells/mL) in 900 µL FCS and transfer to cryovial.
2. Mix cell pellet and FCS by swirling ampoule in ice water bath.
3. Add 100 µL DMSO gradually while holding ampule in the iced water.
4. Close ampules tightly and freeze using the following protocol: 5°C/min to 4°C, 1°C/min to –80°C.
5. Transfer cells directly to liquid nitrogen or ultralow temperature (–135°C) cell freezer bank.

3.5 Thawing Cells

Prior to use, frozen cells are rapidly thawed in a 37°C water bath and then diluted into ≥20 volumes of preheated medium containing 10% FCS and the usual supplements.

1. In the laminar flow cabinet, you should assemble the following items: DMEM (Invitrogen) containing 10% FCS (previously tested for ability to support HBDC growth) and supplements. This should be warmed to room temperature. "Universal container," small beaker, wash bottle containing 70% ethanol, tissue culture Petri dish (100-mm from Sarstedt).
2. Dispense about 10 mL of the medium into the Petri dish and place in the incubator to equilibrate the pH and temperature with culture conditions.
3. Put about 15 mL of the medium into the Universal container—this will be used to wash the cells free of the DMSO.
4. Fill the beaker two-thirds full with warm tap water.
5. Remove the cryovial containing the cells from the liquid nitrogen.
6. Rapidly stir the cryovial in the warm water until the contents have thawed.
7. Swab the outside of the cryovial with 70% ethanol.
8. Open the cryovial and transfer the cell suspension, drop wise, into the Universal container.

9. Mix the closed Universal container by swirling and centrifuge at 250*g* for 2 min to pellet the cells.
10. Remove supernatant and resuspend the cells in 2 mL of medium.
11. Add this suspension to the warmed medium in the incubator and culture as usual.
12. After 24 h, replace the medium with the normal volume of fresh medium. The efficiency of plating obtained after 24 h using this method is typically ≥70%

3.6. Phenotypic Characteristics of Osteoblasts

The phenotypic characterization of HBDCs is described in detail in **ref. 6**. The simplest phenotypic marker to investigate is the enzyme alkaline phosphatase, a widely accepted marker of early osteogenic differentiation. Alkaline phosphatase can be measured by simple enzyme assay or by histochemical staining. Basal activity is initially low, but increases with increasing cell density. Treatment with 1,25(OH)$_2$D$_3$ increases alkaline phosphatase activity.

The most specific phenotypic marker is osteocalcin. This is a protein of M_r 5800 containing residues of the vitamin K–dependent amino acid γ-carboxyglutamic acid. In humans, its synthesis is restricted to mature cells of the osteoblast lineage. It is an excellent late stage marker for cells of this series despite the fact that its precise function in bone has yet to be established. Osteocalcin can be measured by one of the many commercially available kits (*see* **Note 13**). 1,25(OH)$_2$D$_3$ increases the production of osteocalcin in cultures of HBDCs, but not fibroblasts obtained from the same donors.

Reverse transcription–polymerase chain reaction (RT-PCR) can also be used to examine the expression of osteoblastic markers in HBDCs. Whereas qualitative results can be obtained using traditional PCR in conjunction with gel analysis, the availability of real-time fluorescent-based RT-PCR has provided researchers with a powerful tool to assess gene expression quantitatively. There are currently four different principles in common use for real-time PCR, the simplest and cheapest being based on the intercalation of double-stranded DNA-binding dyes. There is no need for any additional fluorescence-labeled oligonucleotide with this principle and as such it can be easily applied to established PCR assays. However, as both specific and non-specific PCR products are detected in these assays, careful optimization of the PCR conditions and a clear differentiation between specific and nonspecific PCR products using melting-curve analysis is crucial. Extensive instructions on how to optimize PCR conditions are normally given by the PCR machine/primer manufacturers, as are specific real-time PCR primer design packages.

3.6.1. Real-Time RT-PCR for Osteoblast Markers

1. Grow HBDC/clonal cell lines as described above. Isolate RNA using Tri Reagent/ Trizol following the maunfacturers' standard protocols (Sigma, Invitrogen).

Particular attention should be paid to avoiding contamination and/or degradation of the RNA.

2. DNase treat RNA, and synthesize cDNA according to standard protocols.

3. Set up 15 μL PCR mixtures (in triplicate) in a 96-well thin-wall plate for serial dilutions (neat, 1:10, 1:100, 1:1000) of the cDNA as follows:

    2 μL cDNA (various concentrations) or RNase/DNase-free water (negative control)

    0.5 μL 100 μmole sense primer

    0.5 μL 10 μmole antisense primer

    7.5 μL Sybr Green Mastermix

    4.5 μL RNase/DNase-free water

4. Perform standard 40-cycle PCR, including 3 min at 95°C followed by 40 cycles 30 s at 95°C, and 30 s at T_m. Fluorescent data are collected during the T_m step (at this point, the intercalating SYBR green will bind to the target sequence and fluorescence will increase). When the PCR enters its exponential phase, the associated fluorescence will rise significantly above the background, this point is known as the threshold, or C_T, cycle (*see* **ref. *33*** for further information). Real-time levels of fluorescence are collected using the iCycler detection system, and the recorded fluorescence is then plotted against cycle number (*see* **Fig. 4A**).

5. Calculation of PCR efficiency. Providing the serial dilutions have been defined within the well series at the plate set-up stage, a standard curve of thermal cycle (C_T cycle) against log starting quantity can be generated (*see* **Fig. 4B**). An efficient reaction with no exogenous contamination will have a slope of the curve of approx –3.2, and a recorded efficiency in the range 95–105%.

6. Melt curve analysis. Once the PCR is complete, the products are heated from 55°C to 95°C at a rate of 1°C/10 s to denature the double-stranded DNA. As the DNA denatures, the recorded fluorescence decreases; plotting temperature against this decrease, $-dF/dT$, generates a melt curve (*see* **Fig. 5A**). This step is critical for confirming the specificity of the PCR product. When designing primers, it is essential that the reaction between the primer and target sequence is thermodynamically stable, and any other byproducts caused by possible primer–dimer formation, nonspecific binding, or secondary structures will be thermodynamically unstable. The presence of any nonspecific products and/or contamination will be represented by the presence of a melt curve at a much lower temperature than that observed for a specific product (*see* **Fig. 5B**).

7. Data are normally expressed as a function of the cycle time C_T of the reaction. The lower the C_T value, the more abundant the message is in the sample (as shown with the serial dilutions and the PCR efficiency plot, **Fig. 4B**). In order to compare expression levels among different samples, or for different genes from one sample, the C_T values must be normalized to a "housekeeping" gene (*see* **ref. *33*** for detailed explanation).

Figure 6 shows real-time RT-PCR data for quiescent SaOS-2 cells using PCR primers for a panel of osteoblastic markers (*see* **Table 1**).

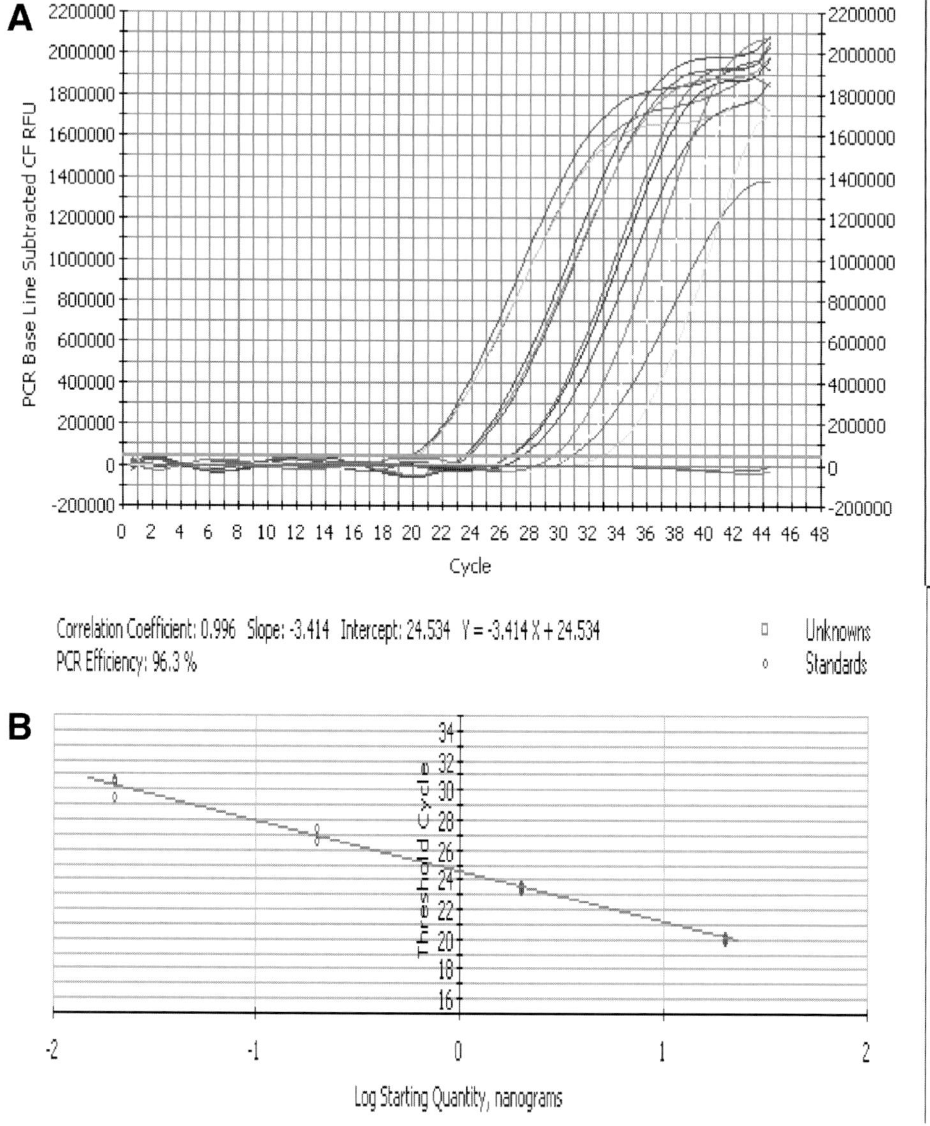

Fig. 4. (**A**) Typical real-time PCR results. Standard reactions were set up using serial dilutions of HBDC cDNA (neat, 1:10, 1:100, 1:1000, 1:10,000) and the same concentration of Ornithine Decarboxylase primers; negative controls (water blanks) reactions were also performed. The C_T cycle was defined automatically using the iCycler software. This graph highlights the differences in serial dilutions and also can be used as an early marker to assess the proximity of replicates within a given reaction. (**B**) Calculation of PCR efficiency based on relative starting quantities and associated C_T values (obtained in **A**). The efficiency of the PCR should be close to 100% when using serial dilutions of the same batch of cDNA, providing no exogenous contaminating factors are present. Efficiencies within the range 95–105% are acceptable.

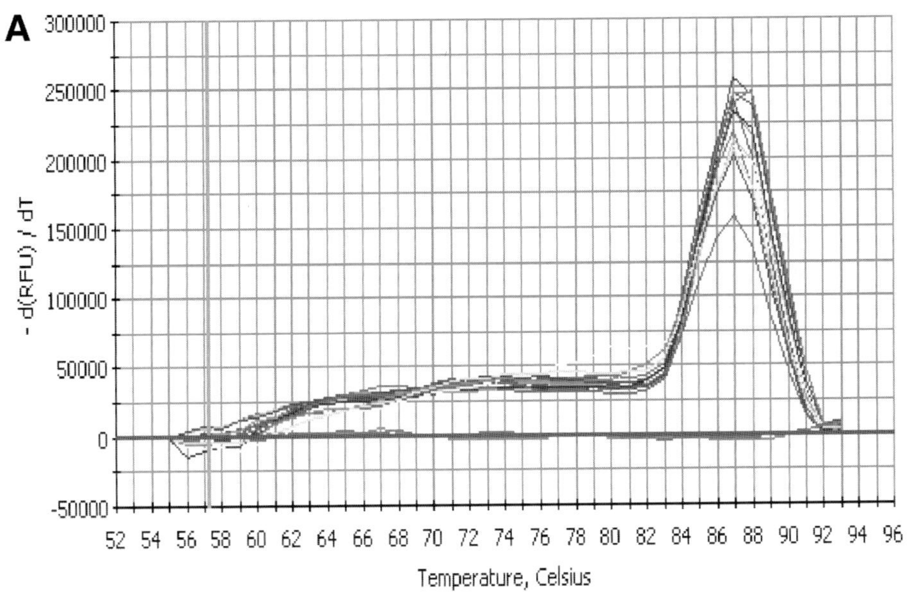

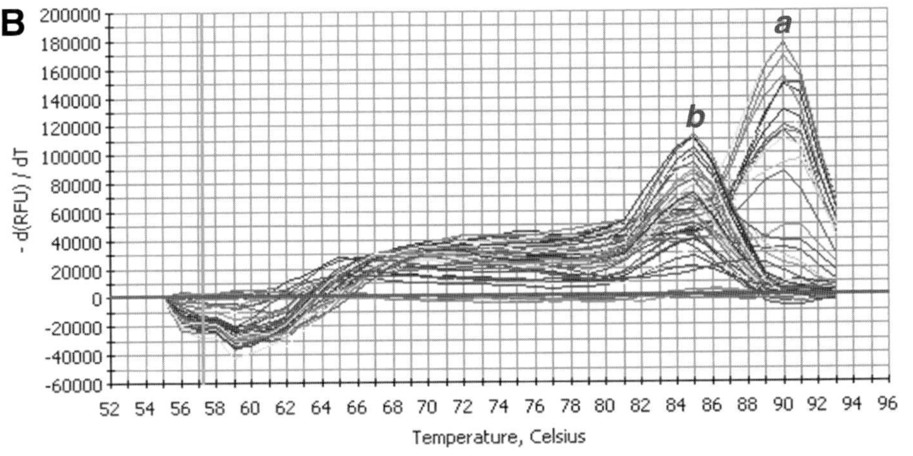

Fig. 5. Melt curve analysis of data obtained using Ornithine decarboxylase primers. In **(A)**, only one peak over a specific high-temperature range is generated, indicating amplification of one specific product in the PCR reaction. **(B)** is an example of melt curve data obtained when more than one product was amplified in the PCR. The specific, thermodynamically stable product is represented by the peak and subsequent sharp decrease in fluorescence at the higher temperature range **(a)**, whereas the formation of primer-dimers **(b)** is represented by the peak and subsequent sharp decrease in fluorescence at a lower temperature range.

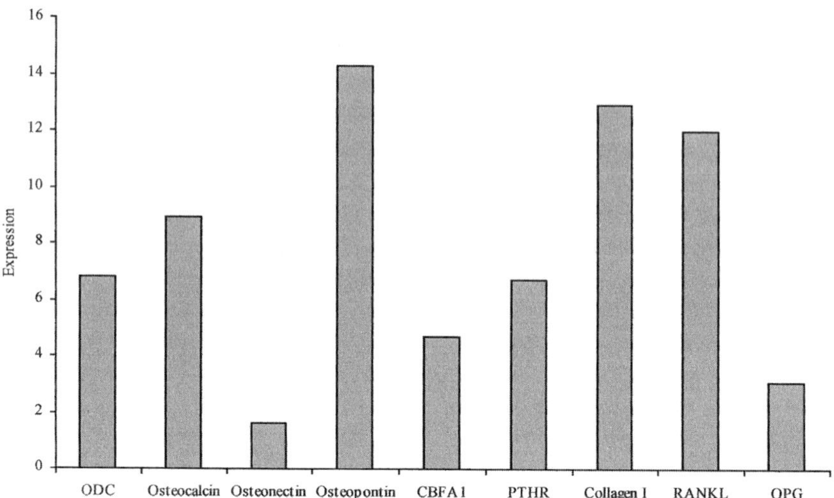

Fig. 6. Expression levels for a panel of osteoblast genes obtained by real-time RT-PCR from cDNA from confluent, quiescent SaOs-2 cells. All values are normalized to the internal house keeping gene β-actin.

3.7. Manipulation of Cultures

3.7.1. In Vitro Mineralization

The major function of the mature osteoblast is to form bone. Despite the overwhelming evidence that cultures of HBDC contain cells of the osteoblast lineage, initial attempts to demonstrate the presence of osteogenic (i.e., bone-forming) cells proved unsuccessful. Subsequently, several authors reported that culture of HBDCs in the presence of ascorbate and millimolar concentrations of the organic phosphate ester β-glycerophosphate (β-GP) led to the formation of mineralized structures resembling the nodules that form in cultures of fetal or embryonic animal bone-derived cells (reviewed in **ref. *34***). These have been extensively characterized and shown by a variety of morphological, biochemical, and immunochemical criteria to resemble embryonic/woven bone formed in vivo. An alternative to the use of β-GP is to provide levels of inorganic phosphate sufficient for supporting the process of cell-mediated mineralization in vitro, and the preferred method when studying HBDCs, is supplementation of the culture medium with 5 m*M* inorganic phosphate.

The protocol for inducing matrix mineralization in cultures of HBDCs is as follows:

1. Fragments of human trabecular bone are prepared as described and cultured in medium additionally supplemented with 100 μ*M* L-ascorbic acid 2-phosphate and either 200 n*M* hydrocortisone or 10 n*M* dexamethasone, which approximates to a

Table 1
Osteoblast-Specific PCR Primers

Primer	Acc. #	Sequence	T_a Opt
β-actin	NM 001101	GGACCTGACTGACTACCTC GCCATCTCTTGCTCGAAG	53.9
Ornithine decarboylase (ODC)	M33764	GCGGATTGCCACTGATGATTC CTCACCTGACACCAACAACATC	53.3
Alkaline phosphatase	J04948	CCTACACGGTCCTCCTATAC TGACTGCTGCCGATACTC	54.9
Collagen I	NM 000088	GCCACTCCAGGTCCTCAG CCACAGCACCAGCAACAC	54.5
Osteopontin	NM 000582	GCGAGGAGTTGAATGGTG CTTGTGGCTGTGGGTTTC	53.9
Osteocalcin	NM 000711	AGCGAGGTAGTGAAGAGAC GAAAGCCGATGTGGTCAG	55.2
Osteonectin	BC 008011	GCTGGATGATGAGAACAACAC AAGAAGTGGCAGGAAGAG	53.4
OPG	NM 002546	GCAGCGGCACATTGGACATG AGGATCTGGTCACTGGGTTTGC	53.7
PTHR1	NM 00316	GCCAACTACAGCGAGTGTG GCCAGGGACACGGAGTAG	53.1
RANKL	ABO37599	CATATCGTTGGATCACAGCAC GGGAACCAGATGGGATGTC	52.2
Core binding Factor α1 (CBFA1)	AH 005498	GGCAGTTCCCAAGCATTTC GCAGGTAGGTGTGGTGTG	54.5

physiological dose of glucocorticoid. For studies of in vitro mineralization, it is prefcrable to obtain trabecular bone from sites containing hematopoietic marrow. In practice, this is usually from the upper femur or iliac crest.

2. When the cells have attained confluence and synthesized, a dense extracellular matrix, typically after 28–35 d, subculture the cells using the sequential collagenase/trypsin-EDTA protocol and plate the cells in 25-cm² flasks at a density of 10^4 viable cells/cm². Change the medium twice weekly.

3. After 14 d, supplement the medium with 5 mM inorganic phosphate. This is achieved by adding 0.01% (v/v) of a 500 mM phosphate solution, pH 7.4, at 37°C prepared by mixing 500-mM solutions of Na_2HPO_4 and NaH_2PO_4 in a 4:1 (v/v) ratio. The filter-sterilized stock solution can be stored at 4°C.

4. After 48–72 h, the cell layers are washed two to three times with 10 mL of SFM prior to fixation with 95% ethanol at 4°C. This can be done *in situ*, for viewing *en face*, or if sections are to be cut following detachment of the cell layer from

the surface of the flask using a cell scraper. Great care is needed if the cell layer is to be harvested intact, particularly when mineralized.

5. For the demonstration of alkaline phosphatase activity (red), Sigma Staining Kit 86-R is used. Per flask, 2.5 mL of staining solution (or sufficient to cover the section) is used. Place specimens in a humidified chamber and incubate for 1 h at 20°C in the dark. Wash under running tap water and counterstain the nuclei for 15 s with hematoxylin (blue). Mineral deposits are then stained using a modification of von Kossa's (black/brown) technique. Prior to examination, mount sections in DPX or cover cell layers in flasks with glycerol.

HBDCs cultured in the continuous presence of glucocorticoid and the long-acting ascorbate analog produce a dense extracellular matrix that mineralizes extensively following the addition of inorganic phosphate. This is the case for the original cell population (E1P1) and that obtained following replating of the trabecular explants (E2P1), which further attests to the phenotypic stability of the cultured cells. Cells cultured in the continuous presence of ascorbate and treated with glucocorticoid at first passage show only a localized and patchy pattern of mineralization, despite possessing similar amounts of extracellular matrix and alkaline phosphatase activity. Cells cultured without ascorbate, irrespective of the presence or absence of glucocorticoid, secrete little extracellular matrix, and do not mineralize.

The ability of the cells to mineralize their extracellular matrix is dependent on ascorbate being present continuously in primary culture. The addition of ascorbate in secondary culture, even for extended periods, cannot compensate for its omission in primary culture. This finding provides further evidence to support the hypothesis that maintenance of adequate levels of ascorbate during the early stages of explant culture is of critical importance for the survival of cells that retain the ability to proliferate extensively and give rise to precursors capable of undergoing osteogenic differentiation. The presence of mineralized matrix is indicative of the true osteogenic nature of the cultured cell population. This is supported by the fact that HBDCs cultured continuously in the presence of ascorbate and glucocorticoid retain the ability to form bone when implanted in vivo within diffusion chambers in athymic mice *(35)*.

3.7.1.1. Quantification of Mineralization

The effect of cytokines and/or novel drugs on the extent of mineralization observed in bone cultures is an excellent indication of their pro- or antiosteogenic nature *(17,36)*. A relatively simple and fast method for quantifying mineralization is as follows. Culture cells as above, with these modifications: Step 2: subculture at a density of 10^4 viable cells/cm^2 in 12- or 24-well plates. Step 4: fix overnight at 4°C in 100% ethanol. Rinse in PBS, and incubate in 2% alizarin red (pH 4.2) (stains mineralized matrix) for 1 h at room temperature with gentle agitation. Wash extensively in 95% ethanol and air-dry. Quantification of alizarin red staining per well can be achieved by scanning the plates with a high-resolution flat bed scanner equipped with a transparency adaptor, "cut out" individual wells using Adobe Photoshop, and then quantify the percentage of alizarin red staining for each well using an image analysis package such as Image J (NIH).

3.7.2. Osteoblasts as Inducers of Osteoclast Formation

Monocytic osteoclast precursor cells differentiate into osteoclasts at bone-resorbing sites. The essential signal for osteoclastogenesis to occur is mediated via receptor activator for nuclear factor κ β ligand (RANKL) *(37)*. This ligand is presented on the osteoblast cell membrane and interacts with the RANK receptor on monocytic osteoclast precursor cells, inducing fusion of these cells to form multinucleated osteoclasts. Osteoprotegerin (OPG) is a soluble decoy receptor to RANKL and is also produced by osteoblasts. Binding of OPG to RANKL prevents osteoclastogenesis from occurring *(38)*. RANKL and OPG expression by osteoblasts may be altered to modulate osteoclast formation, and many regulatory factors control the ratio of expression of these two proteins. Although we found that SaOS-2 cells expressed RANKL mRNA, they did not support osteoclast formation in our experience. One study reported that in response to osteotropic factors such as 1,25-dihydroxy vitamin D_3 and dexamethasone, certain osteoblast cell lines demonstrated increased levels of RANKL expression and decreased OPG expression, whereas other cell lines demonstrated both increased RANKL and OPG expression levels *(39)*. It is this second group of osteoblast-like cells that are unable to support osteoclast formation, owing to an enhancement in the production of the decoy receptor OPG, and it appears that SaOS-2 cells belong to this group. A more recent report by Atkins et al. demonstrates that RANKL expression is related to the differentiation state of human osteoblasts, with expression being highest in immature osteoblast cells *(40)*. We, and others, have found that UMR-106 cells support the generation of functional, resorbing osteoclasts in cocultures with isolated blood monocytes. For a detailed protocol of the osteoclast coculture, please refer to Chapter 4.

3.7.3. Control of Gene Transcription

Alkaline phosphatase activity and in vitro mineralization/nodule formation have been used as indicators of the osteogenic effect of a compound; however, this only tells you the long-term end-point effect. Often a compound will exert its effect as soon as it is exposed to the cell by inducing an intracellular signaling cascade, ultimately resulting in altered gene expression. In the past, Northern blot analysis was the best means to quantitatively measure changes in gene transcription. This is a somewhat tedious process that is reliant on a large amount of

RNA (>10 μg) and a relatively high abundance message, especially if using non-radioactive methods. The advent of real-time RT-PCR has provided the researcher with a means to quickly measure changes in gene expression levels, even of low abundance messages from relatively small amounts of RNA (extracted from as few as 10^4 cells). A basic protocol for SaOS-2 cells is given below:

1. Grow SaOS-2 osteoblast cells to confluence.
2. Remove and discard medium containing 10% FCS.
3. Wash the cell layer three times with 10 mL PBS.
4. Replace with SFM for 24 h.
5. Induce cells with required agonists for required time period to investigate changes in gene expression.
6. Lyse cells, isolate RNA, DNase-treat RNA, and synthesize cDNA according to standard protocols.
7. Perform PCRs containing primers for osteoblast-specific genes (*see* **Table 1**) on a quantitative PCR machine, such as the Bio-Rad icycler thermal cycler.
8. Individual C_T values for the gene of interest are normalized against C_T values for an internal control (such as β-actin) for PCRs performed on the same cDNA sample. Relative expression levels/fold change is then calculated by subtracting the basal level from the stimulated level, i.e.,

$$\text{Fold change} = 2^{-Y}$$

where $Y = X$ stimulated $- X$ control and $X = C_T$ target $- C_T$ β-actin.

3.7.4. MSC and Tissue Engineering

The effect of the surrounding environment, both physical and chemical, on the osteogenic capacity of MSC is of utmost importance in the field of tissue engineering. Investigators are currently concentrating on methods that can be used to ensure that 3D biodegradable scaffolds, loaded with both undifferentiated and predifferentiated MSC, are optimized to enhance the next generation of tissue-engineered bone constructs. These methods include seeding MSC onto scaffolds using agitation *(41)*, and for nonwoven fabrics this method has proved to be effective in increasing homogeneous cell attachment throughout the scaffold without having a detrimental effect on the cells or inhibiting predefined differentiation pathways. A further development of this system involves culturing predifferentiated MSC to osteoblasts in a biodegradable polymer composite of polylactide and polyglycolide (PLGA) scaffold in a rotating oxygen-permeable bioreactor system *(42)*. This study demonstrated that by using various aspects of tissue engineering, i.e., formation of shape-specific scaffolds and increased nutrient and oxygen supply systems, bone formation was obtained throughout the scaffold in vitro, resulting in autologous tissue-engineered scaffolds. Another experimental variable that can be manipulated to enhance the efficiency of potential autologous bone replacements is pressure/ load. A low-pressure force system used during in vitro preparation of a cell-loaded construct (porous HA blocks loaded with undifferentiated MSC) greatly increased in vivo bone formation compared with unloaded controls *(43)*.

MSC are also being used in conjunction with malleable carriers, i.e., fibrin glue, polymers, and hydrogels loaded with osteoconductive particles, i.e., tricalcium phosphate, in order to produce tissue-engineered devices that will fit exactly into any defect *(44,45)*.

Further control, or enhancement, of the osteogenic differentiation of undifferentiated MSC cultured in contact with scaffolds can be achieved by modifying the biomaterial's surfaces with various peptides to drive the osteogenic differentiation pathway of preloaded MSC, and also to enhance the efficiency of attachment of any host preosteogenic stem cells, as demonstrated using an OPF-based (an osteopontin-derived peptide) biomimetic hydrogels *(46)*. Ultimately, this will to help reduce the risk of an inherent defect in the implanted scaffold caused by a population of MSC not undergoing osteogenic differentiation. Tissue-engineered bone constructs are also incorporating genetic engineering with a view to enhancing the efficiency of differentiation of MSC to functional osteoblasts *(47–49)*. Partridge et al. have reported that MSC transfected with adenoviral BMP-2 and loaded onto PLGA polymer scaffolds have an extended maintenance of the osteoblast phenotype in vitro, and production of bone tissue in vivo *(47)*.

Clearly, MSC are providing an exciting avenue of research in the fields of both bone cell biology and tissue engineering.

4. Notes

1. Beresford et al. introduced the more-stable analog L-ascorbic acid 2-phosphate (Wako Pure Chemical Industries, Ltd.) which does not have to be added daily (*see* **ref. 7** for details).
2. Batches of serum vary in their ability to support the growth of HBDCs. It is advisable to screen batches and reserve a large quantity of serum once a suitable batch has been identified. HBDCs will grow in autologous and heterologous human serum, but as yet no comprehensive studies have been performed to identify the effects on growth and differentiation.
3. The authors have obtained consistent results with plasticware from Sarstedt and Becton Dickinson. Smaller flasks or dishes can be used if the amount of bone available is <0.2 g.
4. Several other assays for osteocalcin are commercially available.
5. Gehron-Robey and Termine used prior digestion of minced bone with *Clostridial* collagenase and subsequent culture of explants in medium with reduced calcium concentrations *(50)*. In contrast, Wergedal and Baylink have used collagenase digestion to directly liberate cells *(51)*. Marie et al. have used a method in which

explants are first cultured on a nylon mesh *(52)*. These alternative methods are described in greater detail in **ref. *6***.

6. Bone can be stored for periods of at least 24 h, possibly longer, at 4°C in PBS or SFM prior to culture without any deleterious effect on the ability of the tissue to give rise to populations of osteoblastic cells.

7. Bone cells have been successfully cultured from many anatomical sites including tibia, femur, rib, vertebra, patella, and digits.

8. The available evidence indicates that cultures of HBDCs contain cells of the osteogenic lineage at all stages of differentiation and maturation. This conclusion is consistent with the expression of both early (alkaline phosphatase) and late (osteocalcin, bone sialoprotein) stage markers of osteoblast differentiation. In addition, in ascorbate-treated cultures there is a small subpopulation ($\leq5\%$) of cells that express the epitope recognized by the monoclonal antibody (MAb) STRO-1 *(53)*, which is a cell-surface marker for clonogenic, multipotential marrow stromal precursors capable of giving rise to cells of the osteogenic lineage in vitro. The presence of other cell types, including endothelial cells and those derived from the hematopoietic stem cell, has been investigated using a large panel of MAbs and flow cytometry and/or immunocytochemistry. The results of these studies reveal that at first passage there are no detectable endothelial, lymphoid, or erythroid cells present. A consistent finding, however, is the presence of small numbers of cells ($\leq5\%$) expressing antigens present on cells of the monocyte/macrophage series.

9. If the cells are clumping, resuspend in 2 mL of SFM containing 1 µg/mL DNase I for each dish or flask treated with trypsin-EDTA, and using a narrow bore 2-mL pipet, repeatedly aspirate and expel the medium to generate a cell suspension.

10. In our experience, the minimum plating density for successful subculture is 3500 cells/cm^2. Below this, the cells exhibit extended doubling times and often fail to grow to confluence.

11. Clonal human osteoblasts can be obtained from either the American Type Culture Collection (ATCC) or the European Collection of Cell Culture (ECCAC) (http://www.lgcpromochem.com/atcc/ or http://www.ecacc.org/).

12. Human mesenchymal stem cells can be purchased, cryopreserved, from Cambrex (www.cambrex.com).

13. In the absence of added calcitriol, particularly in SFM or FCS that has been depleted of endogenous calcitriol by charcoal treatment, the amount of ostecalcin produced by HBDCs is below the limits of detection in most assays *(4,5)*. The same applies to the detection of steady-state levels of osteocalcin mRNA. An exception to this general rule is when HBDCs are cultured for extended periods in the presence of L-ascorbate or its stable analog, L-ascorbate-2-phosphate.

References

1. Gallagher, J. A., Beresford, J. N., Caswell, A., and Russell, R. G. G (1987) Subcellular investigations of skeletal tissue, in *Subcellular Pathology of Systemic Disease* (Peters, T. J., ed.), Chapman and Hall, London, pp. 377–397.

2. Bard, D. R., Dickens, M. J., Smith, A. U., and Zarek, J. M. (1972) Isolation of living cells from mature mammalian bone. *Nature* **236,** 314–315.
3. Mills, B. G., Singer, F. R., Weiner, L. P., and Hoist, P. A. (1979) Long term culture of cells from bone affected with Paget's disease. *Calcif. Tiss. Int.* **29,** 79–87.
4. Gallagher, J. A., Beresford, J. N., McGuire, M. K. B., et al. (1984) Effects of glucocorticoids and anabolic steroids on cells derived from human skeletal and articular tissues in vitro. *Adv. Exp. Med. Biol.* **171,** 279–292.
5. Beresford, J. N., Gallagher, J. A., Poser, J. W., and Russell, R. G. G. (1984a) Production of osteocalcin by human bone cells in vitro. Effects of 1,25(OH)$_2$D$_3$, parathyroid hormone and glucocorticoids. *Metab. Bone Dis. Rel. Res.* **5,** 229–234.
6. Gallagher, J. A., Gundle, R., and Beresford, J. N. (1996) Isolation and culture of bone forming cells (osteoblasts) from human bone, in *Human Cell Culture Protocols* (Jones, G. E., ed.), Humana, Totowa, NJ.
7. Gundle, R., Stewart, K., Screen, J., and Beresford, J. N. (1998) Isolation and culture of human bone derived cells, in *Marrow Stromal Cell Culture* (Beresford, J. and Owen, M., eds.) Cambridge University Press, Cambridge.
8. Gallagher, J. A. (2003) Human osteoblast culture, in *Bone Research Protocols* (Helfrich, M. H. and Ralston, S. H., eds.) Humana, Totowa, NJ.
9. Birch, M. A., Ginty, A. F., Walsh, C. A., Fraser, W. D., Gallagher, J. A., and Bilbe, G. (1993) PCR detection of cytokines in normal human and pagetic osteoblast-like cells. *J. Bone. Miner. Res.* **10,** 1155–1162.
10. Chaudhary, L. R., Spelsberg, T. C., and Riggs, B. L. (1992) Production of various cytokines by normal human osteoblast-like cells in response to interleukin-1 beta and tumor necrosis factor-alpha: lack of regulation by 17 beta-estradiol. *Endocrinology* **130,** 2528–2534.
11. Mohan, S. and Baylink, D. J. (1991) Bone growth factors. *Clin. Orthop.* **263,** 30–48.
12. Canalis, E., Pash, J., and Verghese, S. (1993) Skeletal growth factors. *Crit. Rev. Eukaryotic Gene Express.* **3,** 155–166.
13. Prockop, D. J. and Kivivrikko, K. I. (1984) Heritable diseases of collagen. *New Engl. J. Med.* **311,** 376–386.
14. Bowler, W. B., Dixon, C. J., Halleux, C., et al. (1999) Signaling in human osteoblasts by extracellular nucleotides their weak induction of the c-fos proto-oncogene via Ca21 mobilization is strongly potentiated by a parathyroid hormone/cAMP-dependent protein kinase pathway independently of mitogen-activated protein kinase. *J. Biol. Chem.* **274,** 14,315–14,324.
15. Harris, S. A., Tau, K. R., Enger, R. J., Toft, D. O., Riggs, B. L., and Spelsberg, T. C. (1995) Estrogen response in the hFOB 1.19 human fetal osteoblastic cell line stably transfected with the human estrogen receptor gene. *J. Cell Biochem.* **59,** 193–201.
16. Winn, S. R., Randolph, G., Uludag, H., Wong, S. C., Hair, G. A., and Hollinger, J. O. (1999) Establishing an immortalized human osteoprecursor cell line: OPC1. *J. Bone Miner. Res.* **14,** 1721–1733.
17. Rao, L. G., Liu, L. J., Murray, T. M., McDermott, E., and Zhang, X. (2003) Estrogen added intermittently, but not continuously, stimulates differentiation and bone formation in SaOS-2 cells. *Biol. Pharm. Bull.* **26,** 936–945.

18. Minguell, J. J., Erices, A., and Conget, P. (2001) Minireview: mesenchymal stem cells. *Exp. Biol. Med.* **226,** 507–520.

19. Krause, D. S. (2002) Plasticity of marrow derived stem cells. *Gene Therapy* **9,** 754–758.

20. Jiang, Y., Jahagirdar, B. N., Reinhardt, R. L., et al. (2002) Pluripotency of mesenchymal stem cells derived from adult marrow. *Nature* **418,** 41–49.

21. Pittenger, M. F., Mackay, A. M., Beck, S. C., et al. (1999) Multilineage potential of adult human mesenchymal stem cells. *Science* **284,** 143–147.

22. Rosenthal, N. (2003) Prometheus's vulture and the stem cell promise. *N. Engl. J. Med.* **349,** 267–274.

23. Jaiswal, N., Haynesworth, S. E., Caplan, A. I., and Bruder, S. P. (1997) Osteogenic differentiation of purified, culture-expanded human mesenchymal stem cells in vitro. *J. Cell Biochem.* **64,** 295–312.

24. Jones, E. A., Kinsey, S. E., English, A., et al. (2002) Isolation and characterization of bone marrow multipotential mesenchymal progenitor cells. *Arthitis. Rheum.* **46,** 3349–3360.

25. Fibbe, W. E. (2002) Mesenchymal stem cells: A potential source for skeletal repair. *Ann. Rheum. Dis.* **61(Suppl. II),** 29–31.

26. Jaiswal, N., Haynesworth, S. E., Caplan, A. I., and Bruder, S. P. (1997) Osteogenic differentiation of purified, culture-expanded human mesenchymal stem cells in vitro. *J. Cell Biochem.* **64,** 295–312.

27. Chen, D., Harris, M. A., Feng, J. Q., et al. (1998) Differential roles for bone morphogenetic protein (BMP) receptor type IB and IA in differentiation and specification of mesenchymal precursor cells to osteoblast and adipocyte lineages. *J. Cell Biol.* **142,** 295–305.

28. Gronthos, S., Simmons, P. J., Graves, S. E., and Robey, P. G. (2001) Integrin-mediated interactions between human bone marrow stromal precursor cells and the extracellular matrix. *Bone* **28,** 174–181.

29. Bruder, S. P., Jaiswal, N., Ricalton, N. S., Mosca, J. D., Kraus, K. H., and Kadiyala, S. (1998) Mesenchymal stem cells in osteobiology and applied bone regeneration. *Clin. Orthop.* **355,** S247–S256.

30. Bruder, S. P., Jaiswal, N., and Haynesworth, S. E. (1997) Growth kinetics, self-renewal, and the osteogenic potential of purified human mesenchymal stem cells during extensive subcultivation and following cryopreservation. *J. Cell Biochem.* **64,** 278–294.

31. Lennon, D. P., Haynesworth, S. E., Young, R. G., Dennis, J. E., and Caplan, A. I. (1995) A chemically defined medium supports in vitro proliferation and maintains the osteochondal potential of rat marrow-derived mesenchymal stem cells. *Exp. Cell Res.* **219,** 211–222.

32. Bruder, S. P., Fink, D. J., and Caplan, A. I. (1994) Mesenchymal stem cells in bone development, bone repair, and skeletal regeneration therapy. *J. Cell Biochem.* **26,** 283–294.

33. ABI Prism User Bulletin #2, ABI Prism 7700 Sequence Detection System, Dec 2001.

34. Beresford, J. N., Graves, S. E., and Smoothy, C. A. (1993) Formation of mineralised nodules by bone derived cells in vitro: a model of bone formation? *Am. J. Med. Genet.* **45,** 163–178.
35. Gundle, R. G., Joyner, C. J., and Triffitt, J. T. (1995) Human bone tissue formation in diffusion chamber culture *in vivo* by bone derived cells and marrow stromal cells. *Bone* **16,** 597–601.
36. Jones, S. J., Gray, C., Boyde, A., and Burnstock, G. (1997) Purinergic transmitters inhibit bone formation by cultured osteoblasts. *Bone* **21,** 393–399.
37. Yasuda, H., Shima, N., Nakagawa, N., et al. (1998) Osteoclast differentiation factor is a ligand for osteoprotegerin/osteoclastogenesis inhibitory factor and is identical to TRANCE/RANKL. *Proc. Natl. Acad. Sci. USA* **95,** 3597–3602.
38. Tsuda, E., Goto, M., Mochizuki, S., et al. (1997) Isolation of a novel cytokine from human fibroblasts that specifically inhibits osteoclastogenesis. *Biochem. Biophys. Res. Commun.* **234,** 137–142.
39. Nagai, M. and Sato, N. (1999) Reciprocal gene expression of osteoclastogenesis inhibitory factor and osteoclast differentiation factor regulates osteoclast formation. *Biochem. Biophys. Res. Com.* **257,** 719–723.
40. Atkins, G. J., Kostakis, P., Pan, B., et al. (2003) RANKL expression is related to the differentiation state of human osteoblasts. *J. Bone Miner. Res.* **18,** 1088–1098.
41. Takahashi, Y. and Tabata, Y. (2003) Homogeneous seeding of mesenchymal stem cells into nonwoven fabric for tissue engineering. *Tissue Eng.* **9,** 931–938.
42. Abukawa, H., Terai, H., Hannouche, D., Vacanti, J. P., Kaban, L. B., and Troulis, M. J. (2003) Formation of a mandibular condyle in vitro by tissue engineering. *J. Oral Maxillofac. Surg.* **61,** 94–100.
43. Dong, J., Uemura, T., Kikuchi, M., Tanaka, J., and Tateishi, T. (2002) Long-term durability of porous hydroxyapatite with low-pressure system to support osteogenesis of mesenchymal stem cells. *J. Biomed. Mater. Res.* **12,** 203–209.
44. Yamada, Y., Boo, J. S., Ozawa, R., et al. (2003) Bone regeneration following injection of mesenchymal stem cells and fibrin glue with a biodegradable scaffold. *J. Craniomaxillofac. Surg.* **31,** 27–33.
45. Arnold, U., Lindenhayn, K., and Perka, C. (2002) In vitro-cultivation of human periosteum derived cells in bioresorbable polymer-TCP-composites. *Biomaterials* **23,** 2303–2310.
46. Shin, H., Zygourakis, K., Farach-Carson, M. C., Yaszemski, M. J., and Mikos, A. G. (2004) Attachment, proliferation, and migration of marrow stromal osteoblasts cultured on biomimetic hydrogels modified with an osteopontin-derived peptide. *Biomaterials* **5,** 895–906.
47. Partridge, K., Yang, X., Clarke, N. M., et al. (2002) Adenoviral BMP-2 gene transfer in mesenchymal stem cells: in vitro and in vivo bone formation on biodegradable polymer scaffolds. *Biochem. Biophys. Res. Commun.* **22,** 144–152.
48. Peng, H., Wright, V., Usas, A., et al. (2002) Synergistic enhancement of bone formation and healing by stem cell-expressed VEGF and bone morphogenetic protein-4. *J. Clin. Invest.* **110,** 751–759.

49. Mason, J. M., Breitbart, A. S., Barcia, M., Porti, D., Pergolizzi, R. G., and Grande, D. A. (2000) Cartilage and bone regeneration using gene-enhanced tissue engineering. *Clin. Orthop.* **399(Suppl.),** S171–S178.
50. Gehron-Robey, P. and Termine, J. D. (1985) Human bone cells *in vitro*. *Calcif. Tiss. Int.* **37,** 453, 460.
51. Wergedal, J. E. and Baylink, D. J. (1984) Characterisation of cells isolated and cultured from human trabecular bone. *Proc. Soc. Exp. Biol. Med.* **176,** 60–69.
52. Marie, P. J., Sabbagh, A., De Vernejoul, M. C., and Lomri, A. (1988) Osteocalcin and deoxyribonucleic acid synthesis in vitro and histomorphometric indices of bone formation in postmenopausal osteoporosis. *J. Clin. Invest.* **69,** 272–279.
53. Walsh, S., Jefferiss, C., Stewart, K., Jordan, G. R., Screen, J., and Beresford, J. N. (2000) Expression of the developmental markers STRO-1 and alkaline phosphatase in cultures of human marrow stromal cells: regulation by fibroblast growth factor (FGF)-2 and relationship to the expression of FGF receptors 1-4. *Bone* **27,** 185–195.

4

Human Osteoclast Culture
From Peripheral Blood Monocytes

Phenotypic Characterization and Quantitation of Resorption

Katherine A. Buckley, Benjamin Y. Y. Chan, William D. Fraser, and James A. Gallagher

1. Introduction

Research involving osteoclasts has always been difficult to undertake owing to a lack of osteoclast supply. Osteoclasts are terminally differentiated cells and, therefore, they cannot simply divide and be maintained in culture—an osteoclast supply must constantly be replenished. Osteoclasts are large, multinucleated cells formed by the fusion of osteoclast precursors found in the monocyte fraction of blood. The osteoclast is thought to be the only cell capable of excavating authentic resorption lacunae in calcified substrates in vivo and in vitro.

Until relatively recently, the study of osteoclasts in vitro relied on giant cell tumors removed at surgery becoming available, and the subsequent isolation of osteoclasts from these tissues *(1)*. The discovery of receptor activator for nuclear factor kappaB (NF-κB) ligand (RANKL) *(2)* has provided a significant leap forward in understanding the events leading to osteoclast formation, and resulted in the development of techniques to produce routinely functional human osteoclasts *(3)*. These techniques are based on the finding that coactivation of receptors for RANKL and macrophage colony stimulating factor (M-CSF) expressed by osteoclast precursors results in the fusion of these cells to form multinucleated osteoclasts.

Osteoblasts secrete M-CSF, and following stimulation by factors such as 1,25-dihydroxy vitaminD$_3$ and dexamethasone, express elevated levels of

From: *Methods in Molecular Medicine, vol. 107: Human Cell Culture Protocols, Second Edition*
Edited by: J. Picot © Humana Press Inc., Totowa, NJ

RANKL; therefore, co-cultures of these cells and osteoclast precursors isolated from human peripheral blood can be used to generate human osteoclasts. More recently, following commercial availability of a recombinant RANKL protein, another technique of osteoclast generation, negating the need for osteoblasts, has been described and is now routinely used *(4,5)*. These new techniques for osteoclast generation allow the study of both osteoclast formation and resorption. Whereas these procedures are highly valuable in that they provide a constant supply of human osteoclasts, their one disadvantage is that they are time consuming, osteoclasts only forming after approx 7–21 d in culture (*see* **Fig. 1**).

This chapter describes the generation of human osteoclasts from isolated peripheral blood mononuclear cells in both co-cultures and cultures supplemented with recombinant RANKL. Osteoclast characterization and the subsequent assay of osteoclast activity in vitro are also discussed.

2. Materials
2.1. Tissue Culture

1. α-Modification of minimum essential medium (α-MEM) (Gibco), used either alone or supplemented with 10% fetal calf serum (FCS) (Labtech) plus 50 IU/mL of penicillin, 50 µg/mL of streptomycin, and 2 mM L-glutamine (subsequently referred to as complete α-MEM.
2. Phosphate-buffered saline (PBS), prepared without calcium and magnesium for washing cells and cell layers.
3. Trypsin-ethylenediaminetetraacetic acid (EDTA): 0.05% trypsin, 0.03 mM EDTA, 0.85 g NaCl/L (Gibco) for detachment of cells from tissue culture plastic.
4. For human osteoclast cultures Histopaque®-1077 used for blood separation can be purchased from Sigma Diagnostics (Poole, U.K.), 6-mm cover slips can be obtained from Richardson's of Leicester (Leicester, U.K.), and dentine disks can be purchased from IDS (Boldon, U.K.).

2.2. Fixation of Dentine Disks

Use analytical-grade reagents and distilled water for the solutions. The solutions do not need to be sterile, but should be stored in clean containers at 4°C.

1. 0.2 M sodium cacodylate solution: dissolve 8.56 g of sodium cacodylate in 100 mL of water.
2. 8% (v/v) glutaraldehyde solution in water.

Make up the fixative solution by mixing solutions 1 and 2 in equal volumes. The fixative solution should optimally be made up immediately before use, but if necessary, may be stored at 4°C. Discard the solution if discoloration occurs.

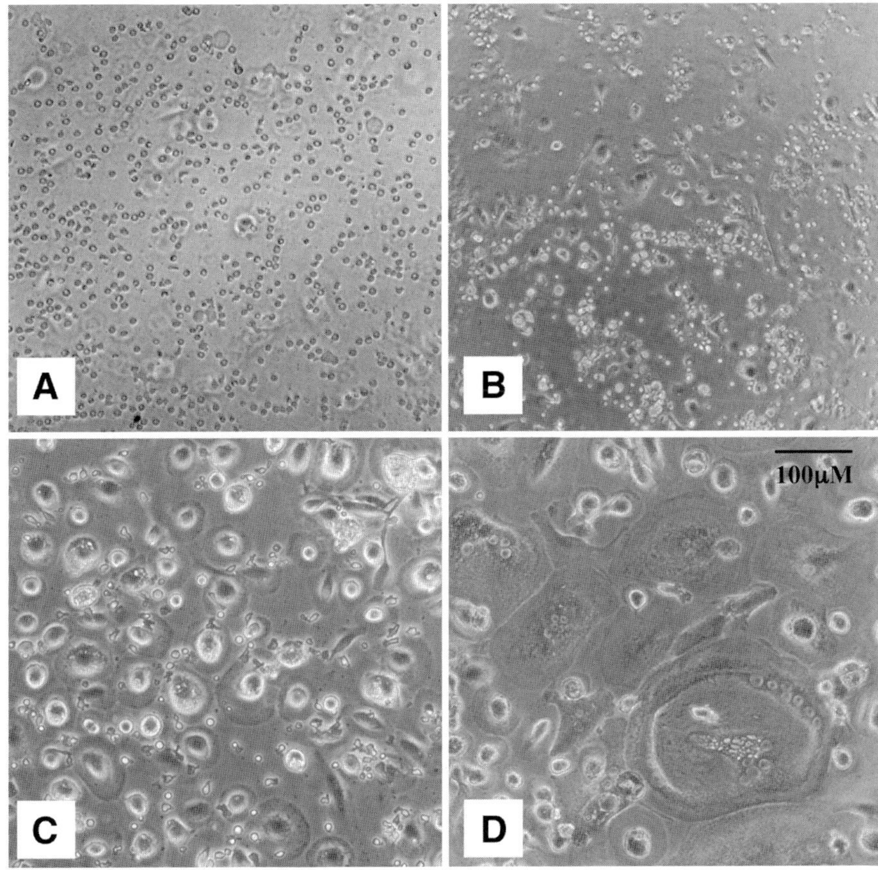

Fig. 1. Osteoclasts form in culture medium supplemented with recombinant RANKL. Images are representative of cell appearance after the indicated number of days in culture using phase-contrast microscopy (all images to same scale). **(A)** D 0: cells are small, spherical and mononuclear. **(B)** D 7: cells are more spread out with larger cytoplasms but are still mononuclear. **(C)** D 14: cells are larger and begin to display an osteoclast-like appearance, some may be multinucleated. **(D)** D 21: cells are larger again, multinucleated, with an osteoclast-like appearance. Some may begin to apoptose after this time-period.

2.3. Staining of Fixed Dentine Disks

Use analytical-grade reagents and distilled water for solutions.

1. 0.5 *M* sodium tetraborate solution: dissolve 19.07 g of disodium tetraborate in 100 mL of water.

2. 1% toluidine blue solution: dissolve 1 g of toluidine blue in 100 mL of 0.5 *M* sodium tetraborate solution (prepared as in **step 1**).

These solutions can be stored at room temperature.

3. Methods

3.1. Dentine Disks and Cover Slips

Sterile, precut 5-mm dentine disks can be obtained from IDS or alternatively prepared in the following way:

1. Cut thin wafers of dentine (200 μm) using a water-cooled low-speed diamond saw.
2. Soak the wafers for 1 h in distilled water, then punch disks from these wafers using a circular punch (5 mm diameter). Dentine disks may be marked with a pencil on one side for later identification of the surface not containing cells.
3. Sonicate dentine disks in double-distilled water (ddH$_2$O) for three 15-min washes. Wash the disks in 70% ethanol, allow to air-dry, then arrange the disks in a Petri dish, pencil-marked-side face down. Place the Petri dish over a mirror and sterilize under ultraviolet (UV) light in a flow cabinet for 2 h.
4. Cover slips are similarly sterilized by UV irradiation for 2 h.

3.2. Human Osteoclast Generation

Human osteoclasts can be generated from their precursors, found in the monocyte fraction of blood or bone marrow. These cultures may be set up either on dentine disks or cover slips, to study both osteoclast resorption and osteoclast formation, respectively.

3.2.1. Co-Culture

The principle of human co-cultures for osteoclast formation is that osteoblasts in these cultures express RANKL, which induces the fusion of monocytic osteoclast precursors also present in the culture, although these osteoblasts may be expressing factors in addition to RANKL to influence the process of osteoclastogenesis that we are as yet unaware of. The co-culture described here utilizes the rat osteosarcoma cell line UMR-106, but it cannot be assumed that all osteoblast cell lines will be able to induce osteoclast formation in these cultures (*see* **Note 1**).

1. The dentine disks or cover slips to be used should be prewetted by soaking in 100 μL complete α-MEM in wells of a 96-well plate, for not less than 1 h at 37°C, but preferably overnight (*see* **Fig. 2**).
2. UMR-106 cells should be grown to confluence, then washed twice with warm PBS and incubated for 5 min (*see* **Note 2**) with 1 mL of warm trypsin-EDTA.

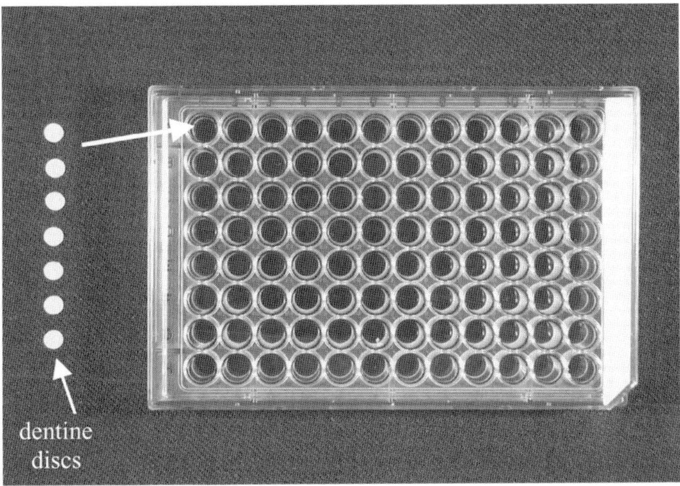

Fig. 2. Dentine disks cut using a low-speed diamond saw and circular punch are sterilized, then placed in wells of a 96-well plate where osteoclasts may be cultured on their surface.

Wash the detached cells and resuspend them in complete α-MEM. Perform a cell count and add the appropriate volume of cell suspension containing 2×10^4 UMR-106 cells to wells of the 96-well plate, containing the prewetted dentine disks or cover slips, then incubate overnight at 37°C in a humidified atmosphere of 95% air and 5% CO_2.

3. The following day, collect 10–100 mL human peripheral blood into lithium-heparin monovettes, or some other heparinized, sterile container. Mix the blood with an equal volume of nonsupplemented α-MEM, then layer 8 mL of this mixture over 5 mL of Histopaque®-1077 in 15-mL Falcon tubes and centrifuge at 400g for 30 min at 10°C.

4. After centrifugation, remove the clear upper layer and collect the opaque interface containing the monocyte and lymphocyte fraction of blood, using a fine pipet, and dilute into 20 mL of non-supplemented α-MEM.

5. Centrifuge these cells at 300g for 20 min, then wash the cell pellet a second time in nonsupplemented α-MEM. Resuspend the cell pellet in an appropriate volume of complete α-MEM (*see* **Note 3**), take an aliquot of cell suspension, mix 1:1 with 10% acetic acid to lyse contaminating red cells, and determine the cell number by performing a cell count.

6. Add an appropriate volume of cell suspension containing 1×10^5 cells to wells of the 96-well plate containing UMR-106 cells and dentine disks or cover slips.

7. Incubate the cells at 37°C for a minimum of 1 h to allow the blood mononuclear cells to adhere, then remove the nonadherent cells by removing the dentine disks or cover slips from the 96-well plate (*see* **Note 4**), and washing in a Petri dish

containing nonsupplemented α-MEM. This should remove the lymphocytes because these cells do not adhere.

8. Transfer the dentine disks or cover slips to a 24-well plate, taking care to ensure the surface containing the cells is facing upward. Three cover slips or dentine disks can be placed in the same well. Incubate in 1 mL complete α-MEM containing 25 ng/mL M-CSF, 10^{-7} M 1,25-dihydroxy vitaminD$_3$ (*see* **Note 5**) and 10^{-8} M dexamethasone at 37°C in a humidified atmosphere of 95% air and 5% CO$_2$ for approx 3 wk.

9. Replace the complete α-MEM, containing all of the above added factors, every 2–3 d.

10. After 2 wk in culture, it is advisable to transfer cover slips or dentine disks to a fresh 24-well plate, because by this stage, the UMR-106 cells will have grown to cover the surface of the well. By the end of the culture period, the UMR-106 cell layer should have detached from the dentine disk or cover slip surface to reveal large multinucleated osteoclasts beneath.

3.2.2. Recombinant RANKL Cultures

This procedure negates the need for osteoblasts in the culture by utilizing commercially available recombinant RANKL. In our experience, it is an easier and more reliable method of osteoclast production than the co-culture.

1. The dentine disks or cover slips to be used should be prewetted by soaking in 100 μL complete α-MEM in wells of a 96-well plate for no less than 1 h at 37°C, but preferably overnight.

2. For blood collection, monocyte isolation, washing, and counting follow the method for **Subheading 3.2.1., steps 3–5**.

3. In this culture, add an appropriate volume of cell suspension containing 5×10^5 cells to the prewetted cover slips or dentine disks in the 96-well plate (*see* **Note 6**).

4. Incubate the cells at 37°C for a minimum of 1 h to allow the blood monocytes to adhere, then remove the nonadherent cells (i.e., the lymphocytes) by washing the dentine disks or cover slips by repeated pipetting (*see* **Note 7**), while they are still in the wells, in nonsupplemented α-MEM.

5. Incubate the remaining adherent cells in 100 μL complete α-MEM containing 25 ng/mL M-CSF and 30 ng/mL recombinant RANKL, at 37°C in a humidified atmosphere of 95% air and 5% CO$_2$ for approx 2–3 wk. In our experience, osteoclasts first appear between 7–14 d, but larger multinucleated osteoclasts are not seen until d 14 onward.

6. Replace the complete α-MEM containing 25 ng/mL M-CSF and 30 ng/mL RANKL every 2–3 d.

3.2.3. Characterization of Osteoclasts Produced in Human Cultures

The cell with the greatest similarity to the osteoclast is thought to be the macrophage polykaryon. This cell has a similar lineage and it is often difficult to distinguish between the two. The fundamental difference, however, is the inabil-

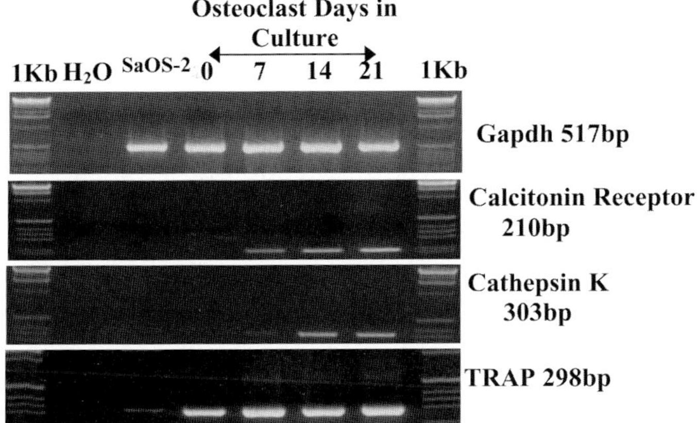

Fig. 3. Development of osteoclast markers by cells generated in recombinant RANKL-supplemented cultures of isolated blood monocytes. PCRs using primers designed specifically for calcitonin receptor, cathepsin K, and TRAP mRNA were performed for 35 cycles on cDNA generated from RNA isolated from recombinant RANKL-supplemented cultures of peripheral blood mononuclear cells after 0, 7, 14, and 21 d. Primer sequences: Calcitonin receptor, Forward 5′-tgc ggt ggt att atc tct tgg-3′, Reverse 5′-ttc cct cat ttt ggt cac aag-3′, product = 210 bp. Cathepsin K, Forward 5′-acc gca tgg ttc aga tta tc-3′, Reverse 5′-cag tca tct tct gaa cca cc-3′, product = 303 bp. TRAP, Forward 5′-ctg tcc tgg ctc aag aaa cag-3′, Reverse 5′-cat agt gga agc gca gat agc-3′ product = 298 bp. PCR for glyceraldehyde-3-phosphate dehydrogenase (gapdh) was performed on the same cDNA samples for 30 cycles to confirm the presence of equal amounts of cDNA between samples. SaOS-2 cell (an osteosarcoma cell line) cDNA was used as a negative control. (Reprinted from *Bone*, **31(5)** 582–590. Buckley, K. A., Hipskind, R. A., Gartland, A., Bowler, W. B., and Gallagher, J. A. (2002) Adenosine triphosphate stimulates human osteoclast activity via upregulation of osteoblast-expressed receptor activator of nuclear factor-kappaβ ligand. With permission from Elsevier Science.)

ity of macrophage polykaryons to resorb calcified substrates, osteoclasts being the only cells considered capable of resorption. The multinucleated cells produced in these cultures may also be identified as osteoclasts by the detection of certain osteoclast markers expressed by the cells. In addition, these markers can allow the differentiation of monocytes into osteoclasts to be monitored over time.

1. Cathepsin K: This enzyme is thought to be the predominant cysteine protease expressed by osteoclasts *(6)*. Following RNA isolation and cDNA synthesis, messenger RNA encoding cathepsin K using specifically designed primers, can be identified from cells in recombinant RANKL-supplemented cultures by PCR (*see* **Fig. 3**) (*see* **Notes 8** and **9**). In addition, a polyclonal antibody specific for cathepsin K used in immunocytochemistry will stain the cytoplasm of multinucleated cells grown in these cultures (obtained from IDS, Boldon, UK).

2. Calcitonin receptor: The polypeptide hormone calcitonin is a potent inhibitor of osteoclastic bone resorption. Again, PCR can be performed, using specifically designed primer sequences to identify calcitonin receptor mRNA in cells grown in recombinant RANKL-supplemented cultures (*see* **Fig. 3**) (*see* **Note 8**). Cells produced in this manner have displayed desensitization to calcitonin following exposure, owing to downregulation of receptor expression *(7)*, therefore confirming that the behavior of these osteoclasts is identical to mature osteoclasts isolated from tissue samples.

3. Tartrate-resistant acid phosphatase (TRAP) (antibodies available from Novocastra): Enzyme cytochemical studies have indicated that osteoclasts *in situ* and giant cell tumors express TRAP in abundance *(8,9)*. This expression, however, extends to other cells of the monocyte/macrophage lineage; therefore, this marker does not distinguish between osteoclasts and their precursors. We have found that multinucleated cells in RANKL-supplemented cultures express this enzyme (*see* **Fig. 3**).

4. Other markers indicative of an osteoclast phenotype include type II carbonic anhydrase and the vitronectin receptor (integrin $\alpha v\beta 3$) (antibodies available from IDS). Type II carbonic anhydrase is abundantly expressed by osteoclasts *(10)*, where it catalyzes the production of protons. The vitronectin receptor is functionally involved in adhesion and possibly resorption, but is also expressed by many other cells of the monocyte/macrophage lineage.

These phenotypic markers can still only be regarded as suggestive of osteoclastic differentiation; the only definitive marker for osteoclasts remains the ability to form resorption lacunae on calcified substrates.

3.3. Fixation and Staining of Dentine Disks

1. Following the culture of osteoclasts on dentine disks, remove the culture medium, wash the dentine disks in PBS, then transfer the disks to clean wells containing enough fixative solution (*see* **Subheading 2.2.**) to cover the disks. Fix for 5 min.

2. Wash with distilled water, then stain for 3 min with a 1% solution of toluidine blue buffered with sodium tetraborate (*see* **Subheading 2.3.**). Remove excess stain by washing in 70% ethanol for 1 min.

3. Rinse in tap water and air-dry before examining the wafers under the microscope. Wafers can then be stored indefinitely at room temperature.

3.4. Quantitation of Resorption

Osteoclast activity in vitro can be assessed both by scanning electron microscopy (SEM) and reflected light microscopy (RLM) *(11)*. In our lab, we routinely use the latter technique. Although SEM produces images of excellent quality and stereophotogrammetry of scanning electron micrographs enables the area and volumes of resorption lacunae to be determined, this procedure requires expertise and expensive equipment, and is impractical for the assessment of a large number of wafers because it is highly time-consuming.

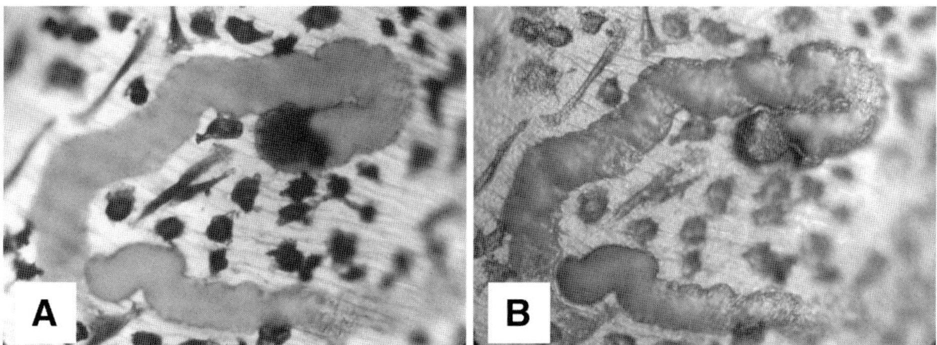

Fig. 4. Recombinant RANKL-generated osteoclasts form authentic resorption lacunae on dentine disks. Blood monocytes were grown on dentine disks in the presence of recombinant RANKL. After 3 wk, dentine disks were fixed and stained with toluidine blue. (**A**) A trail of resorption excavated by an osteoclast which can be seen lying at the end of the resorption pit, viewed by transmitted light microscopy. (**B**) The identical image viewed by reflected light microscopy.

RLM avoids many of the problems associated with SEM, providing accurate information on the area, volume, and depth of resorption lacunae using an adapted standard microscope.

3.4.1. Area Measurements

1. Using reflected light and the ×10 objective lens, resorption lacunae can be clearly identified even when the dentine disk is covered by a dense layer of cells. Larger pits can also be seen easily by transmitted light microscopy. **Figure 4** shows a resorption lacuna excavated by osteoclasts generated in recombinant RANKL-supplemented medium visualized by transmitted and reflected light microscopy.
2. Using a microscope fitted with a CCTV camera or a drawing-arm, overlay a grid of points onto the image of the dentine disk. Quantitation can be done manually by point counting, or can be automated using custom-designed counting software *(12)*. The magnification should be adjusted to ensure that the grid covers an area of 1 mm² on the surface of the dentine disk.
3. Area of resorption can be measured over the whole disk, or alternatively random fields can be selected and resorption expressed as a percentage of the area. When selecting random fields repeat until the measurements between fields become consistent to ensure an accurate representation of resorption on the whole wafer *(11)*.
4. To determine resorption area on the dentine disk, record the number of points of the grid falling onto resorption lacunae.
5. Resorption can be expressed as a relative measurement compared to resorption produced in vehicle wells where no compounds in addition to RANKL and M-CSF are introduced.

6. In recombinant RANKL-supplemented cultures we have found that approx eight dentine disks per treatment are required to provide significant results. The coefficient of variation (%CV) of our data from these assays is approx 20. In co-cultures, the variation is greater (%CV = approx 25–30) and it is advised that as many as 12 dentine disks are used per treatment.

3.4.2. Volume Measurements

The quantitation of mean plan-area of resorption in most cases provides a good indicator of osteoclast activity. If a more accurate measurement of osteoclast activity is required, however, the volume of resorption lacunae may be determined.

1. Bring the surface of the dentine disk into focus using the ×50 objective lens (numerical aperture 0.75). The position of the fine-focus knob of the microscope at the surface of the dentine disk should be noted.
2. Using the Olympus BH2 microscope, move the fine-focus knob one division at a time, representing a change in depth of 2 µm. For other instruments, the manufacturer should be consulted. At each depth, record the area of the lacuna by counting the number of points of the grid falling within the part of the lacuna that is in focus, or has yet to come into focus.
3. Repeat this procedure for 10 resorption lacunae chosen at random.
4. Calculate the volume of resorption lacunae by application of the Cavalieri estimator of volume *(13)*. This states that the volume of a sectioned object is "the sum of the products of the sectioned areas × the section separation."

3.4.3. β-CTX Measurement

The organic matrix of bone consists of 90% of type I collagen, which is preferentially synthesized in bone. During bone resorption, mature type I collagen is degraded and small degradation products are generated. In vivo these will eventually be excreted, whereas in vitro they remain in the cell culture medium bathing the cells. C-terminal telopeptide (CTX) is one such collagen degradation product. CTX is an aspartic acid 1211 residue of the 1209AHDGGR1214 sequence *(14)*, which can undergo spontaneous post-translational modifications as bone ages, resulting in the α-aspartic acid form converting to the β-form *(15)*. Because dentine disks used in our experiments are obtained from mature animals, β-CTX is likely to be the predominant bone-degradation product.

An electrochemiluminescence immunoassay (ECLIA) (Roche® Elecsys β-CrossLaps Serum assay) can be used to quantify β-CTX. This is a sandwich immunoassay with two monoclonal antibodies specific for the β-isomerized amino acid sequence (EKAHD-β-GGR) of the C-terminal telopeptide of type I collagen. Samples can be measured by automated assay on the Elecsys analyzer

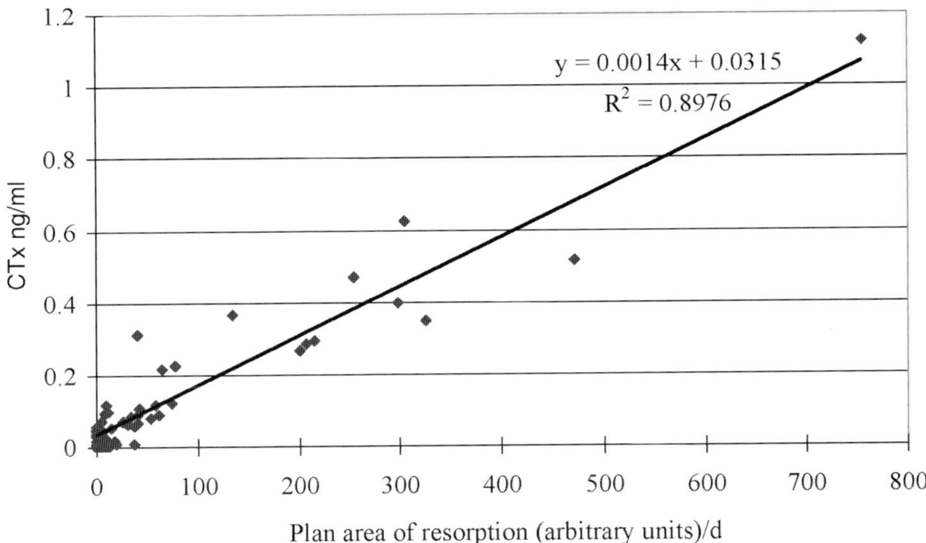

Fig. 5. Area of resorption by recombinant-RANKL generated osteoclasts correlates with concentration of CTX in the culture medium.

(Roche® Diagnostics, Lewes, U.K.). Samples are prepared in the following manner:

1. Collect cell culture media every 2–3 d when changing cell culture medium, and store at –70°C.
2. On the day of performing the β-CTX assay, thaw all samples at room temperature.
3. Centrifuge samples at 400g for 5 min before loading into the Roche® Elecsys automated analyzer.
4. In addition measure internal standard samples of cell culture medium alone cultured on dentine disks (*see* **Note 10**).

Following quantification of area of resorption by RLM, and measurement of β-CTX by ECLIA, a significant correlation is observed between these two measurements of resorption activity (*see* **Fig. 5**). Our data are in agreement with findings reported by Breuil et al. *(16)*, thus we suggest that the use of β-CTX as a marker for bone resorption activity could replace the more time-consuming method of point counting area of resorption lacunae using RLM.

4. Notes

1. It has become apparent that this osteogenic potential of osteoblasts is not universal to all osteoblast cell lines, some supporting and others not supporting osteoclast formation. This difference is due to the ratio of expression of RANKL and

osteoprotegerin (OPG) (a soluble decoy receptor to RANKL) by the osteoblasts following stimulation with osteotropic factors such as 1,25dihydroxy vitaminD$_3$ and dexamethasone *(17)*. Binding of OPG to RANKL prevents osteoclastogenesis from occurring *(18)*. In our experience, the human osteosarcoma cell line SaOS-2 is unable to support osteoclast formation.

2. UMR-106 cells should be incubated with trypsin-EDTA for the minimum length of time possible, as prolonged exposure impairs the osteogenic capacity of this cell line.

3. We recommend that for each 20 mL of blood used, the isolated monocyte fraction should be finally resuspended in 4 mL of complete α-MEM to obtain a cell concentration that will allow the addition of volumes of this suspension that will be held by wells of a 96-well plate.

4. To remove dentine disks or cover slips from wells of 96-well plates, use a fine needle to lever the disks or cover slips into an upright position, then pick up with fine forceps.

5. Although osteoblasts produce M-CSF, the rat M-CSF produced by UMR-106 cells is not recognized by the human M-CSF receptor on the osteoclast precursors, therefore, human recombinant M-CSF must be added. 1,25dihdroxy vitaminD$_3$ should be stored under nitrogen, at −70°C and in the absence of light to maintain its activity.

6. In this case, for each 20 mL of blood used, the isolated monocyte fraction should be finally resuspended in 1 mL of complete α-MEM to obtain a cell concentration that will allow the addition of volumes of this suspension that will be held by wells of a 96-well plate (20 mL of blood provides approximately enough cells for 50 wells, depending on the donor).

7. In our experience, this washing step is very important in the resulting number and quality of osteoclasts produced in the culture. Ensure the whole surface of the dentine disk or cover slip is washed evenly and take care not to flip the dentine disks over. A rough guide to washing would be to use two changes of medium, and with each medium change pipet up and down four times, firmly but not vigorously.

8. We have observed that after 7 d in RANKL-supplemented cultures, cells begin to exhibit an osteoclast-like appearance, but are mostly still mononuclear (*see* **Fig. 1**). Despite this, we have found that messenger RNAs encoding the osteoclast markers cathepsin K and calcitonin receptor are present in cells at this stage of the culture period (*see* **Fig. 3**). This suggests that the expression of markers typical of osteoclast cells may precede cell fusion in the development of osteoclasts from their precursors, consistent with a recent report from other investigators *(19)*.

9. We have found that osteoclasts do not grow very well on tissue culture plastic surfaces, instead preferring glass. Therefore, for RNA isolation we still place sterile glass cover slips in the base of wells of a 96-well plate. Isolating RNA from 30 or more wells is sufficient for cDNA synthesis.

10. Because this assay is designed to measure β-CTX in serum, some detection will be found in cell culture medium that has not been exposed to collagen degradation if

it contains some quantity of FCS. Set up an internal standard by placing three (or more) dentine disks in wells of a 96-well plate and culturing these in the absence of cells. Collect the culture medium at the same time point as for the wells containing cells.

References

1. Walsh, C. A., Carron, J. A., and Gallagher, J. A. (1996) Isolation of osteoclasts from human giant cell tumour and long term marrow cultures, in *Human Cell Culture Protocols* (Jones, T. L., ed.), Humana, Totowa, NJ, pp. 263–276.
2. Yasuda, H., Shima, N., Nakagawa, N., et al. (1998) Osteoclast differentiation factor is a ligand for osteoprotegerin/osteoclastogenesis inhibitory factor and is identical to TRANCE/RANKL. *Proc. Natl. Acad. Sci. USA* **95,** 3597–3602.
3. Quinn, J. M. W., Fujikawa, Y., McGee, O. D., and Athanasou, N. A. (1997) Rodent osteoblast-like cells support osteoclastic differentiation of human cord blood monocytes in the presence of M-CSF and 1,25-dihydroxyvitamin D_3. *Int. J. Biochem. Cell Biol.* **29,** 173–179.
4. Neale, S. D., Smith, R., Wass, J. A., and Athanasou, N. A. (2000) Osteoclast differentiation from circulating mononuclear precursors in Paget's disease is hypersensitive to 1,25-dihydroxyvitamin D_3 and RANKL. *Bone* **27,** 409–416.
5. Buckley, K. A., Hipskind, R. A., Gartland, A., Bowler, W. B., and Gallagher, J. A. (2002) Adenosine triphosphate stimulates human osteoclast activity via upregulation of osteoblast-expressed receptor activator of nuclear factor-kappaB ligand. *Bone* **31,** 582–590.
6. Kamiya, T., Kobayashi, Y., Kanaoka, K., et al. (1998) Fluorescence microscopic demonstration of cathepsin K activity as the major lysosomal cysteine proteinase in osteoclasts. *J. Biochem.* **123,** 752–759.
7. Samura, A., Wada, S., Suda, S., Iitaka, M., and Katayama, S. (2000) Calcitonin receptor regulation and responsiveness to calcitonin in human osteoclast-like cells prepared in vitro using receptor activator of nuclear factor-κB ligand and macrophage colony stimulating factor. *Endocrinol.* **141,** 3774–3782.
8. Minkin, C. (1982) Bone acid phosphatase: tartrate-resistant acid phosphatase as a marker of osteoclast function. *Calcif. Tiss. Int.* **34,** 285–290.
9. Scheven, B. A., Kawilarang-De-Haas, E. W., Wassenaar, A. M., and Nijweide, P. J. (1986) Differentiation kinetics of osteoclasts in the periosteum of embryonic bones in vivo and in vitro. *Anat. Rec.* **214,** 418–423.
10. Hall, G. E. and Kenny, A. D. (1985) Role of carbonic anhydrase in bone resorption induced by 1,25 dihydroxyvitaminD3 in vitro. *Calcif. Tiss. Int.* **37,** 134–142.
11. Walsh, C. A., Beresford, J. N., Birch, M. A., Boothroyd, B., and Gallagher, J. A. (1991) Application of reflected light microscopy to identify and quantitate resorption by isolated osteoclasts. *J. Bone Miner. Res.* **6,** 661–671.
12. Armour, K. J., van 't Hof, R. J., Armour, K. E., Torbergsen, A. C., Del Soldato, P., and Ralston, S. H. (2001) Inhibition of bone resorption in vitro and prevention of ovariectomy-induced bone loss in vivo by flurbiprofen nitroxybutylester (HCT1026). *Arthritis Rheum.* **44,** 2185–2192.

13. Sterio, D. C. (1983) The unbiased estimation of number and sizes of arbitrary particles using the dissector. *J. Microscopy* **134,** 127–136.
14. Okabe, R., Nakatsuka, K., Inaba, M., et al. (2001) Clinical evaluation of the elecsys β-crosslaps serum assay, a new assay for degradation products of Type I collagen C—telopeptides. *Clin. Chem.* **47,** 1410–1414.
15. de la Piedra, C., Calero, J. A., Traba, M. L., Asensio, M. D., Argente, J., and Munoz, M. T. (1999) Urinary alpha and beta C—telopeptides of collagen I: clinical implications in bone remodeling in patients with anorexia nervosa. *Osteoporosis Int.* **10,** 480–486.
16. Breuil, V., Cosman, F., Stein, L., et al. (1998) Human osteoclast formation and activity in vitro: effects of alendronate. *J. Bone Miner. Res.* **13,** 1721–1729.
17. Nagai, M. and Sato, N. (1999) Reciprocal gene expression of osteoclastogenesis inhibitory factor and osteoclast differentiation factor regulates osteoclast formation. *Biochem. Biophys. Res. Com.* **257,** 719–723.
18. Tsuda, E., Goto, M., Mochizuki, S., et al. (1997) Isolation of a novel cytokine from human fibroblasts that specifically inhibits osteoclastogenesis. *Biochem. Biophys. Res. Commun.* **234,** 137–142.
19. Blair, H. C., Sidonio, R. F., Friedberg, R. C., Khan, N. N., and Dong, S. S. (2000) Proteinase expression during differentiation of human osteoclasts in vitro. *J. Cell. Biochem.* **78,** 627–637.

5

Human Chondrocyte Cultures as Models of Cartilage-Specific Gene Regulation

Mary B. Goldring

1. Introduction

The mature articular chondrocyte embedded in the cartilage matrix is a resting cell with no detectable mitotic activity and a very low synthetic activity *(1)*. The markers of mature articular chondrocytes are type II collagen (COL2A1), other cartilage-specific collagens IX (COL9) and XI (COL11), the large aggregating proteoglycan aggrecan, and link protein (*see* **Table 1**). Chondrocytes also synthesize a number of small proteoglycans such as biglycan and decorin and other specific and nonspecific matrix proteins both in vivo and in vitro. As the single cellular constituent of adult articular cartilage, chondrocytes are responsible for maintaining the cartilage matrix in a low turnover state of equilibrium. In mature cartilage, the chondrocytes synthesize matrix components very slowly. The turnover of collagen in normal adult articular cartilage has been estimated to occur with a half-life of longer than 100 yr *(2,3)* and the half-life of aggrecan subfractions is in the range of 3–24 yr, whereas the glycosaminoglycan constituents on the aggrecan core protein are more readily replaced *(4)*. Furthermore, normal chondrocyte metabolism *in situ* occurs in low oxygen tension and is remote from a vascular supply. Thus, it is not surprising that changes in expression of these cartilage matrix constituents occur when the chondrocytes are isolated and placed in monolayer culture, where they increase synthetic activity by several orders of magnitude.

Primary cultures of articular chondrocytes isolated from various animal and human sources have served as useful models for studying the mechanisms controlling responses to growth factors and cytokines (*see*, for review, **refs. 5–7**). Early attempts to culture chondrocytes were frustrated by the tendency of these

From: *Methods in Molecular Medicine, vol. 107: Human Cell Culture Protocols, Second Edition*
Edited by: J. Picot © Humana Press Inc., Totowa, NJ

Table 1
Proteins Synthesized by Mature Chondrocytes

Collagens
    Type II
    Type IX
    Type XI
    Type VI
    Types XII, XIV
Proteoglycans
    Aggrecan
    Versican
    Link protein
    Biglycan (DS-PGI)
    Decorin (DS-PGII)
    Epiphycan (DS-PGIII)
    Fibromodulin
    Lumican
    PRELP (proline/arginine-rich and leucine-rich repeat protein)
    Chondroadherin
    Perlecan
    Lubricin (SZP)
Other noncollagenous proteins (structural)
    Cartilage oligomeric matrix protein (COMP; thrombospondin-5))
    Thrombospondin-1 and -3
    Cartilage matrix protein (matrilin-1); matrilin-3
    Fibronectin
    Tenascin-c
    Cartilage intermediate layer protein (CILP)
    Fibrillin
    Elastin
Other noncollagenous proteins (regulatory)
    S-100
    Chondromodulin-I (SCGP) and -II
    Glycoprotein (gp)-39, YKL-40
    Matrix Gla protein (MGP)
    CD-RAP (cartilage-derived retinoic acid-sensitive protein)
    Growth factors
Membrane-associate proteins
    Integrins ($\alpha1\beta1$, $\alpha2\beta1$, $\alpha3\beta1$, $\alpha5\beta1$, $\alpha6\beta1$, $\alpha10\beta1$, $\alpha v\beta3$, $\alpha v\beta5$)
    Anchorin CII (annexin V)
    CD44
    Syndecan-3

The collagens, proteoglycans, and other noncollagenous proteins in the cartilage matrix are synthesized by chondrocytes at different stages during development and growth of cartilage and the mature articular chondrocyte may have a limited capacity to maintain and repair some of the matrix components, particularly proteoglycans. Proteins that are associated with chondrocyte cell membranes are also listed because they permit specific interactions with extracellular matrix proteins.

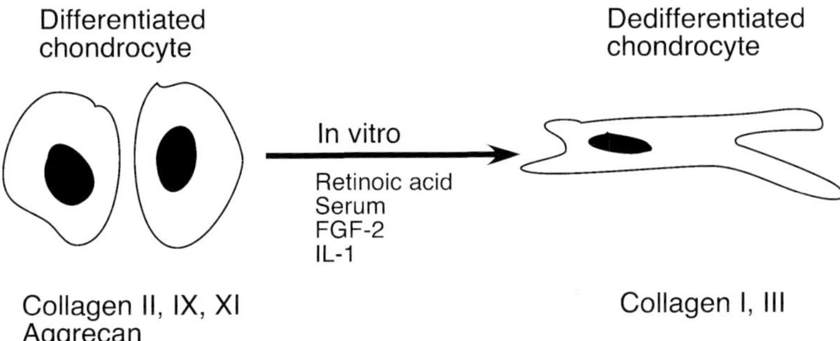

Fig. 1. Schematic representation of the "switch" from the differentiated to the dedifferentiated chondrocyte phenotype that occurs during culture and in response to certain cytokines and growth factors in vitro. The extracellular matrix genes that are differentially expressed are indicated.

cells to "dedifferentiate" in monolayer culture and their inability to proliferate in suspension culture where cartilage-specific phenotype was maintained *(8–10)*. In monolayer culture, chondrocytes maintain a rounded, polygonal morphology *(see* **Fig. 1**), but there may be a progressive change to a fibroblast-like morphology with passage of time, especially after subculture. High-density monolayer cultures maintain the cartilage-specific phenotype until they are subcultured, although gene expression of type II collagen is generally more labile than that of aggrecan *(11–13)*. Concomitant with the loss of differentiated phenotype, which can be accelerated by plating the cells at low densities or by treatment with cytokines such as interleukin 1 (IL-1) *(14,15)* or retinoic acid, the chondrocytes acquire some, but not all, characteristics of the fibroblast phenotype, i.e., type I collagen.

Because the stability of the phenotype of isolated chondrocytes is critically dependent on cell shape and cell density *(16,17)*, high-density micromass cultures are useful if sufficient numbers of chondrocytes are isolated *(18,19)*, particularly for studying proteoglycan biosynthesis *(20)*. It is also possible to expand the cultures through a limited number of subcultures and "redifferentiate" the cells in fluid or gel suspension culture systems, where the chondrocytes regain morphology and the cessation of proliferation is associated with increased expression of cartilage-specific matrix proteins *(21)*. Culture systems that support chondrocyte phenotype include suspension culture in spinner flasks *(22)*, in dishes coated with a nonadherent substrates *(23–25)*, in pellets *(26)*, and in three-dimensional matrices such as collagen gels *(27)*, agarose *(12,28,29)*, alginate *(30,31)*, or collagen sponges *(32,33)*. Serum-free defined media of varying compositions, but usually including insulin, have also been used, frequently in combination with the other culture systems mentioned above *(34)*.

The use of chondrocytes of human origin has been problematical, because the source of the cartilage cannot be controlled, sufficient numbers of cells are not readily obtained from random operative procedures, and the phenotypic stability and proliferative capacity of adult human chondrocytes are lost more quickly upon expansion in serial monolayer cultures than in cells of juvenile human *(11)* or embryonic or postnatal animal origin *(35,36)*. Alternatively, explant cultures of human, but usually bovine, articular cartilage where the chondrocytes remain encased within their own extracellular matrix, have been used as in vitro models to study cartilage biochemistry and metabolism *(37)*. However, many experimental manipulations are done more easily using isolated chondrocytes. Recent studies have focused on adult human articular chondrocytes as target cells for immortalization, using immortalizing antigens such as SV40-TAg *(38–40)*, human papilloma virus type 16 (HPV-16) early function genes E6 and E7 *(41)* and telomerase *(42)*. Strategies that maintain high cell density and decrease cell proliferation must also be applied to immortalized chondrocyte cell lines, since stable integration of immortalizing genes disrupts normal cell cycle control but does not stabilize expression of the type II collagen gene *(43,44)*.

This chapter will focus on strategies for isolation, culture, and characterization of isolated human articular chondrocytes and will also describe approaches for using different chondrocyte culture systems for evaluating chondrocyte phenotype and studying the regulation of chondrocyte functions.

2. Materials

2.1. Isolation and Culture of Human Chondrocytes

1. Growth medium for chondrocytes: Mix equal volumes of Dulbecco's modified Eagle's medium (DMEM), high glucose (4.5 g/L), and Ham's F-12, 1:1 mixture (Cambrex, Walkersville, MD), both with L-glutamine and without HEPES, to give final concentrations of 3.151 g/L glucose, 365 mg/L L-glutamine/7.36 mg/L L-glutamic acid, and 110 mg/L sodium pyruvate. Add 10% FCS immediately before use.
2. Dulbecco's phosphate-buffered Ca^{2+}- and Mg^{2+}-free saline phosphate-buffered serum (PBS).
3. Trypsin-ethylenediaminetetraacetic acid (EDTA) solution: 0.05% trypsin and 0.02% EDTA in Hanks' balanced salt solution (HBSS) without Ca^{2+} and Mg^{2+} (Life Technologies, Gaithersburg, MD).
4. Hank's balanced salt solution (HBSS) with Ca^{2+} and Mg^{2+} (Life Technologies, Inc., Gaithersburg, MD).
5. Serum substitutes for experimental incubations: Nutridoma-SP (Roche Applied Science) is provided as sterile concentrate (100X, pH 7.4; storage at 15–25°C protected from light). Dilute 1:100 (v/v) with sterile DMEM/Ham's F-12, without FCS, and use immediately. ITS+ (BD Biosciences) is an alternative serum substitute.

   Except where specified, cell culture reagents can be obtained from a number of different suppliers, including Life Technologies, Gaithersburg, MD; Cambrex, Inc.,

Walkersville, MD; Sigma, St. Louis, MO; or Intergen, Purchase, NY. Reserve serum testing, which is offered by these suppliers, is recommended for selection of lots of FCS that maintain chondrocyte phenotype (*see* **Note 1**). FCS and trypsin-EDTA are stored at –20°C, but should not be refrozen after thawing for use.

6. Enzymes for cartilage digestion:
   a. Hyaluronidase (Sigma): 1 mg/mL in PBS: Prepare freshly and filter through a sterile 0.22-μm filter.
   b. Trypsin (Life Technologies): 0.25% in HBSS without Ca^{2+} and Mg^{2+}.
   c. Collagenase (bacterial, clostridiopeptidase A; Worthington Biochemical Corp., Freehold, NJ) or Collagenase D (Roche Applied Science): 3 mg/mL in DMEM with 10% FCS for articular cartilage or serum-free for costal cartilage: Prepare freshly in ice-cold DMEM and filter immediately through a sterile 0.22-μm filter.

2.2. Suspension Culture Systems for Chondrocytes

1. Agarose-coated dishes *(24)*: Weigh out 10 mL high-melting-point agarose in an autoclavable bottle and add 100 mL of dH_2O. Autoclave with cap tightened loosely, allow to cool to ~approx 55°C, and pipet quickly into culture dishes (1 mL/3.5-cm well of 6-well plate, 3 mL/6-cm dish, or 9 mL/10-cm dish). Allow the gel to set at 4°C for 30 min and wash the surface two or three times with PBS. Plates may be used immediately or wrapped tightly with plastic or foil to prevent evaporation and stored at 4°C.

2. Poly-HEMA (poly-2-hydroxyethyl-methacrylate)-coated dishes *(25)*: Prepare a 10% (w/v) solution by dissolving 5 g of polyHEMA (BD Biosciences) in 50 mL of ethanol in a sterile capped bottle or centrifuge tube. Leave overnight at 37°C with gentle shaking to dissolve polymer completely. Centrifuge the viscous solution for 30 min at 2000*g* to remove undissolved particles. Layer the polyHEMA solution on dishes at 0.3 mL/well of 6-well plate or 0.9 mL/6-cm dish and leave with lids in place to dry overnight in a tissue-culture hood. Expose open dishes to bactericidal ultraviolet light for 30 min to sterilize.

3. Agarose: Autoclave 2% (w/v) low-gelling-temperature agarose in dH_2O, cool to 37°C, and dilute with an equal volume of 2X DMEM containing 20% FCS either without cells or with a chondrocyte suspension.

4. Alginate (Keltone LVCR, NF; ISP Alginates Inc., 6605 Nancy Ridge Dr., San Diego, CA). Low viscosity (LV) alginate is used generally. Request LVCR for more highly purified preparation.
   a. Prepare 1.2% (w/v) solution of alginate in 0.15 *M* NaCl.
   b. Dissolve alginate in a 0.15 *M* NaCl solution, heating the solution in a microwave oven until it just begins to boil. Swirl and heat again two or three times until the alginate is dissolved completely. (**Caution:** Do not autoclave.) Allow the solution to cool to about 37°C and sterile filter. Filtering when warm permits the viscous solution to pass through the filter.
   c. Prepare 102 m*M* $CaCl_2$ and 0.15 *M* NaCl solutions in tissue culture bottles and autoclave.

d. Prepare 55 m*M* Na citrate, 0.15 *M* NaCl, pH 6.0, sterile filter, and store at 4°C. Make fresh weekly.

5. Three-dimensional (3D) scaffolds: Several types of scaffolds are available commercially, including the following:

   a. Gelfoam® (Pharmacia and Upjohn, Kalamazoo, MI), sterile absorbable collagen sponge, purchased as sponge-size 12–7 mm (2 cm × 6 cm × 7 mm; cat. no. 09-0315-03, box of 12).

   b. BD™ Three Dimensional Collagen Composite Scaffold (BD Biosciences): Contains a mixture of bovine type 1 and type III collagens and is provided as 3D scaffolds with 48-well plates.

   c. BD™ Three Dimensional OPLA® Scaffold (BD Biosciences): Contains a synthetic polymer synthesized from D,D-L,L polylactic acid and is provided as 3D scaffolds with 48-well plates.

6. Cell lysis solution for recovery of cells from scaffolds: 0.2% v/v Triton X-100, 10 m*M* Tris-HCl, pH 7.0, 1 m*M* EDTA.

7. Collagenase solution for recovery of cells from scaffolds: 0.03% (w/v) collagenase (Worthington, Freehold, NJ) in HBSS.

2.3. Analysis of Matrix Protein Synthesis

2.3.1. Alcian Blue Staining

1. 2.5% glutaraldehyde (diluted from 50% solution; Sigma) in 0.4 *M* $MgCl_2$ and 25 m*M* sodium acetate, pH 5.6.
2. Alcian blue 8GX (Sigma). Dissolve in the 2.5% glutaraldehyde solution to give final concentration of 0.05%. Filter the solution through Whatmann paper (or coffee filter).
3. Washes: 3% acetic acid solutions without and with 25 and 50% ethanol.

2.3.2. Collagen Typing

1. L-[5-^{3}H]proline (1 mCi/mL; specific activity >20 Ci/mmol) at 25 μCi/mL in serum-free culture medium supplemented with 50 μg/mL ascorbate and 50 μg/mL β-aminoproprionitrile fumarate (β-APN). Sterile filter 10X solution of ascorbic acid and β-APN (5 mg of each dissolved in 10 mL serum-free culture medium), dilute in medium at 1/10 (v/v) to give the volume required for the incubation, and add 25 μL of [^{3}H]proline per mL using a sterile pipet tip.
2. Pepsin/acetic acid solution: Dissolve 2 mg of pepsin in 1 mL dH_2O, then add 58 μL glacial acetic acid per each milliliter of solution and cool on ice.
3. Gel sample buffer (GSB): 0.1 M Tris-HCl, pH 7.6, 3% (w/v) SDS, and 16% (v/v) glycerol. Reagents for sodium dodecyl sulfate-polyacrylamide gel electrophoresis (SDS-PAGE) *(45)* and autoradiography.
4. Loading dye: 1% (w/v) bromophenol blue (BPB; sodium salt; Sigma).
5. Tris-glycine/SDS-5% gradient polyacrylamide or 7–15% gradient gels and Laemmli buffer system *(45)*.

2.3.3. Proteoglycan Synthesis

1. Biosynthetic labeling of proteoglycans: [^{35}S]sodium sulfate (2 mCi/mL; specific activity >1000 Ci/mmol). Add to culture medium at 50 µCi/mL.
2. 7 *M* urea.
3. DE52 columns, 3.5 × 12-cm containing 4 mL of DEAE-cellulose (Whatman, Hillsboro, OR).
4. Guanidine extraction buffer: 4.0 *M* guanidine-HCl, buffered with 50 m*M* sodium acetate, and containing 10 m*M* disodium EDTA. Add immediately before use 100 m*M* 6-aminocaproic acid, 2.5 m*M* benzamidine HCl, 5 m*M* N-ethylmaleimide, and 0.25 m*M* phenylsulfonyl fluoride (PMSF) from 100X stock solutions in absolute ethanol. The 2X guanidine extraction buffer is prepared at twice the concentrations above.
5. SDS-PAGE: Pre-cast Tris-glycine SDS-polyacrylamide 4–20% gradient gels (Bio-Rad). Use the GSB and BPB solutions for loading the samples and the Lemmli buffer system *(45)* (*see* **Subheading 2.3.2., steps 3** to **5**).

2.3.4. Immunocytochemistry

1. Fixative: 2% paraformaldehyde in 0.1 *M* cacodylate buffer, pH 7.4. Dissolve 10 g paraformaldehyde in 150 mL dH$_2$O in Ehrlenmeyer flask on hot plate in fume hood (do not exceed 65°C). Add ~approx 2 mL of 1 *N* NaOH while stirring, and stir until solution is clear. Let solution cool for 15 min. Add 250 mL 0.2 *M* cacodylate buffer, pH 7.4, and adjust pH if necessary.

2.3.5. Western Blotting

1. RIPA buffer: 50 m*M* Tris-HCl, pH 7.4, 150 m*M* NaCl, 0.5% sodium deoxycholate, 0.1% SDS, and 1% NP-40. At time of use, add Complete™, EDTA-free proteinase inhibitor cocktail (Roche Applied Science).
2. Tris-glycine/SDS-polyacrylamide gels at polyacrylamide concentration appropriate for the size of matrix protein to be analyzed (*see* **Subheadings 2.3.2.** and **2.3.3.**).
3. Nylon-supported nitrocellulose membranes, Immobilon-P, 0.45-µm (Millipore).
4. Transfer buffer: 25 m*M* Tris, pH 7.6, 192 m*M* glycine, 20 % (v/v) methanol.
5. Tris-buffered saline (TBS)/Tween (TBST): 20 m*M* Tris-HCl, 137 m*M* NaCl, 0.1% (v/v) Tween-20, pH 7.6. Add 5% (w/v) nonfat dry milk (Carnation) as required.
6. Primary antibody and horseradish peroxidase-conjugated secondary antibody: Dilute according to the supplier's instructions in TBST containing 5% (w/v) nonfat dry milk.
7. Enhanced Chemiluminescence ECL Western blotting analysis system (Amersham, Arlington Height, IL).

2.3.6. Antibodies

Antibodies that detect human collagens and proteoglycans that are specific to cartilage are available commercially from Southern Biotechnology Associates, Inc. (Birmingham, Alabama; http://southernbiotech.com), IBEX Technologies, Inc. (Montreal, Quebec, Canada; http://www.ibex.ca/), and Chemicon International (Temecula, CA; http://www.chemicon.com). Some antibody preparations are useful for developing quantitative enzyme-linked immunosorbent assay (ELISA) assays.

2.4. Analysis of mRNA

1. RNA extraction kit: The TRIzol® reagent (Invitrogen) or RNeasy® Mini Kit (Qiagen) are suitable for extraction of total RNA from chondrocytes.
2. Sterile, RNase-free solutions, polypropylene tubes, and other materials.

2.5. Analysis of Gene Expression by Transfection of Regulatory DNA Sequences

1. EndoFree Plasmid Maxi Kit (Qiagen).
2. Lipid-based transfection reagent such as LipofectAMINE PLUS™ Reagent (Life Technologies, Inc.) or FuGENE 6 (Roche Molecular Biochemical, Indianapolis, IN).
3. Serum-free culture medium for transfections: Opti-MEM (Life Technologies, Inc.) or DMEM/F-12 (test for optimal transfection efficiency).
4. Passive Lysis Buffer (Promega).
5. Coomassie Plus Protein Assay Reagent (Pierce Chemical Company, Rockford, IL).
6. Luciferase Assay System (Promega, Madison, WI).
7. Dual-Luciferase Reporter Assay (Promega) with the pRL-TK Renilla luciferase control vector.
8. Adenovirus producer cell line: 293 (ATTC CRL 1573; transformed primary human embryonic kidney).

3. Methods

The methods are described as follows: (1) the isolation of human chondrocytes from cartilage and their primary culture in monolayer and passaging, (2) suspension culture systems for maintaining chondrocyte phenotype, analysis of the (3) synthesis and (4) mRNA expression of cartilage-specific matrix proteins, and (5) analysis of gene expression by transfection of regulatory DNA sequences.

3.1. Isolation and Culture of Human Chondrocytes in Monolayer

1. Human adult articular cartilage is obtained, after Institutional Review Board approval, from the knee joints or hips after orthopaedic surgery for joint replacement or reconstruction, or at autopsy, and dissected free from underlying bone and any adherent connective tissue.
2. Place slices of cartilage in a 10-cm dish and wash several times with PBS. Incubate slices at 37°C in hyaluronidase for 10 min followed by 0.25% trypsin for 30–45 min with two or three washes in PBS after each enzyme treatment. Use approx 10 mL of proteinase solution for digestion of each gram of tissue.

Table 2
Culture Vessel Area vs Chondrocyte Number Required
for Plating Density of approx 2.5 × 10⁴ cells/cm²

Diameter	Area (cm²)	No. of cells plated
16-mm well (24-well)	2	50,000
2.2-cm well (12-well)	3.8	100,000
3.5-cm well (6-well)	10	250,000
6-cm plate	28	750,000
10-cm plate	79	2×10^6

3. Add collagenase solution, chop the cartilage in small pieces using a scalpel blade, and incubate at 37°C overnight (18–24 h) for articular cartilage and up to 48 h for costal cartilage until the cartilage matrix is completely digested and the cells are free in suspension (*see* **Note 2**). Break up any clumps of cells by repeated aspiration of the suspension through a 10-mL pipet or a 12-cc syringe without a needle.

4. Transfer cell suspension to a sterile 50-mL conical polypropylene tube and wash the plate with PBS to recover remaining cells and combine in tube. Centrifuge cells at 1000*g* in a bench-top centrifuge for 10 min at room temperature and wash the cell pellet three times with PBS, resuspending cells each time and centrifuging.

5. Resuspend the final pellet in DMEM/F12 containing 10% FCS, perform cell count with a Coulter counter or hemacytometer, and bring up to volume with culture medium to give 1×10^6 cells per mL. For monolayer culture, plate cells at 2.5×10^4/cm² (*see* **Table 2**) in dishes or wells containing culture medium, and agitate without swirling to distribute the cells evenly. Incubate at 37°C in an atmosphere of 5% CO_2 in air with medium changes after 2 d and every 3 or 4 d thereafter, as described (*see* **Note 3**). Primary of adult articular chondrocytes incubated in the absence or presence of IL-1β are shown in **Fig. 2**.

6. Preparation of subcultured cells (*see* **Note 4**): Remove culture medium by aspiration with a sterile Pasteur pipet attached to a vacuum flask and wash with PBS. Add trypsin-EDTA (1 mL/10-cm dish) and incubate at room temperature for 10 min with periodic gentle shaking of dish and observation through microscope to ensure that cells have come off the plate. If significant numbers of cells remain attached, continue the incubation for a longer time (≤20 min) or at a higher temperature (37°C) and/or scrape the cell layer with a sterile plastic scraper or syringe plunger. Repeatedly aspirate and expel the cell suspension into the plate using a 5- or 10-mL pipet containing culture medium, and then transfer to a sterile conical 15- or 50-mL polypropylene tube. Perform cell counts or determine the split ratio required (usually one dish into two, or 1:2, for adult articular chondrocytes, or 1:5 for more rapidly growing, denser juvenile chondrocytes). Distribute equal volumes of the cell suspension in dishes or wells that already contain culture medium, rocking plates back-and-forth (not swirling) immediately after each addition to ensure uniform plating density (*see* **Note 5**).

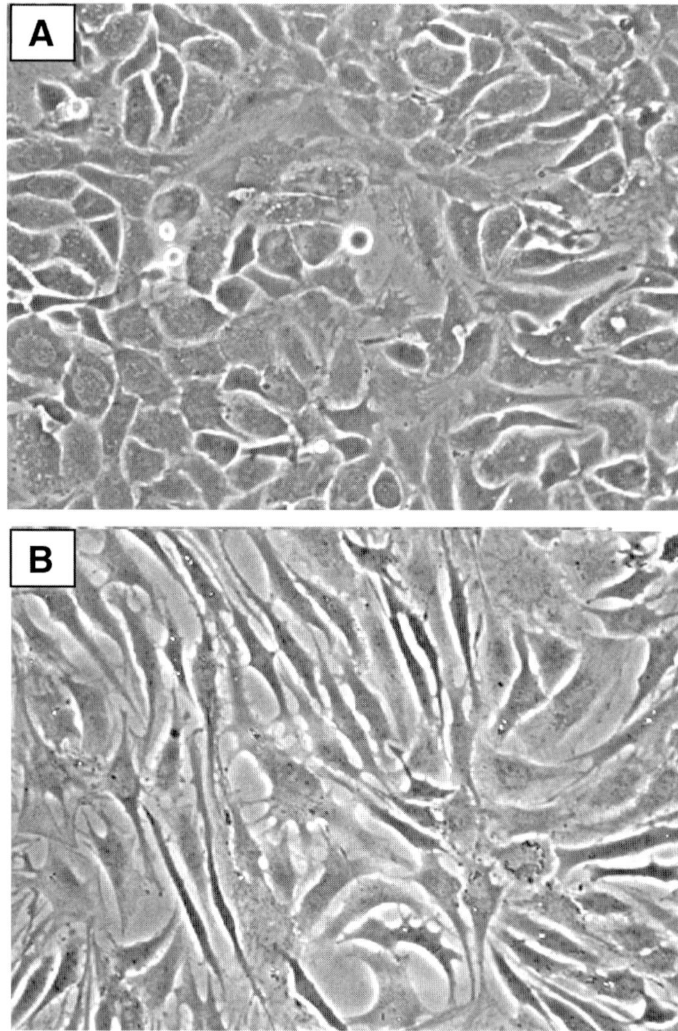

Fig. 2. Morphology of human articular chondrocytes grown in monolayer culture on plastic. Chondrocytes were isolated from articular cartilage and cultured in DMEM/F12 containing 10% FCS until confluent and then subcultured and grown again to confluence. (**A**) Primary chondrocytes display the characteristic cobblestone morphology. (**B**) Passaged chondrocytes display features of a dedifferentiated phenotype, including a portion of cells with fibroblast-like morphology.

7. For experiments, plate cells in dishes or wells at a concentration of 2×10^6 cells/cm² in DMEM/F12 containing 10% FCS. Remove the growth medium, when the cultures are confluent. Add serum-free medium containing a serum substitute (*see* **Note 6**), such as Nutridoma-SP, followed 18–24 h later by the test agent of

interest. Continue incubation at 37°C for short-term time courses of 0.25–24 h or longer time courses of several days.

3.2. Suspension Culture Systems for Chondrocytes

Chondrocytes in monolayer culture are susceptible to loss of phenotype during prolonged culture and particularly after subculture. Thus, it is necessary to use culture conditions that maintain differentiated chondrocyte features or that permit redifferentiation (*see* **Note 7**). Freshly isolated chondrocytes may be cultured immediately in suspension, where they do not proliferate, or they may be expanded in monolayer culture and placed in suspension culture after several passages.

3.2.1. Fluid Suspension Cultures on Agarose- or polyHEMA-Coated Dishes (see **Note 8**)

1. Trypsinize monolayer cultures, spin down cells, wash with PBS, centrifuge, and resuspend in culture medium containing 10% FCS at 1×10^6 cells/mL.
2. Transfer chondrocyte suspension to dishes that have been coated with 1% agarose or with 0.9% polyHEMA and culture for 2–4 wk. The cells first form large clumps that begin to break up after 7–10 d and eventually form single-cell suspensions.
3. Change the medium weekly by carefully removing the medium above settled cells while tilting the dish, centrifuging the remaining suspended cells, and replacing them in the dish after resuspension in fresh culture medium.
4. To recover cells for direct experimental analysis, for redistribution in agarose- or polyHEMA-coated wells, or for culture in monolayer, transfer the cell suspension to 15- or 50-mL conical tubes, gently washing the agarose surface at least twice with culture medium to recover remaining cells, and spin down and resuspend cells in an appropriate volume of culture medium for plating or in extraction buffer for subsequent experimental analysis.

3.2.2. Suspension Culture within Agarose

1. Precoat plastic tissue culture dishes with cell-free 1% agarose in culture medium (0.5 mL/3.5-cm, 1.5 mL/ 6-cm, or 4.5 mL/10-cm dish) and allow to gel at room temperature. Add the same volume of 1% agarose in medium containing chondrocytes at a density of $1–4 \times 10^6$ cells/mL of gel, incubate at 37°C for 20–30 min to allow the cells to settle, and leave at room temperature until the agarose forms a gel. Add culture medium containing 10% FCS and incubate at 37°C with medium changes every 3–4 d.
2. After incubations with test reagents and/or radioisotopes in minimal volumes of appropriate culture medium, the whole cultures may be stored frozen or medium and gel treated separately. For subsequent analysis, add appropriate guanidine extraction buffer directly to the gel (for proteoglycan or RNA extraction) or whole cultures may be adjusted to 0.5 *M* acetic acid, treated with pepsin, and neutralized,

as described below for analysis of collagens. To remove agarose and debris, the samples are centrifuged in a high-speed centrifuge at >10,000g at 4°C.

3.2.3. Alginate Bead Cultures (see **Note 9**)

1. Trypsinize several 10-cm plates and wash the cells with PBS. Determine the cell count with a hemacytometer and pellet the cells.
2. Resuspend the pellet in a 1.2% solution of alginate in 0.15 M NaCl at a concentration of 1–4 × 10^6 cells per mL. Slowly express the alginate suspension in a dropwise manner through a 10-cc syringe equipped with a 22-gauge needle into a 50-mL polypropylene centrifuge tube containing 40 mL of 102 mM CaCl$_2$. Allow the beads to polymerize in the CaCl$_2$ solution for 10 min and wash twice with 25 mL of 0.15 M NaCl. The alginate beads should *not* be washed in PBS, as they will become cloudy.
3. Resuspend the beads at 7–15 beads per mL in growth medium supplemented with 25 µg/mL Na ascorbate (*see* **Note 10**) and decant to a culture dish or flask. Culture in DMEM/F12 with medium changes every 3 d, carefully pipetting the spent culture medium from the top of the settled beads.
4. At the end of the culture period (one to several weeks), add the appropriate guanidine extraction buffer or centrifuge at 500g for 10 min to recover the chondrocytes with pericellular matrix.
5. Alternatively, to recover cells from alginate, carefully aspirate the medium from the cultures and wash twice with PBS. Depolymerize the alginate by adding three volumes of a solution of 55 mM Na citrate/0.15 M NaCl and incubate at 37°C for 10 min. Aspirate the solution over the surface of the dish several times to dislodge adherent cells (the cells are sticky) and transfer the suspension to a 50-mL centrifuge tube. Because the solution is quite viscous, centrifuge the cells at 2000g for a minimum of 10 min to completely pellet the cells. Wash the cells twice with PBS before using them for further analysis.

3.2.4. Culture on 3-D Scaffolds (see **Note 11**)

1. Gelfoam®: Use sterile scalpel blade to cut into pieces of 1 × 1 × 0.5 cm^3 and place in wells of sterile six-well plates. Inoculate by dropping 50 µL of growth medium containing 10^6 cells on each sponge. Place in incubator for 1.5–2 h, then add 100 µL medium and culture for an additional 1–3 h. Add medium to cover and continue incubation overnight or longer.
2. BD™ 3D Collagen Composite or OPLA® Scaffolds: Place scaffolds (0.5 cm^3) in the 48-well plates provided, in 96-well plates, or other plate as required. Seed scaffolds by dropping 100 µL of growth medium containing 1–5 × 10^4 cells. Incubate for 1 h, add 150 µL of medium to each scaffold, and incubate for 1.5–3 h. Add medium as required for further culture and experimental conditions.
3. Recovery of cells from scaffolds for analysis:
   a. Prepare cell lysates for DNA analysis using 250–500 µL of cell lysis solution [0.2% (v/v) Triton X-100, 10 mM Tris-HCl, pH 7.0, 1 mM EDTA] per scaffold in 1.5-mL tube. Freeze samples at –70°C and subject to two freeze/thaw cycles,

thawing at room temperature for 45–60 min. Break up scaffolds with pipet tip, centrifuge, and transfer lysates to fresh tubes. The cell lysates may be analyzed using the Picogreen Assay Kit according to the manufacturer's protocol (Molecular Probes).

b. Recover cells for RNA extraction and other analyses by treatment of minced scaffolds with 0.03% (w/v) collagenase in HBSS for 10–15 min at 37°C. Collect cells by centrifugation, wash with PBS, and add appropriate extraction buffer to the final pellet.

3.3. Analysis of Matrix Protein Synthesis

Chondrocyte culture models are used to examine the effects of cytokines and growth/differentiation factors on the synthesis of chondrocyte phenotypic markers by staining the glycosaminoglycans with alcian blue, characterizing the collagens and proteoglycans synthesized, and perform immunohistochemistry using specific antibodies against these proteins.

3.3.1. Alcian Blue Staining

1. Monolayer cultures: Remove culture medium from confluent cultures that have been incubated in the presence of 50 mg/mL ascorbic acid for at least 4 h and wash with PBS. Add alcian blue/glutaraldehyde solution at room temperature for several hours, remove excess stain by washing with 3% acetic acid, and store cultures in 70% ethanol for subsequent examination by light photomicrography.
2. Alginate bead cultures:
   a. Using a 25-mL pipet, transfer five beads to a 12 × 15 cm tube, and wash twice with 2 mL PBS. Add 1 mL alcian blue stain, and 50 µL of a 50% glutaraldehyde solution. Store for 24 h at 4°C.
   b. Aspirate the stain from the beads, and wash twice with 2 mL of 3% (v/v) acetic acid. Destain the beads with rocking at room temp for 5 min sequentially with 2 mL of each of the following solutions: (1) 3% acetic acid, (2) 3% acetic acid/25% ethanol, and (3) 3% acetic acid/50% ethanol. Store the beads in 70% ethanol.
   c. Before photographing the stained beads, gently flatten beneath a glass cover slip, taking care not to disrupt the alginate matrix. Alternatively, the beads may be embedded and sectioned prior to photography.

3.3.2. Collagen Typing

1. Biosynthetic labeling of collagens (*see* **Note 12**): Remove serum-containing culture medium, wash with serum-free medium, and add [^{3}H]proline at 25 µCi/mL for a further 24 h in serum-free culture medium supplemented with 50 µg/mL ascorbate and 50 µg/mL β-APN (or without β-APN to retain collagen in the pericellular matrix). Remove culture medium and store at –20°C. Wash cell layer with PBS and solubilize by adding equal volumes of serum-free culture medium and 1 *M* ammonium hydroxide (an aliquot may be analyzed for DNA).

2. Collagen typing: To analyze pepsin-resistant collagens, add pepsin/acetic acid solution to an equal volume of either labeled culture medium or solubilized cell solution for 16 h at 4°C, lyophilize, redissolve in 2X SDS sample buffer, and neutralize with 1 μL additions of 2 *M* NaOH to titrate the color from yellow-green to blue (but not to violet). To analyze procollagens and fibronectin, add 2X SDS sample buffer containing 0.2% β-ME to an equal volume of the culture medium. Heat samples to boiling for 10 min and load on SDS gels (5% acrylamide running gels or 7–15% gradient gels) that include a radiolabeled rat tail tendon collagen standard in one lane. Perform delayed reduction with 0.1% β-ME on pepsinized samples to distinguish α1(III) from α1(I or II) collagens. Absence of the α2(I) collagen band generally indicates the absence of type I collagen synthesis. In cultures containing a mixture of type I and type II collagen, definitive identification of these collagens requires Western blotting using specific antibodies (*see* **Subheading 3.3.4.**).

3.3.3. Proteoglycan Synthesis

1. Monolayer cultures: Aspirate the culture medium and wash the cell layer with PBS. Add DMEM/F12 containing 10% FCS supplemented with [^{35}S]sulfate at 50 μCi/mL and 25 μg/mL ascorbate. Incubate at 37°C for 18 h. Remove the conditioned medium to a 15-mL polypropylene tube and wash the cell layer three times with PBS.
   a. Medium extraction and purification: Add an equal volume of 7 *M* urea to the medium. Mix and count 10 μL in a liquid scintillation counter. Pass up to 4 mL through DEAE- Cellulose (DE52, Whatman) column, which has been pre-equilibrated with 7 *M* urea, to remove unincorporated cpm. Elute proteoglycans (PGs) with 2 mL of 4 *M* guanidine extraction buffer. To 500 μL of column eluate, add two volumes of 100% ethanol and precipitate PGs for 2 h at –20°C. Spin at 10,000*g* for 20 min at 4°C. Wash final pellet with 70% ethanol.
   b. Cell layer extraction and purification: Add 4 *M* guanidine extraction buffer to cell layer e.g., 2 mL/25-cm^2 flask) and extract at 4°C for 24 h with rocking. Transfer extract to a 2-mL screw cap microcentrifuge tube and spin at 10,000*g* for 20 min at 4°C to pellet particulate material from the sample. Remove supernatant to a fresh 2-mL tube and count 10 μL in a liquid scintillation counter. (Also, 10–50 μL aliquots may be taken at this point for DNA analysis.) Take 250 μL for precipitation of PGs by addition of three volumes of 100% ethanol at –20°C for 2 h (or overnight for alginate extracts), spin at 10,000*g* for 20 min at 4°C, and wash final pellet with 70% ethanol. Dry pellet at room temperature for 30 min.
2. Alginate cultures: Aspirate the medium from a culture containing 50 beads in a 25-cm^2 flask cultured on end. Add 4 mL of growth medium supplemented with [^{35}S]sulfate at 50 μCi/mL and 25 μg/mL ascorbate. Incubate at 37°C for 18 h. Remove the conditioned medium to a 12-mL polypropylene tube and wash the beads three times with 5 mL of PBS (5 min/wash).

   a. Medium extraction and purification: Perform as described for monolayer cultures in **Subheading 3.3.3., step 1a**.

   b. Extraction from alginate beads and purification: Transfer the beads to a 15-mL polypropylene tube, extract radiolabeled PGs with 2 mL of 4 M guanidine extraction buffer at 4°C for 24 h with rocking. Transfer extract to a 2-mL screw cap microcentrifuge tube and proceed as described for monolayers in **Subheading 3.3.3., step 1b**.

3. Characterization of proteoglycans in monolayer and alginate cultures:

   a. Sulfate incorporation into proteoglycans: Dissolve ethanol-precipitated pellet in 100 µL of 1X GSB. Count 3 µL in a liquid scintillation counter and calculate the cpm incorporated, after accounting for dilutions, against the concentration of DNA in the cell extract. The [^{35}S]sulfate incorporation may also be determined, after passing the guanidine extracts over Sephadex G-25M in PD 10 columns and eluting under dissociative conditions, by scintillation counting.

   b. SDS-PAGE analysis: Take a volume of cell or medium extract in GSB that corresponds to 0.25 µg of DNA in the cell extract and add DTT to a final concentration of 0.5 mM and 1% bromophenol blue to a final concentration of 0.1%. Heat 5 min at 100°C and store remaining sample at –20°C. Electrophorese ~approx 20,000 cpm on a 4–20% polyacrylamide gradient gel. Fix the gel in acetic acid/methanol for 1 h, dry, and expose to film at –80°C. Visualize radiolabeled PGs by autoradiography, as shown in **Fig. 3** *(46–48)* (*see* **Note 13**).

3.3.4. Immunocytochemistry

1. Plate cells in plastic Lab-Tech four-chamber slides (Nunc, Inc., Naperville, IL) at 6×10^4 cells/chamber in culture medium containing 10% FCS. Add 25 µg/mL ascorbic acid with the first medium change and daily thereafter (*see* **Note 10**).

2. When the cultures have reached confluence, add the desired test reagents. At the end of the incubation period, carefully wash the chambers three times with PBS, and fix the cells with 2% paraformaldehyde in 0.1 M cacodylate buffer, pH 7.4, for 2 h at 4°C.

3. Rinse twice with 0.1 M cacodylate buffer. Add antibodies that recognize human type II collagen, aggrecan, and so on (*see* **Subheading 2.3.6.**) to different chambers at concentrations recommended by the supplier. Incubate separate chamber slides with chondroitinase ABC for 30 min at 37°C prior to addition of monoclonal antibodies in order to expose epitopes.

4. Visualize the staining by incubation with a gold-conjugated secondary antibody (Auroprobe LM, Amersham) followed by silver enhancement (e.g., IntenSE Kit, Amersham).

3.3.5. Western Blotting

1. Plate cells in six-well tissue culture plates at a density of 0.25×10^6 cells per well in culture medium containing 10% FCS. Add 25 µg/mL ascorbic acid with

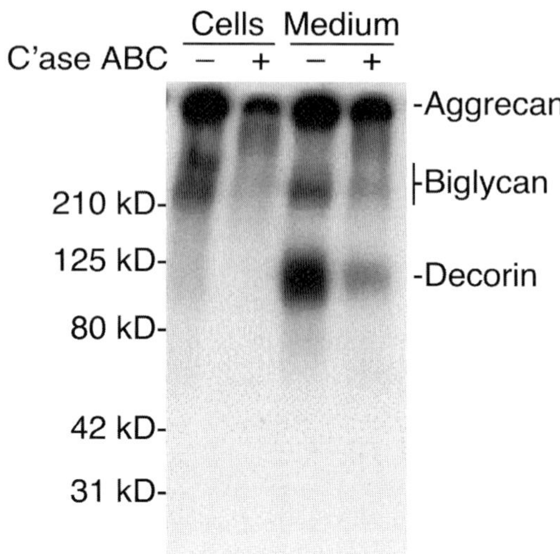

Fig. 3. SDS-PAGE analysis of proteoglycans synthesized by human articular chondrocytes. Primary cultures were grown to confluence in DMEM/F12 containing 10% FCS. The cultures were incubated with [^{35}S]sulfate in the presence of 25 µg/mL ascorbate during the final 18 h. Radiolabeled proteoglycans were purified by ethanol precipitation of 4 *M* guanidine extracts of cell monolayers or by DEAE ion-exchange chromatography and ethanol precipitation of 7 *M* urea extracts of the medium prior to SDS-PAGE on a 4–20% gradient gel. Equal volumes of cell and medium extracts representing equivalent proportions of the total culture were loaded on the gel. Molecular weight standards are indicated to the left of the panel. The migration of aggrecan, biglycan, and decorin is indicated to the right. Treatment with chondroitinase (C′ase) ABC resulted in the expected loss of sulfation.

the first medium change and daily thereafter (*see* **Note 10**). When the cultures have reached confluence, add the desired test reagents.

2. At the end of the incubation, remove medium, wash cell layers with 3 mL/well of PBS, and add 0.2 mL of ice-cold RIPA buffer containing freshly added proteinase inhbitors. Extract for 15 min at 4°C with gentle rocking, scrape and transfer cell lysate to a microcentrifuge tube. Incubate 30–60 min on ice, microcentrifuge at 10,000*g* for 10 min at 4°C, and transfer supernatant to new tube, discarding pellet.

3. Mix cell lysate (20–40 µg of protein) with equal volume of 2X GSB, boil for 2–3 min, and electrophorese on a Tris-glycine/SDS polyacrylamide pre-cast gel, and transfer to Immobilon-P membrane by electroblotting with a transfer current of 100 V for 1 h.

4. Block membrane with 10 mL TBST containing 5% nonfat powdered milk at 4°C for 1 h with rocking, and wash three times (10 min/wash) in 10 mL TBST.

5. Incubate membrane with the antibody at the appropriate dilution in 5 mL of TBST containing 5% nonfat powdered milk for 1 h at room temperature with rocking. Wash membrane three times (5 min/wash) with TBS, 0.05% Tween-20. Incubate

with secondary antibody diluted in TBST for 45 min and repeat washes as above, followed by one wash with TBS for 5 min.

6. Detect bound antibody by enhanced chemiluminescence according to the manufacturer's protocol and expose to film for autoradiography.

3.4. Extraction and Analysis of RNA

1. Recovery of cells for extraction: Trypsinize monolayer cultures as described in **Subheading 3.1.3., step 6**, and add medium containing 10% FCS to the suspension to inactivate the trypsin. Depolymerize the alginate cultures as described in **Subheading 3.2.3., step 3**. Transfer cell suspension to sterile, RNase-free polypropylene tube of appropriate size, centrifuge at 1000g at 4°C, washing two or three times with ice-cold PBS.

2. Extract total RNA using TRIzol® reagent (Invitrogen), RNeasy® Mini Kit (Qiagen), or other method as preferred. Add extraction buffer to final pellet, vortex vigorously, and transfer to RNase-free 1.2-mL centrifuge tube. Continue according to the manufacturer's instructions. The final pellets are usually washed with 75% ethanol and centrifuged in a speed-vac apparatus, but not to dryness. Dissolve pellets in nuclease-free water.

3. Read ODs at 240, 260, and 280. Use OD 260 to calculate RNA concentration. The final preparations should give yields of approx 10 µg of RNA per 1 × 10^6 cells with the appropriate A_{260}:A_{280} ratio of approx 2.0. Store at –20°C in nonself-defrosting freezer or at –80°C.

4. Analyze mRNAs by Northern blotting *(15)*, semiquantitative RT-PCR *(40,49)*, or real-time PCR *(50)* by published methods. The Northern blots shown in **Fig. 4** indicate that type II collagen gene expression is relatively stable in primary cultures of human chondrocytes, at least through the first 12 d of culture. Type I collagen gene expression is low, but TGF-β is a potent stimulant. This experiment also illustrates that the effects of TGF-β on type II collagen gene expression may be time-dependent with inhibition at the early time point, but stimulation later on.

3.5. Analysis of Gene Expression by Transfection
of Regulatory DNA Sequences

Transfection studies may be performed to analyze the DNA sequences and transcription factors involved in the regulation of gene expression using plasmid vectors, in which the expression of reporter genes such as CAT *(51)* or luciferase *(40,52,53)* is driven by gene regulatory sequences, such as those regulating type II collagen gene (COL2A1) transcription (*see* **Fig. 5**). Coexpression of wild-type or dominant-negative mutants of transcription factors, protein kinases, and other regulatory molecules, mediated by plasmid or adenoviral vectors may be performed to further dissect the mechanisms involved.

3.5.1. Transient Transfections Using Luciferase Reporter Plasmids

1. Prepare plasmids using the EndoFree Plasmid Maxi Kit, according the manufacturer's instructions (Qiagen), to generate endotoxin-free DNA (*see* **Note 14**).

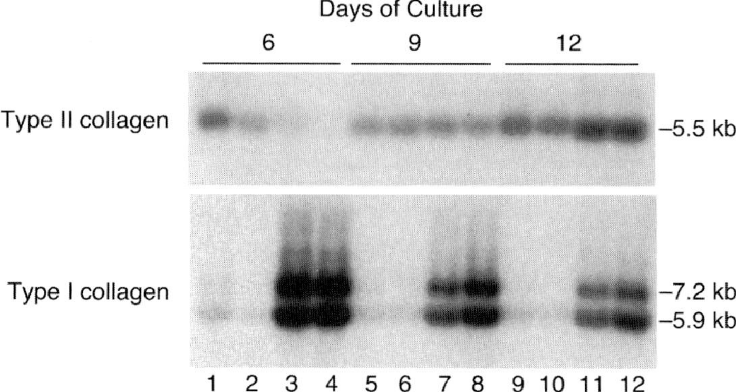

Fig. 4. Northern blotting analysis of type II collagen and type I collagen mRNAs expressed by adult human articular chondrocytes in primary culture. Freshly isolated chondrocytes in primary culture were left untreated (lanes 1, 5, 9), or treated with 1 µ*M* indomethacin (lanes 2, 6, 10), which blocks the effects of endogenous prostaglandins *(51)*, 10 ng/mL TGF-β (lanes 3, 7, 11), or TGF-β +indomethacin (lanes 4, 8, 12) with medium changes on d 2 and every 3 d thereafter. The cells were harvested 24 h after medium change on d 6, 9, and 12. Total RNA was extracted and the mRNAs encoding type II and type I collagens were analyzed on Northern blots. The transcript sizes are indicated on the right.

2. On the day before transfection, seed cells in six-well tissue-culture plates at $2.5–5 \times 10^5$ cells/well in DMEM/F-12 containing 10% FCS. (Determine optimal cell number to ensure that cultures are greater than 50% confluent at the time of transfection.) Hyaluronidase may be added during these medium changes to facilitate the infection efficiency (*see* **Note 15**).

3. Change the medium to fresh growth medium 24 h and 3 h before the transfection to ensure that the cells are actively dividing. Prepare lipid/DNA complexes in serum-free DMEM/F-12 or Opti-MEM using LipofectAMINE+ or FuGENE 6, according to the manufacturer's protocol. Prepare in bulk for multiple transfections. Do not vortex at any step:

   a. LipofectAMINE+: For each well, add 92 µL of serum-free medium to a small sterile polypropylene tube, add 1 µL of plasmid DNA (maximum of 1 µg; *see* **Note 16**), and tap gently to mix. Add 6 µL of PLUS reagent, mix and incubate for 15 min at room temperature. Dilute 4 µL of LipofectAMINE+ reagent into 100 µL of serum-free medium, mix and add to each reaction mixture. Mix and leave at room temperature for an additional 15–30 min at room temperature.

   b. FuGENE 6: For each well, add 96 µL of serum-free medium to a small sterile polypropylene tube, add 3 µL of FuGENE 6 reagent, and tap gently to mix. Add 1 µL plasmid of DNA (maximum of 1 µg) to the prediluted FuGENE 6 reagent and incubate for 15 min at room temperature.

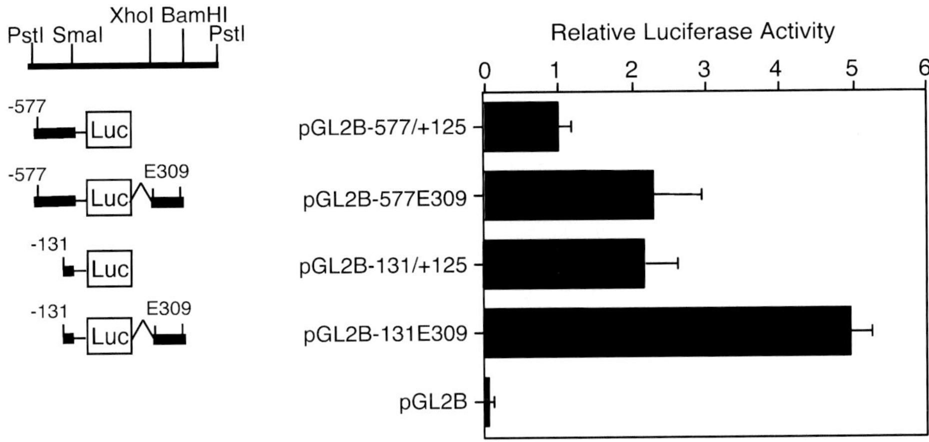

Fig. 5. Expression of the *COL2A1* promoter human articular chondrocytes. Luciferase reporter constructs containing COL2A1 sequences were transfected in primary chondrocytes using LipofectAMINE+ and incubated in DMEM/F-12 containing 1% FCS for 24 h prior to harvest for luciferase assay. The promoter constructs, pGL2B-577/+125, and pGL2B-131/+125 without or with the E309 enhancer region (+2388/+2696 bp) from intron 1 were compared. Luciferase activity was normalized to the amount of protein and expressed as relative activity to that of untreated cells transfected with pGL2B-577/+125E309. Each value is calculated as the mean ± SD of the results from 3 to 6 wells and results are representative of at least three experiments. Note that the empty vector, pGL2-Basic, expressed at 1 to 2% of the levels of the *COL2A1* promoter constructs.

4. While the lipid/DNA complexes are forming, replace culture medium on cells with serum-free medium to give a final volume of 1 mL. Add the lipid/DNA complex mixture dropwise to the well and incubate for 4 h at 37°C.
5. Dilute the transfection medium by adding to the wells an equal volume of DMEM/F-12 containing 2% Nutridoma-SP (or 20% FCS), and incubate 2 h to overnight. Add test agent without medium change and incubate further for 18 h (up to 48 h) (*see* **Note 17**).

3.5.2. Cotransfections using Plasmid Vectors for Expression of Recombinant Proteins

1. Prepare plasmids as described above.
2. Titrate each expression vector and its corresponding empty vector, at amounts ranging from 10 to 200 ng per well, against a fixed amount of reporter vector. Equalize the total amount of reporter plus expression plasmid in each well (<1 µg/well) by adding empty vector and maintain equal volumes (*see* **Note 16**).
3. After cotransfection, incubate the cells for 18–24 h to permit expression of recombinant protein prior to treatment with test reagents.

3.5.3. Adenoviral-Mediated Expression of Recombinant Proteins

1. Infect the 293 producer cell line with adenoviral vector containing the cDNA encoding the wild-type or mutant protein to be coexpressed and determine the titer (moi) by standard techniques.
2. Incubate immortalized chondrocytes in DMEM/F-12 containing 10% FCS for 18 h following transfection of the reporter construct.
3. Remove medium and wash cells with PBS.
4. Add 1 mL of serum-free medium containing adenovirus at 1:125 moi. Incubate at 37°C for 90 min.
5. Add 1 mL of DMEM/F-12 containing 20% FCS and continue incubation for 18 h.
6. Change medium to fresh DMEM/F-12 containing 10% FCS or 1% Nutridoma-SP, incubate for 1 h, and treat with test agent for 18 h.

3.5.4. Luciferase Assay

1. Prepare cell lysates by extraction with 200 µL of passive lysis buffer, which will passively lyse cells without the requirement of a freeze-thaw cycle. Scrape cells with policeman and transfer solubilized cells to 1.5-mL microcentrifuge tube. Microcentrifuge 5 min at maximum speed at 4°C, transfer supernatant to clean microcentrifuge tube, and store on ice.
2. Determine the protein content using the Coomassie Plus Protein Assay Reagent (Pierce Chemical Company, Rockford, IL).
3. Determine luciferase activities using the Luciferase Assay System or equivalent, according to manufacturer's protocol. Mix, manually or automatically, 20 µL of cell lysate with 100 µL of Luciferase Assay Reagent and read in a luminometer. Normalize to the amount of protein (or internal control such as β-galactosidase) and express as relative activity against the empty vector or untreated control. Perform each treatment in triplicate wells and each experiment at least three times to ensure reproducibility and significance.
4. The Dual-Luciferase Reporter Assay (Promega), in which 20 ng of the pRL-TK Renilla luciferase control vector is included in the transfections, may be used routinely or as necessary to check the purity of new plasmid preparations or relative activities of mutant and wild-type constructs.

4. Notes

1. Batches of serum should be tested and selected on the basis of the capacity to support expression of chondrocyte-specific matrix gene expression. High capacity to induce cell proliferation is not necessarily associated with the ability to maintain phenotype.
2. Chondrocytes are quite resilient and tolerate the prolonged incubation times required for complete dissociation of the matrix and the absence of serum in the costal cartilage digestion. If the digestion is not complete by the end of the allotted time, then more collagenase solution may be added, or the suspension may be recovered and the fragments left behind for further digestion. These conditions

result in suspensions that are essentially single cell, and therefore, it is not necessary to resort to filtration through a nylon mesh, as has been done by others when shorter digestion times are used *(12)*. These considerations are important for decreasing the loss of chondrocytes during their isolation from valuable human cartilage specimens.

3. After initial plating of the primary cultures, the chondrocytes require 2–3 d before they have settled down and spread out completely. Culture for ~approx 4–7 d is required before reasonable amounts of total RNA may be extracted. Although the cultures may continue to express chondrocyte phenotype (e.g., type II collagen and aggrecan mRNAs) for several weeks, expression of nonspecific collagens I and III may begin as early as d 7 after isolation. Adult articular chondrocytes are strongly contact-inhibited and they may lose phenotype within 1–2 wk of monolayer culture. Juvenile costal chondrocytes continue to express chondrocyte phenotype (e.g., type II collagen mRNA) for several weeks and will form multilayer cultures. After they are subcultured, both types of chondrocytes cease the expression of chondrocyte matrix proteins, but this loss of phenotype is reversible and the cells may be redifferentiated in suspension culture within or on top of a nonadherent matrix.

4. Because chondrocytes adhere strongly to tissue culture plastic, possibly because of the presence of calcium ion-binding glycosaminoglycans in the pericellular matrix and cell membrane, a trypsin-EDTA solution rather than trypsin alone should be used for full recovery of chondrocytes from tissue culture plastic during passaging. It is preferable not to use any antibiotics in order that any contamination that arises becomes apparent immediately. If necessary, standard concentrations of penicillin–streptomycin, gentamycin, and so on, that are suggested for fibroblast cultures are acceptable for use in chondrocyte cultures.

5. Primary chondrocyte cultures should be used for experimental analyses immediately before or just after confluence is reached to permit optimal matrix synthesis and cellular responsiveness. If the cells are not used or subcultured, they may be left at confluence for several weeks with weekly medium changes as long as the volume of the culture medium is maintained. If long-term culture results in the deposition of excessive matrix that is not easily digested with trypsin-EDTA, then a single-cell suspension may be obtained by using a dilute solution of collagenase (0.25%) and trypsin (0.25%) in PBS.

6. The synthetic activities of chondrocytes in monolayer culture are inversely related to proliferative activities. Thus, the expression of genes encoding matrix proteins and their deposition into the extracellular matrix increase compared to cell growth-associated genes. For experiments, the growth medium should not be changed within 3 d before addition of the test agent. Alternatively, the cells should be made quiescent by changing to serum-free medium supplemented with an insulin-containing serum substitute such as Nutridoma-SP or ITS+, followed 18 to 24 h later by the addition (without medium change) of the test agent of interest. Confluent cultures may tolerate serum-free medium containing 0.3% BSA for up to 48 h or longer.

7. Although the growth and maintenance of chondrocytes in primary culture or after subculture requires the use of 10% FCS, the loss of phenotype that occurs under these conditions may be delayed if the cells are plated at 4- to 10-fold higher density. Because high cell yields are not usually attainable from human cartilage sources, the reversibility of the loss of phenotype may be exploited by expanding the chondrocyte populations in monolayer cultures, redifferentiating the cells in fluid suspension culture and replating them in monolayer immediately before performing the experimental procedure.

8. After several passages in monolayer, chondrocytes may be redifferentiated by 2 wk or more of culture in alginate beads or in suspension over agarose or poly-HEMA.

9. The method for culture of chondrocytes in alginate beads has been adapted from previously published methods *(30,31)*. For long-term alginate cultures, high viscosity alginate may provide more stable beads. Concentrations of serum as low as 0.5%, serum substitutes, or combinations of growth and differentiation factors or hormones have been used successfully, depending upon the experimental protocol, to permit chondrocyte phenotypic expression. Note that articular chondrocytes do not proliferate when cultured in fluid or gel suspension.

10. Ascorbate, which is required for synthesis and secretion of proteoglycans and collagens, is added daily to alginate or other 3D cultures to permit secretion and deposition of extracellular matrix, particularly when staining techniques are to be used. Add 25 µg/mL of ascorbate during the final 24–72 h of incubation when radiolabeling proteoglycans with ^{35}S-sulfate or collagens with ^{3}H-proline for characterization by SDS-PAGE. It is not necessary to maintain ascorbate in cultures if analysis of type II collagen or aggrecan mRNA is the endpoint, since effects on gene transcription may vary according to the time of incubation *(54)*.

11. Culture of immortalized chondrocytes in 3D scaffolds is a useful approach for tissue engineering applications. The commercially available methods are recommended because of their ease of use. Published methods are available for fabricating collagen sponges *(32)* and other 3D scaffolds, where the composition may be manipulated, for example, by using type II collagen and/or adding proteoglycans and other cartilage-specific matrix components. The biodegradable scaffolds are particularly useful if the cell-seeded scaffolds are to be implanted in animals. For studies entirely in vitro, where incubation periods of more than a few days are required, it is recommended that cultures be performed in wells that fit the size of the scaffolds. Otherwise, the culture surface of the well or dish should be coated with a nonadherent substrate or treated in such a way as to prevent attachment of cells that may migrate out from the sponges. Additional analytical methods have been described using Gelfoam® *(55)* and BD™ 3D scaffolds (see website at http://www.bdbiosciences.com/discovery_labware/Products/tissue_engineering/).

12. Biosynthetic labeling and immunocytochemistry procedures are readily performed on chondrocytes in a solid suspension system such as alginate, agarose, or collagen gels. Alginate culture may be the method of choice, since the chondrocytes are easily recovered by depolymerization of the alginate with a calcium chelator.

13. Various methods are available for analysis and characterization of proteoglycans. We have found the described methods to be a convenient and rapid approach for screening the relative amounts and molecular sizes of newly synthesized proteoglycans. Specific antibodies are available for more precise identification by either Western blotting or immunocytochemistry, as described in **Subheading 2.3.6.**

14. Although chondrocytes are generally less susceptible than monocyte/macrophages and other immune cells to endotoxin, it is possible that the transfection conditions, the proliferative state of the cells, or other factors may sensitize the cells to low concentrations of endotoxin *(56)*. Endotoxin itself induces and activates transcription factors that are common to inflammatory responses and may thus up- or downregulate the promoter of interest, thereby masking the response to a cytokine or growth factor.

15. Hyaluronidase added before and/or during transfections has been shown to increase transfection efficiencies in chondrocytes *(57–59)*.

16. The total amount of plasmid to be transfected, including reporter, expression, and internal control plasmids, should not exceed 1 µg and the optimal amount for the culture system should be tested empirically. If variable amounts of expression vector, for example, are included, the total amount of plasmid in each well should be equalized by the addition of the empty vector. Wells transfected with appropriate empty vector controls, without and with treatment with test agent, should also be included.

17. The times of incubation following transfection before addition of the test agent may vary according to the cell density and culture condition and should be tested empirically. Nutridoma-SP or other serum substitute may be used for experiments that require quiescent cells for the reasons indicated in **Note 6**. Note also that test agents should be added without medium change to avoid induction of pathways by serum growth factors or other medium constituents.

Acknowledgment

Dr. Goldring's research was supported in part by NIH Grants AR45378 and AG22021 and a Biomedical Science Grant from the Arthritis Foundation. The author is grateful to Lujian Tan, Haibing Peng, Lii-Fang Suen, James Birkhead, Pamela Kaplan, Michael Byrne, Merrilee Flannery, James Robbins, Makoto Osaki, and Bob Choy for supplying technical expertise and data and to Dr. Benjamin Bierbaum and his staff at the New England Baptist Hospital for supplying cartilage specimens.

References

1. Goldring, M. B. Chondrocytes, in *Kelley's Textbook of Rheumatology, 7th Edition* (2004) (Harris, E. D. J., Ruddy, S., Sledge, C. B., Sergent, J. S., and Budd, R. C., eds.). Elsevier, Philadelphia, Ch. 13, pp. 50–81.

2. Maroudas, A., Palla, G., and Gilav, E. (1992) Racemization of aspartic acid in human articular cartilage. *Connect. Tiss. Res.* **28,** 161–169.

3. Verzijl, N., DeGroot, J., Thorpe, S. R., et al. (2000) Effect of collagen turnover on the accumulation of advanced glycation end products. *J. Biol. Chem.* **275,** 39,027–39,031.

4. Maroudas, A., Bayliss, M. T., Uchitel-Kaushansky, N., Schneiderman, R., and Gilav, E. (1998) Aggrecan turnover in human articular cartilage: use of aspartic acid racemization as a marker of molecular age. *Arch. Biochem. Biophys.* **350,** 61–71.

5. Goldring, M. B. (1993) Degradation of articular cartilage in culture: regulatory factors, in *Joint Cartilage Degradation: Basic and Clinical Aspects* (Woessner, J. F., Jr. and Howell, D. S., eds.), Marcel Dekker, New York, pp. 281–345.

6. Goldring, M. B. (2000) The role of the chondrocyte in osteoarthritis. *Arthritis Rheum.* **43,** 1916–1926.

7. Goldring, M. B. (2000) Osteoarthritis and cartilage: the role of cytokines. *Curr. Rheumatol. Rep.* **2,** 459–465.

8. Holtzer, J., Abbott, J., Lash, J., and Holtzer, A. (1960) The loss of phenotypic traits by differentiated cells in vitro. I. Dedifferentiation of cartilage cells. *Proc. Natl. Acad. Sci. USA* **46,** 1533–1542.

9. Ham, R. G. and Sattler, G. L. (1968) Clonal growth of differentiated rabbit cartilage cells. *J. Cell. Physiol.* **72,** 109–114.

10. Green, W. T., Jr. (1971) Behavior of articular chondrocytes in cell culture. *Clin. Orthopaed. Rel. Res.* **75,** 248–260.

11. Goldring, M. B., Sandell, L. J., Stephenson, M. L., and Krane, S. M. (1986) Immune interferon suppresses levels of procollagen mRNA and type II collagen synthesis in cultured human articular and costal chondrocytes. *J. Biol. Chem.* **261,** 9049–9056.

12. Aulthouse, A. L., Beck, M., Friffey, E., et al. (1989) Expression of the human chondrocyte phenotype in vitro. *In Vitro Cell. Devel. Biol.* **25,** 659–668.

13. Kolettas, E., Buluwela, L., Bayliss, M. T., and Muir, H. I. (1995) Expression of cartilage-specific molecules is retained on long-term culture of human articular chondrocytes. *J. Cell Sci.* **108 (Pt 5),** 1991–1999.

14. Goldring, M. B., and Krane, S. M. (1987) Modulation by recombinant interleukin 1 of synthesis of types I and III collagens and associated procollagen mRNA levels in cultured human cells. *J. Biol. Chem.* **262,** 16724–16729.

15. Goldring, M. B., Birkhead, J., Sandell, L. J., Kimura, T., and Krane, S. M. (1988) Interleukin 1 suppresses expression of cartilage-specific types II and IX collagens and increases types I and III collagens in human chondrocytes. *J. Clin. Invest.* **82,** 2026–2037.

16. von der Mark, K., Gauss, V., von der Mark, H., and Muller, P. (1977) Relationship between cell shape and type of collagen synthesised as chondrocytes lose their cartilage phenotype in culture. *Nature* **267,** 531–532.

17. Watt, F. M. (1988) Effect of seeding density on stability of the differentiated phenotype of pig articular chondrocytes in culture. *J. Cell. Sci.* **89 (Pt 3),** 373–378.

18. Kuettner, K. E., Pauli, B. U., Gall, G., Memoli, V. A., and Schenk, R. K. (1982) Synthesis of cartilage matrix by mammalian chondrocytes in vitro. I. Isolation, culture characteristics, and morphology. *J. Cell Biol.* **93,** 743–750.

19. Bassleer, C., Gysen, P., Foidart, J. M., Bassleer, R., and Franchimont, P. (1986) Human chondrocytes in tridimensional culture. *In Vitro Cell. Devel. Biol.* **22,** 113–119.

20. Thonar, E. J., Buckwalter, J. A., and Kuettner, K. E. (1986) Maturation-related differences in the structure and composition of proteoglycans synthesized by chondrocytes from bovine articular cartilage. *J. Biol. Chem.* **261,** 2467–2474.

21. Hauselmann, H. J. and Hedbom, E. (1999) In vitro models of cartilage metabolism, in *Dynamics of Bone and Cartilage Metabolism* (Seibel, M. J., Robins, S. P., and Bilezekian, J. P., eds.), Academic, New York, pp. 325–338.

22. Norby, D. P., Malemud, C. J., and Sokoloff, L. (1977) Differences in the collagen types synthesized by lapine articular chondrocytes in spinner and monolayer culture. *Arthrit. Rheum.* **20,** 709–716.

23. Glowacki, J., Trepman, E., and Folkman, J. (1983) Cell shape and phenotypic expression in chondrocytes. *Proc. Soc. Exp. Biol. Med.* **172,** 93–98.

24. Castagnola, P., Moro, G., Descalzi-Cancedda, F., and Cancedda, R. (1986) Type X collagen synthesis during in vitro development of chick embryo tibial chondrocytes. *J. Cell Biol.* **102,** 2310–2317.

25. Reginato, A. M., Iozzo, R. V. and Jimenez, S. A. (1994) Formation of nodular structures resembling mature articular cartilage in long-term primary cultures of human fetal epiphyseal chondrocytes on hydrogel substrate. *Arthrit. Rheum.* **37,** 1338–1349.

26. Croucher, L. J., Crawford, A., Hatton, P. V., Russell, R. G., and Buttle, D. J. (2000) Extracellular ATP and UTP stimulate cartilage proteoglycan and collagen accumulation in bovine articular chondrocyte pellet cultures. *Biochim. Biophys. Acta.* **1502,** 297–306.

27. Gibson, G. J., Schor, S. L., and Grant, M. E. (1982) Effects of matrix macromolecules on chondrocyte gene expression: Synthesis of a low molecular weight collagen species by cells cultured within collagen gels. *J. Cell Biol.* **93,** 767–774.

28. Benya, P. D. and Shaffer, J. D. (1982) Dedifferentiated chondrocytes reexpress the differentiated collagen phenotype when cultured in agarose gels. *Cell* **30,** 215–224.

29. Aydelotte, M. B. and Kuettner, K. E. (1988) Differences between sub-populations of cultured bovine articular chondrocytes. I. Morphology and cartilage matrix production. *Conn. Tiss. Res.* **18,** 205–222.

30. Guo, J., Jourdian, G. W., and MacCallum, D. K. (1989) Culture and growth characteristics of chondrocytes encapsulated in alginate beads. *Conn. Tiss. Res.* **19,** 277–297.

31. Hauselmann, H. J., Aydelotte, M. B., Schumacher, B. L., Kuettner, K. E., Gitelis, S. H., and Thonar, E. J.-M. A. (1992) Synthesis and turnover of proteoglycans by human and bovine adult articular chondrocytes cultured in alginate beads. *Matrix* **12,** 116–129.

32. Mizuno, S., Allemann, F., and Glowacki, J. (2001) Effects of medium perfusion on matrix production by bovine chondrocytes in three-dimensional collagen sponges. *J. Biomed. Mater. Res.* **56,** 368–375.

33. Xu, C., Oyajobi, B. O., Frazer, A., Kozaci, L. D., Russell, R. G., and Hollander, A. P. (1996) Effects of growth factors and interleukin-1 alpha on proteoglycan and

type II collagen turnover in bovine nasal and articular chondrocyte pellet cultures. *Endocrinology* **137,** 3557–3565.

34. Adolphe, M., Froger, B., Ronot, X., Corvol, M. T., and Forest, N. (1984) Cell multiplication and type II collagen production by rabbit articular chondrocytes cultivated in a defined medium. *Exp. Cell Res.* **155,** 527–536.

35. Gerstenfeld, L. C., Kelly, C. M., Von Deck, M., and Lian, J. B. (1990) Comparative morphological and biochemical analysis of hypertrophic, non-hypertrophic and 1,25(OH)$_2$D$_3$ treated non-hypertrophic chondrocytes. *Conn. Tiss. Res.* **24,** 29–39.

36. Adams, S. L., Pallante, K. M., Niu, Z., Leboy, P. S., Golden, E. B., and Pacifici, M. (1991) Rapid induction of type X collagen gene expression in cultured chick vertebral chondrocytes. *Exp. Cell Res.* **193,** 190–197.

37. Poole, A. R. (1989) Honor Bridgett Fell, Ph.D., D.Sc. F.R.S., D.B.E., 1900–1986. The scientist and her contributions. *In Vitro Cell Dev. Biol.* **25,** 450–453.

38. Benoit, B., Thenet-Gauci, S., Hoffschir, F., Penformis, P., Demignot, S., and Adolphe, M. (1995) SV40 large T antigen immortalization of human articular chondrocytes. *In Vitro Cell. Dev. Biol.* **31,** 174–177.

39. Steimberg, N., Viengchareun, S., Biehlmann, F., et al. (1999) SV40 large T antigen expression driven by col2a1 regulatory sequences immortalizes articular chondrocytes but does not allow stabilization of type II collagen expression. *Exp. Cell. Res.* **249,** 248–259.

40. Robbins, J. R., Thomas, B., Tan, L., Choy, B., Arbiser, J. L., Berenbaum, F., and Goldring, M. B. (2000) Immortalized human adult articular chondrocytes maintain cartilage-specific phenotype and responses to interleukin-1β. *Arthrit. Rheum.* **43,** 2189–2201.

41. Grigolo, B., Roseti, L., Neri, S., Gobbi, P., Jensen, P., Major, E. O., and Facchini, A. (2002) Human articular chondrocytes immortalized by HPV-16 E6 and E7 genes: Maintenance of differentiated phenotype under defined culture conditions. *Osteoarthr. Cartilage* **10,** 879–889.

42. Piera-Velazquez, S., Jimenez, S. A., and Stokes, D. (2002) Increased life span of human osteoarthritic chondrocytes by exogenous expression of telomerase. *Arthr. Rheum.* **46,** 683–693.

43. Goldring, M. B., Birkhead, J. R., Suen, L.-F., et al. (1994) Interleukin-1β-modulated gene expression in immortalized human chondrocytes. *J. Clin. Invest.* **94,** 2307–2316.

44. Goldring, M. B. (2004) Immortalization of human articular chondrocytes for generation of stable, differentiated cell lines, in *Methods in Molecular Medicine. Osteoarthritis: Methods and Protocols.* (Sabatini, M., Pastoureau, P., and de Ceuninck, F., eds.), Humana, Totowa, NJ, pp. 23–25.

45. Laemmli, U. K. (1970) Cleavage of structural proteins during the assembly of the head of bacteriophage T4. *Nature* **227,** 680–685.

46. Vogel, K. G., Sandy, J. D., Pogany, G., and Robbins, J. R. (1994) Aggrecan in bovine tendon. *Matrix Biol.* **14,** 171–179.

47. Robbins, J. R. and Vogel, K. G. (1997) Mechanical loading and TGF-β regulate proteoglycan synthesis in tendon. *Arch. Biochem. Biophys.* **342,** 203–211.
48. Kokenyesi, R., Tan, L., Robbins, J. R., and Goldring, M. B. (2000) Proteoglycan production by immortalized human chondrocyte cell lines cultured under conditions that promote expression of the differentiated phenotype. *Arch. Biochem. Biophys.* **383,** 79–90.
49. Dell'Accio, F., De Bari, C., and Luyten, F. P. (2001) Molecular markers predictive of the capacity of expanded human articular chondrocytes to form stable cartilage in vivo. *Arthr. Rheum.* **44,** 1608–1619.
50. Bau, B., Gebhard, P. M., Haag, J., Knorr, T., Bartnik, E., and Aigner, T. (2002) Relative messenger RNA expression profiling of collagenases and aggrecanases in human articular chondrocytes in vivo and in vitro. *Arthr. Rheum.* **46,** 2648–2657.
51. Goldring, M. B., Fukuo, K., Birkhead, J. R., Dudek, E., and Sandell, L. J. (1994) Transcriptional suppression by interleukin-1 and interferon-γ of type II collagen gene expression in human chondrocytes. *J. Cell. Biochem.* **54,** 85–99.
52. Osaki, M., Tan, L., Choy, B. K., et al. (2003) The TATA-containing core promoter of the type II collagen gene (COL2A1) is the target of interferon-γ-mediated inhibition in human chondrocytes: requirement for Stat1α, Jak1, and Jak2. *Biochem. J.* **369,** 103–115.
53. Tan, L., Peng, H., Osaki, M., et al. (2003) Egr-1 mediates transcriptional repression of COL2A1 promoter activity by interleukin-1β. *J. Biol. Chem.* **278,** 17,688–17,700.
54. Hering, T. M., Kollar, J., Huynh, T. D., Varelas, J. B., and Sandell, L. J. (1994) Modulation of extracellular matrix gene expression in bovine high-density chondrocyte cultures by ascorbic acid and enzymatic resuspension. *Arch. Biochem. Biophys.* **314,** 90–98.
55. Wu, Q. Q. and Chen, Q. (2000) Mechanoregulation of chondrocyte proliferation, maturation, and hypertrophy: ion-channel dependent transduction of matrix deformation signals. *Exp. Cell Res.* **256,** 383–391.
56. Cotten, M., Baker, A., Saltik, M., Wagner, E., and Buschle, M. (1994) Lipopolysaccharide is a frequent contaminant of plasmid DNA preparations and can be toxic to primary human cells in the presence of adenovirus. *Gene Ther.* **1,** 239–246.
57. Lu Valle, P., Iwamoto, M., Fanning, P., Pacifici, M., and Olsen, B. R. (1993) Multiple negative elements in a gene that codes for an extracellular matrix protein, collagen X, restrict expression to hypertrophic chondrocytes. *J. Cell. Biol.* **121,** 1173–1179.
58. Viengchareun, S., Thenet-Gauci, S., Steimberg, N., Blancher, C., Crisanti, P., and Adolphe, M. (1997) The transfection of rabbit articular chondrocytes is independent of their differentiation state. *In Vitro Cell Dev. Biol.* **33,** 15–17.
59. Madry, H. and Trippel, S. B. (2000) Efficient lipid-mediated gene transfer to articular chondrocytes. *Gene Ther.* **7,** 286–291.

6

Human Myoblasts and Muscle-Derived SP Cells

Grace K. Pavlath and Emanuela Gussoni

1. Introduction

Skeletal muscle cells can be used in vitro for the study of myogenesis, as well as in vivo as gene-delivery vehicles for the therapy of muscle and non-muscle diseases. These skeletal muscle cells are derived from muscle satellite cells that lie between the basal lamina and the sarcolemma of differentiated muscle fibers *(1)*. Normally quiescent after the period of muscle development and growth during fetal life and the early postnatal period, these cells are induced to proliferate upon muscle damage and fuse with existing muscle fibers. Satellite cells isolated and grown in vitro are called myoblasts. Myoblasts proliferate in mitogen-rich media, but upon reaching high cell density followed by exposure to mitogen-poor media, are induced to differentiate and become postmitotic. Muscle differentiation is characterized by the fusion of myoblasts to form multinucleated myotubes that express differentiation-specific proteins. In this chapter, methods are given for the isolation of myoblasts from human muscle tissue using two different techniques: (a) flow cytometry *(2)* and (b) cell cloning *(3,4)*. Recent reports have also highlighted the existence of highly primitive cells within mouse skeletal muscle, whose relationship with satellite cells is still under study *(5–7)*. These primitive cells have been purified using different methods and techniques, including the preplating technique *(8–10)* and the fluorescence-activated cell sorter (FACS) *(11–14)*. Depending on the isolation technique, these cells have been named differently. Muscle SP cells have been isolated from mouse skeletal muscle by staining the dissociated primary muscle cells with the vital DNA dye Hoechst 33342, followed by FACS purification *(11–14)*. Mouse muscle SP cells have demonstrated hematopoietic and myogenic differentiation potential both in vitro and

From: *Methods in Molecular Medicine, vol. 107: Human Cell Culture Protocols, Second Edition*
Edited by: J. Picot © Humana Press Inc., Totowa, NJ

in vivo *(11–14)* and these studies are being extended to human-derived SP cells. Methods to isolate human muscle SP cells from fetal and from adult skeletal muscle are also given. The methods in this chapter are applicable to muscle tissue from both fetal and postnatal donor as well as from normal and diseased individuals.

2. Materials

All solutions are prepared using double-distilled water (ddH$_2$O). All solutions and materials are sterile. Whenever possible, reagents that have been tissue-culture tested by the manufacturer are recommended. Specific vendors are indicated for certain items. Alternate sources for these particular items have not been tested.

2.1. Cells/Tissue

1. MRC-5 (ATCC Accession Number CRL 171): These cells are used in preparation of the conditioned media required at the early stages of cell growth in vitro (*see* **Note 1**). The preparation of conditioned media is described in **Subheading 3.1.**
2. Muscle tissue:
   a. Biopsy material: Muscle biopsies should be 0.5 cm^3 to 1.0 cm^3 (approx 0.5–1 g).
   b. Autopsy material: Collect muscle sample as soon as possible within the first 12–24 h after death.

2.2. Transport, Tissue Dissociation, Growth, and Freezing of Human Myoblasts

1. Transport medium: Dulbecco's phosphate-buffered saline without calcium and magnesium (PBS-CMF),1% glucose, 100 U/mL penicillin, 100 µg/mL streptomycin (store at 4°C). Place the muscle sample in a plastic 50-mL centrifuge tube in transport medium. Dissociate immediately or store at 4°C for up to 24 h. If shipping muscle sample, fill the tube completely with transport medium to prevent desiccation of the sample.
2. Dispase II (Boehringer Mannheim cat. no. 295 825): Comes as sterile 2.4 U/mL stock solution. Aliquot and store frozen at –20°C.
3. Collagenase D (Boehringer Mannheim cat. no. 1088 874): 1 mg/mL stock solution prepared using PBS-CMF with 5 mM CaCl$_2$. Aliquot and store frozen at –20°C.
4. Digestion medium: Prepare a 1 : 1 mixture of dispase II and collagenase D prior to use. Final concentration is 1.2 U/mL dispase and 0.5 mg/mL collagenase D. Use 200 µL per 100 mg of tissue.
5. Basic fibroblast growth factor (bFGF) (Promega cat. no. G-5071): Prepare by dissolving 25 µg powder in 1 mL of PBS-CMF containing 1% bovine serum albumin (BSA). Aliquot and store frozen at –20°C. Do not freeze–thaw aliquots. FGF is added directly to the media in the culture dish at each feeding.

6. Growth media:
   a. MRC-5 maintenance medium (MM; store at 4°C, use within 2 wk). Fetal bovine serum (FBS) is directly added to Dulbecco's modified Eagle's medium (DMEM) (formulation containing 1000 mg/L glucose) to give a final concentration of 10%.
   b. Human muscle growth medium (HuGM; store at 4°C, use within 2 wk) (*see* **Note 2**): The following are added to Ham's F10 to give the indicated final concentrations: 15% FBS, 100 U/mL penicillin, 100 µg/mL streptomycin.
7. Human muscle fusion medium (FM) (*see* **Note 3**): The following are added to DMEM (formulation containing 1000 mg/L glucose) to give the indicated final concentration: 2% horse serum, 100 U/mL penicillin, 100 µg/mL streptomycin.
8. Freezing medium: 90% calf serum, 10% dimethylsulfoxide (DMSO).
9. Trypsin: 0.05% trypsin/0.53 m*M* ethylenediaminetetraacetic acid (EDTA).

2.3. Flow Cytometry of Human Myoblasts

1. 5.1H11 monoclonal antibody against muscle-specific isoform of neural cell adhesion molecule *(15)* (*see* **Note 4**). Use hybridoma supernatant neat or purified antibody at 1 µg/10^6 cells.
2. Biotinylated anti-mouse IgG (cat. no. BA 2000, Vector Labs, Burlingame, CA). Use water to reconstitute the contents of the vial. Use at 7 µg/mL in PBS + 0.5% BSA for immunostaining.
3. Streptavidin conjugated to either FITC (cat. no. 43-4311, Zymed, South San Francisco, CA) or Texas Red (cat. no. 43-4317, Zymed). These reagents are light sensitive. Use water to reconstitute the contents of the vial. Use at 0.5 µg/mL in PBS + 0.5% BSA for immunostaining.
4. PBS-CMF containing 0.5% BSA (sterile filtered through 0.22-µm filter; store at 4°C and use cold in the immunostaining procedure).
5. Propidium iodide (PI): 1 mg/mL stock solution prepared in water. Aliquot and store at –20°C; light sensitive. Use at 1 µg/mL.
6. Gentamicin: use at 50 µg/mL.
7. Sterile 1.5-mL microcentrifuge tubes.
8. Sterile polystyrene tubes (12 × 75 mm) (*see* **Note 5**).
9. 70-µm cell strainers (Falcon cat. no. 2350).

2.4. Purification of Human Muscle-Derived SP Cells

1. Transport medium: PBS-CMF, 1% glucose, 100 U/mL penicillin, 100 µg/mL streptomycin (store at 4°C). Place the muscle sample in a plastic 50-mL centrifuge tube in transport medium. Dissociate immediately or store at 4°C overnight.
2. Dispase II (Worthington LS02100 or Roche 295825). 2.4 U/mL stock solution prepared in PBS-CMF. Aliquot stock solution in 5-mL aliquots and store at –20°C.
3. Collagenase 4 (Worthington CLS-4). 4 mg/mL stock solution prepared in PBS-CMF. Aliquot stock solution in 5-mL aliquots and store at –20°C.

4. Digestion medium for fetal muscle tissue: Each gram of fetal tissue is digested using 4 mL of the following solution: 1 mL dispase II, 0.5 mL collagenase 4, 2.5 mL PBS-CMF. Final enzyme concentration is 0.6 U/mL dispase II and 0.5 mg/mL collagenase 4.

5. Digestion medium for adult muscle tissue: Each gram of adult tissue is digested using 4 mL of the following solution: 1.6 mL dispase II, 2 mL collagenase 4, 0.4 mL PBS-CMF. Final enzyme concentration is 1 U/mL dispase II and 2 mg/mL collagenase 4.

6. Hoechst 33342 (Sigma B2661). 1 mg/mL stock solution prepared in water (light sensitive). Aliquot stock solution in 1-mL aliquots and store at –20°C.

7. Verapamil (Sigma V4629): 5 mM stock solution prepared in water. Aliquot stock solution in 1-mL aliquots and store at –20°C. Verapamil is added at a final concentration of 50 µM.

8. Reserpine (Sigma R0875): 5 mM stock solution prepared in chloroform. Aliquot stock solution in 10-mL aliquots and store at –20°C. Reserpine is added at a final concentration of 5 µM.

9. PI (Sigma P4170): 1 mg/mL stock solution prepared in water; aliquot and store at –20°C; light sensitive.

10. 40-µm cell strainers (Falcon 352340).

11. 100-µm cell strainers (Falcon 352360).

12. PBS-CMF containing 0.5% BSA (sterile filtered through 0.22-µm filter; store at 4°C).

3. Methods

3.1. Preparation of Conditioned Medium (see Note 6)

MRC-5 cells are grown in MRC-5 Maintenance Medium (MM) in a humidified 37°C incubator containing 10% CO_2.

1. Remove a cryogenic vial from liquid nitrogen storage and rapidly thaw it in a 37°C water bath.

2. Transfer the contents of the vial to a 100-mm dish containing 10 mL of MM.

3. On the following day, aspirate the old media and refeed the cells with 10 mL of fresh MM.

4. Culture the cells in 100-mm dishes in 10 mL MM per dish. Feed the cultures with fresh medium every 2–3 d, subculture the cells using trypsin when cells reach approx 80% confluency and dilute them 1:5.
   To subculture MRC-5 cells:
   a. Aspirate the old media and gently rinse the cells twice with PBS-CMF.
   b. Add 1 mL of trypsin per 100-mm dish. Incubate for a few minutes at room temperature until the cells round up or lift off the dish.
   c. Add 9 mL of MM to the dish and gently pipet the cells to disperse.
   d. Seed the cells into 100-mm dishes in 10 mL of MM.

5. When the cells in the desired number of 100-mm dishes (usually 5–10) reach 70–80% confluency, refeed the cultures with 10 mL of HuGM.

6. Leave HuGM on the MRC-5 cells overnight.
7. Collect the conditioned HuGM (CM) and store at 4°C. Refeed the cultures with 10 mL of fresh HuGM and incubate overnight again.
8. Collect the CM again. Pool all the CM, filter through 0.45-µm filter units (*see* **Note 7**), and freeze at –20°C in 5-mL aliquots.
9. When needed, thaw aliquots of CM at 37°C and mix 1:1 with fresh HuGM prior to use.

3.2. Tissue Preparation/Dissociation/Initial Plating of Human Myoblasts

Dissociates of human muscle tissue are grown initially in CM (*see* **Subheading 3.1.**) but after adaptation to culture conditions are grown in HuGM. All culturing is done in a humidified 37°C incubator containing 5% CO_2.

1. Add approx 10 mL of F10 into each of four 100-mm dishes.
2. Wash the muscle sample four times by gently agitating with forceps in each of the four dishes containing F10.
3. In another 100-mm dish containing F10, remove any obvious connective or fatty tissue (*see* **Note 8**).
4. Transfer the tissue sample to a 100-mm dish.
5. Add the 1:1 mixture of dispase II and collagenase D.
6. Mince the tissue finely using sterile razor blades until the pieces are approx 1 mm³ in size.
7. Use a wide bore pipet tip to transfer the contents into a sterile tube.
8. Rinse the dish with an equal volume of 1:1 dispase II/collagenase D and add this mixture to the tube above.
9. Place the tube in a 37°C water bath.
10. Incubate for 2 × 15 min. Triturate after the first 15-min interval by pipetting up and down.
11. Remove the contents of the tube and place the liquid through the 70-µm cell strainer.
12. Wash the tube by adding 1 mL of PBS-CMF and add to the cell strainer.
13. Wash the cell strainer with PBS-CMF.
14. Centrifuge cells at 800–900*g* for 5 min.
    If proceeding with myoblast isolation using flow cytometry, continue as follows:
15. Pool and resuspend the cell pellet(s) in 4 mL of 1:1 HuGM /CM. Transfer the cell suspension (consisting of myofiber debris, red blood cells, myoblasts, and fibroblasts) to a 60-mm dish. Proceed to **Subheading 3.3.**
    If proceeding with myoblast isolation using cloning procedures, continue as follows:
16. Pool and resuspend the cell pellet(s) in 10 mL of 1:1 HuGM /CM. Assuming the following viable cell yields (**2–4,16**):
    5×10^3 cells/0.1 g tissue from normal donors
    200-cells/0.1 g tissue from Duchenne muscular dystrophy patients
    5×10^5 cells/0.1 g tissue from fetal sources

Plate 2 rows each of a 96-well plate at 1.0, 0.5, 0.25, and 0.125 cells/well. Pellet and freeze the remaining cells at 0.1 g tissue/mL freezing medium in 0.2-mL aliquots. Skip to **Subheading 3.6.**

3.3. Growth and Expansion of Primary Dissociates Containing Mixed Cell Populations (Bulk Cultures)

1. Feed with 1:1 HuGM/CM at d 1 or 2 if the cells are 30–40% confluent. Otherwise, feed with 1:1 HuGM/CM at d 4 or 5 (*see* **Note 9**) or when cells reach 40% confluency (whichever is sooner).
2. Feed the cells every 2–3 d with HuGM. If the cells are less than 40% confluent, feed the cultures with 1:1 HuGM/CM.
3. When cells are about 70–80% confluent, subculture as follows:
   To subculture human muscle cells (*see* **Note 10**):
   a. Aspirate the old media and gently rinse the cells twice with PBS-CMF.
   b. Add 0.5 mL of trypsin (0.05% trypsin/0.53 m*M* EDTA) per 60-mm dish. Incubate for a few minutes at room temperature until the cells round up or lift off the dish.
   c. Add 3–4 mL of HuGM to the dish and gently pipet the cells to disperse.
   d. Seed the cells into 100-mm dishes at 5–10 × 10^5 cells per dish in 10 mL of HuGM. Dishes should be approximately 20% confluent once the cells attach to the dish.
4. Feed cells every 2–3 d with HuGM, subculturing when the cells reach 70–80% confluency (*see* **Note 11**).
5. Continue feeding and subculturing until there are at least four 100-mm dishes at 70–80% confluency. At this point, one half of the cells should be frozen (*see* **Note 12**) and the other half used for flow cytometric purification of the myoblasts.

3.3.1. To Freeze Cells

1. Rinse and trypsinize the cells as described in **step 3**, but use 1 mL of Trypsin/ EDTA per 100-mm dish. Collect into a 15-mL tube.
2. Determine the number of cells using a hemocytometer.
3. Centrifuge the cells at 800–900*g* for 2 min.
4. Gently aspirate the media and resuspend in freezing media to give approx 2 × 10^6 cells/mL.
5. Aliquot the cell suspension at 0.5 mL per 2 mL cryogenic freezing vial.
6. Place the freezing vials in a foam-filled box (*see* **Note 13**) and transfer to –70°C freezer for up to 1 wk. Transfer to liquid nitrogen cryogenic unit for long-term storage.

3.4. Flow Cytometry (Bulk Cultures)

1. Trypsinize the cells and transfer to a 15-mL centrifuge tube (*see* **Note 14**). Add PBS-CMF containing 0.5% BSA (*see* **Note 15**). Centrifuge at 800–900*g* for 2 min at room temperature.

2. Mix the pellet by tapping the tube with your finger several times, then resuspend in 1 mL of PBS-CMF + 0.5% BSA. Transfer the cells to a 1.5-mL microcentrifuge tube and centrifuge for 3 s in a microcentrifuge. Gently aspirate the supernatant.

3. Loosen the pellet by tapping the tube as above, then add either 5.1H11 hybridoma supernatant neat or 1 µg of 5.1H11 antibody per 10^6 cells in PBS-CMF + 0.5% BSA. Mix by gently pipetting up and down.

4. Incubate for 20 min on ice.

5. Wash two or three times with 1 mL of ice-cold PBS-CMF + 0.5% BSA, centrifuge for 3 s in a microcentrifuge at room temperature, loosening the pellet each time.

6. Add 1 mL of biotinylated anti-mouse IgG and gently mix the cells by pipetting up and down.

7. Incubate for 20 min on ice.

8. Wash two to three times as in **step 5**.
   Working with the light in the tissue-culture hood OFF from here to the end:

9. Loosen the pellet. Add 1 ml of streptavidin conjugated to desired fluorophore (*see* **Note 16**) in PBS-CMF + 0.5% BSA. Add PI stock to 1 µg/mL (*see* **Note 17**).

10. Incubate for 10–15 min on ice.

11. Wash two times as in **step 5**.

12. Resuspend in 1 mL of ice-cold PBS-CMF + 0.5% BSA and filter through a 70-µm cell strainer to remove cell clumps (*see* **Note 18**). Wash the filter with several ml of PBS + 0.5% BSA and centrifuge the tube for 2 min at 800–900*g* at room temperature.

13. Resuspend the cell pellet in approx 0.2 mL of ice-cold PBS-CMF + 0.5% BSA containing 0.5 µg/mL PI. Cells should be at a density of 10^7/mL for flow cytometry. Transfer to a 5-mL polystyrene tube.

13. Keep the cells on ice and in the dark prior to flow cytometry.

14. Collect the purified myoblasts (*see* **Fig. 1**) into a 5-mL tube containing HuGM. Keep the cells on ice after collection (*see* **Note 19**).

15. Centrifuge the myoblasts at 800–900*g* for 2 min at room temperature, resuspend in fresh HuGM + 50 µg/mL gentamicin (*see* **Note 20**), seed into 100 dishes at $5–10 \times 10^5$ per dish.

3.5. Culturing of Purified Human Myoblasts After Flow Cytometry

1. Feed myoblast cultures every 2–3 d with HuGM, subculturing when the cells reach 70–80% confluency (*see* **Note 21**). Remove gentamicin from the HuGM media at the first refeeding after the flow cytometry step.

2. Continue feeding and subculturing until the desired number of myoblasts is obtained (*see* **Note 22**).

3. After several weeks in culture, test the following for each muscle sample:
   a. Purity of culture. Restain a small number of cells (approx 5×10^5) with the NCAM antibody and determine the percentage of myoblasts after several weeks in culture (*see* **Note 23**). The purity of the culture should be maintained at greater than 98% *(2)*.

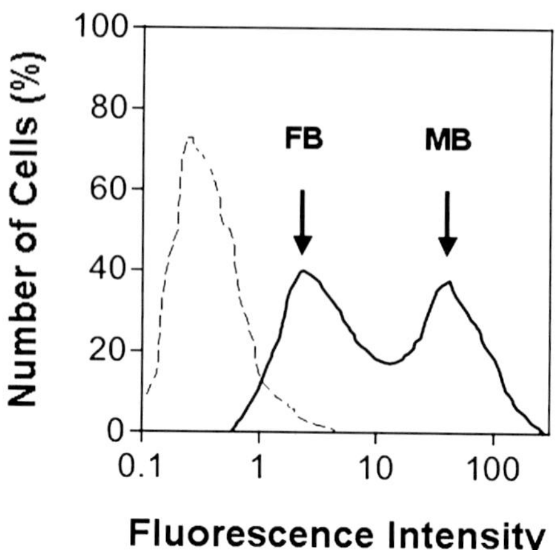

Fig. 1. Schematic of a typical FACS analysis of a mixed primary culture of myoblasts (MB) and fibroblasts (FB) labeled with an αN-CAM antibody (5.1H11). Dotted line: control staining in which the 5.1H11 antibody was omitted from the staining procedure and the cell mixture was exposed to biotinylated anti-mouse IgG antibody and Texas Red-streptavidin. Solid line: the cell mixture was labeled with 5.1H11, biotinylated anti-mouse IgG antibody and Texas Red-streptavidin.

    b. Differentiative capacity.
       i. Collagen coat some 35-mm dishes (*see* **Note 24**).
      ii. Grow myoblasts to approx 70% confluency in HuGM.
     iii. Aspirate the old medium and refeed with FM (*see* **Note 25**).
     iv. Refeed cultures every day with fresh FM. Multinucleated myotubes will form typically within 2–3 d.
      v. Determine the fusion index:
Count the number of nuclei inside and outside myotubes in a number of different fields. The fusion index is defined as:

$$\frac{\text{number of nuclei inside myotubes}}{\text{total number of nuclei}} \times 100$$

3.6. Growth of Clonal Cultures

1. Feed the 96-well plates with 1:1 HuGM/CM at 4–5 d intervals. Clones should be ready to transfer to 24-well plates by d 11. Myoblasts and fibroblasts can be distinguished by morphologic criteria *(3)*.
2. Keep feeding with 1:1 HuGM/CM every 2–3 d. Every 3 d if the cultures are sparse, every 2 d when denser. Subculture the cells when they reach 70–80% confluency and transfer to a 60-mm dish containing HuGM.

3. At the next subculture, plate some cells in one well of a collagen-coated (*see* **Note 24**) 24-well dish containing HuGM to test the fusion index of the clone (*see* **Subheading 3.5.** and **Note 26**) and the remainder in an uncoated 100-mm dish.

4. Feed the cells every 2–3 d with HuGM, freezing (*see* **Subheading 3.3.**) when the cells reach 70–80% confluency.

3.7. Isolation and Purification of Human Muscle-Derived SP Cells

3.7.1. Tissue Dissociation

1. Weigh tissue.
2. Under sterile conditions in a tissue-culture hood, place the tissue in a 10-cm dish with 2–4 mL PBS-CMF to prevent tissue from drying out. Remove all visible fat and connective tissue.
3. Transfer the cleaned muscle tissue to a new 10-cm dish with 2 mL PBS-CMF.
4. Mince the tissue finely using sterile razor blades until the pieces are approx 1 mm^3 in size.
5. Digest each gram of tissue using 4 mL of either Fetal Muscle Dissociation Buffer or Adult Muscle Dissociation Buffer. Tissue is digested in a 15- or 50-mL conical tube at 37°C in an EnviroGenie nutator 20/40 stir/rpm for 45–90 min or until solution can be easily drawn up with a 1-mL pipet.
6. Filter solution through cell strainers (100 μm followed by 40 μm).
7. Centrifuge cells at 365g for 10 min at 4°C.

3.7.2. Cell Staining with Hoechst 33342

1. Resuspend pelleted cells in prewarmed (37°C) PBS-CMF-0.5% BSA at a concentration of 1×10^6 cells/mL (live and dead cells, *see* **Note 27**).
2. Transfer 0.5 mL of resuspended cells to either a 15- or 50-mL tube depending on cell number (*see* **Note 28**). These cells will be stained in the presence of verapamil or reserpine as a negative control for SP cells.
   Working with the light in the tissue-culture hood OFF from here to the end:
3. For SP cell tube: Hoechst 33342 is added at a final concentration of 3.5–9 μg/mL for fetal tissue and 7.5–12.5 μg/mL for adult human tissue (*see* **Note 29**).
4. For verapamil or reserpine negative control tube (*see* **Note 30**): Verapamil is added at a final concentration of 50 μM. If using reserpine instead of verapamil, it is added at a final concentration of 5 μM. Hoechst is added at the same final concentration as for the SP cell tube.
5. Wrap tubes with aluminum foil and incubate in a 37°C water bath for 60 min (adult tissue) or 90 min (fetal tissue).
6. Add 5–10 volumes of ice-cold PBS-0.5% BSA and pellet cells at 365g for 10 min at 4°C.
7. Resuspend negative control tube in 500 μL PBS-0.5% BSA, add PI to a final concentration of 2 μg/mL.

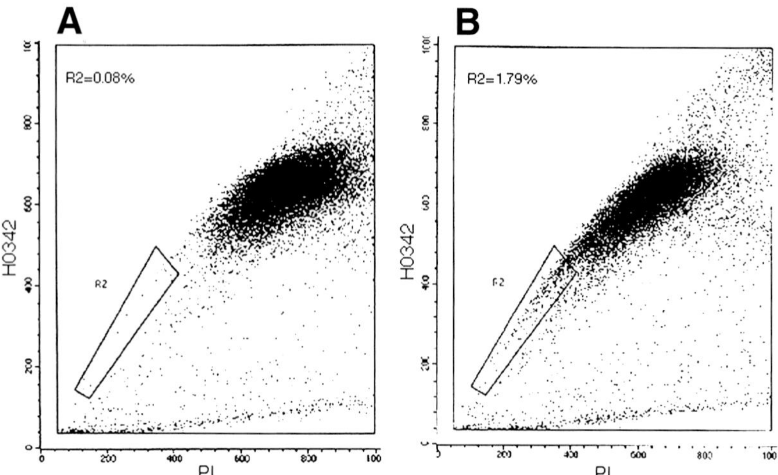

Fig. 2. FACS profile of muscle SP cells detected in human fetal skeletal muscle. SP cells are within the gated areas. SP cells are visible in B (sample stained in the absence of reserpine), but absent in A (negative control sample stained in the presence of reserpine).

8. Samples containing SP cells are resuspended in 1–2 mL PBS–0.5% BSA (cell concentration is up to 10^7 cells/mL). PI is also added to a final concentration of 2 µg/mL.
9. Samples are stored on ice protected from the light until FACS analysis.
10. For FACS analysis, Hoechst and PI are excited at 350 nm and their fluorescence measured at 450 and 600 nm, respectively (*see* **Note 31**). Muscle SP cells are collected in a sterile 1.5-mL microcentrifuge tube containing PBS-CMF (*see* **Fig. 2** and **Note 32**).

4. Notes

1. Initially, one can use conditioned medium prepared from MRC-5 cultures, but later, for simplicity, one may want to collect the spent HuGM directly from cultures of either the primary human muscle dissociates or the purified human myoblasts.
2. Serum-free medium may be substituted if desired *(17)*.
3. Serum-free medium may be substituted: DMEM (formulation containing 1000 mg/L glucose), 1% insulin-transferrin-selenium supplement (GIBCO 51300-044), 100 U/mL penicillin, 100 µg/mL streptomycin.
4. The monoclonal antibody 5.1H11 may be substituted with a monoclonal antibody against Leu19 (Becton-Dickinson, Mountain View, CA) that crossreacts with the NCAM isoform expressed by human myoblasts and regenerating fibers *(18–20)*. This Leu19 antibody has been used for immunofluorescence analyses of cultured

human myoblasts, but not for flow cytometry. In theory, the anti-Leu19 antibody should also work in the flow cytometric procedure outlined here.

5. Check with the operator of the FACS machine to be used to determine which manufacturer's brand of tube fits the particular instrument. Slight differences among manufacturers can render the 12 × 75-mm polystyrene tube unable to maintain an adequate seal with the FACS machine.

6. Whether toxic compounds from the media are removed or whether stimulatory compounds are added to the media during the conditioning process is not known.

7. The filtration rate through the 0.45-μm filter is very slow. Only about 25–50 mL will be collected before the filter clogs up and prevents further filtration.

8. Removing as much of the connective tissue as possible is advantageous because the yield of contaminating fibroblasts is reduced in the primary dissociates.

9. There is a lag phase of several days before the cells start dividing in vitro.

10. Bulk cultures should never be plated at lower than 20% confluent.

11. Cells should be evenly plated in the dishes. Otherwise, the cells will get denser in certain areas and the myoblasts may start to fuse with each other to form differentiated myotubes that are postmitotic. One should maintain the cultures with spaces between the cells, never approaching close to confluency. At high cell densities, the cells will start to become elongated indicating that differentiation has been triggered. During the isolation and expansion process in culture, one wants small, mononucleated cells with a good mitotic index. If this is not observed, then something is wrong with the culture conditions.

12. If there are problems with the immunostaining or the FACS machine or the purified myoblasts become microbially contaminated after collection, then the remainder of the cells can be thawed, expanded and used for flow cytometric purification again.

13. The vials must be frozen while wrapped in insulating material. Without insulation, the viability of the cells decreases dramatically.

14. Start with as many cells as possible. Approximately $5–6 \times 10^6$ is a good starting point, but scale up or down as needed. The number of myoblasts obtained after FACS depends on how much of the total starting cell population was expressing NCAM (e.g., myoblasts). Fetal muscle tissue has a much higher proportion of fibroblasts than myoblasts *(2)*.

15. Do not substitute serum for the BSA anywhere in this protocol. The use of serum during the staining procedure leads to clumping of cells.

16. Texas Red is excited at 590 nm by a tunable dye laser. Not all FACS machines are outfitted with this type of laser. The more commonly found machines are equipped with an argon-ion laser that excites FITC at 488 nm. For further details about the FACS machine itself, *see* Webster et al. *(2)*.

17. PI stains only dead cells. Dead cells will nonspecifically react with the antibodies in this procedure resulting in cells with high fluorescence levels. Unless the dead cells (propidium iodide positive) are gated out electronically on the FACS machine from the analysis, these dead cells will mistakenly be identified as 5.1H11 positive myoblasts. Less than 5% of the starting cell population should be stained with propidium iodide.

18. Cell clumps clog up the nozzle of the FACS machine.
19. The viability of the cells decreases if they are not kept on ice.
20. Occasionally, one encounters problems with microbial contamination after the FACS purification step. Even though HuGM contains penicillin and streptomycin, the addition of gentamicin to the cultures immediately after plating in culture helps to prevent the loss of cultures to bacterial contamination.
21. The doubling time of human myoblasts is approx 24 h.
22. Freeze a number of vials at low passage number after sorting. Myoblasts from normal postnatal donors are capable of approx 40 doublings, whereas fetal myoblasts can undergo 60 doublings. Myoblasts from individuals with neuromuscular disease have a limited proliferative capacity in culture *(4,21)*. As myoblasts begin to senescence in culture, they become bigger, contain many stress fibers, the doubling time decreases and the fusion index decreases.
23. If the fibroblasts were not sufficiently eliminated during the cell sorting procedure owing to either inadequate setting of sort parameters or to cell clumping, the contaminating fibroblasts will rapidly overgrow and the percentage of myoblasts will decrease dramatically.
24. Vitrogen100 (Cohesion) is diluted to give a final concentration of 0.1 mg/mL collagen in 0.1 *N* acetic acid. The solution is added to tissue-culture dishes to cover the bottom and placed overnight at 37°C. The next morning, remove the collagen solution (store at 4°C, can reuse up to 6 wk or 10 uses, whichever comes first) and put the dish back at 37°C to dry for a few hours. When dry, the dishes can be taped shut and stored in a drawer at room temperature indefinitely. Rinse the dishes once with PBS-CMF before adding any cells.
25. Two to three days is the average time for myotubes to form. Some muscle samples take a shorter time, some longer. Some muscle samples form long, thin myotubes, whereas others tend to form shorter, fatter ones.
26. The fusion rate and the morphology of the myotubes derived from different myoblast clones can vary dramatically within the same muscle sample. If the clone does not fuse, it is most likely a fibroblast clone but it could be a poor fusing myoblast clone. To definitively prove the muscle origin of clones, one can immunostain clones in small dishes with a muscle-specific antibody (5.1H11, Leu19, desmin).
27. Both dead and live cells must be counted, as both will uptake the Hoechst dye.
28. If the total number of cells to be stained (dead and alive) is 3 million or less, use a 15-mL Falcon tube. For 4 million cells or more, use a 50-mL Falcon tube for staining. This is to allow enough room to add 5–10 volumes cold PBS-0.5% BSA in the later wash step (**step 3.7.6**).
29. If <20% of the cells are dead, use the lower concentration of Hoechst indicated. If >30% of the cells are dead, use the higher concentration of Hoechst.
30. Verapamil and reserpine inhibit the transporter responsible for Hoechst efflux in SP cells. The transporter responsible for Hoechst dye exclusion has been recently identified (ABCG2) *(22,23)*. As a result, SP cells are not visible in Verapamil or reserpine-treated samples (negative control). Verapamil or reserpine is added to the

control tube first, followed by the Hoechst dye. The negative control tube is used to set the SP cell gate during FACS analysis and prior to SP cell purification.

31. More details on FACS setting can be found in the book chapter by Goodell *(24)* or at http://www.bcm.tmc.edu/genetherapy/goodell/new_site/protocols.html.

32. The expected yield of SP cells is 1–4% of live mononuclear cells in fetal tissue, 0.03–0.3% of live mononuclear cells in adult tissue.

References

1. Mauro, A. (1961) Satellite cells of skeletal muscle fibers. *J. Biophys. Biochem. Cytol.* **9,** 493–495.
2. Webster, C., Pavlath, G. K., Parks, D. R., Walsh, F. S., and Blau, H. M. (1988) Isolation of human myoblasts with the fluorescence-activated cell sorter. *Exp. Cell Res.* **174,** 252–265.
3. Blau, H. M., Webster, C., and Pavlath, G. K. (1983) Defective myoblasts identified in Duchenne muscular dystrophy. *Proc. Natl. Acad. Sci. USA* **80,** 4856–4860.
4. Webster, C. and Blau, H. M. (1990) Accelerated age-related decline in replicative life-span of Duchenne muscular dystrophy myoblasts: implications for cell and gene therapy. *Somat. Cell Mol. Genet.* **16,** 557–565.
5. Zammit, P. and Beauchamp, J. (2001) The skeletal muscle satellite cell: stem cell or son of stem cell? *Differentiation* **68,** 193–204.
6. Seale, P. and Rudnicki, M. A. (2000) A new look at the origin, function, and "stem-cell" status of muscle satellite cells. *Dev. Biol.* **218,** 115–124.
7. Seale, P., Sabourin, L. A., Girgis-Gabardo, A., Mansouri, A., Gruss, P., and Rudnicki, M. A. (2000) Pax7 is required for the specification of myogenic satellite cells [In Process Citation]. *Cell* **102,** 777–786.
8. Qu, Z., Balkir, L., van Deutekom, J. C., Robbins, P. D., Pruchnic, R., and Huard, J. (1998) Development of approaches to improve cell survival in myoblast transfer therapy. *J. Cell Biol.* **142,** 1257–1267.
9. Qu-Petersen, Z., Deasy, B., Jankowski, R., et al. (2002) Identification of a novel population of muscle stem cells in mice: potential for muscle regeneration. *J. Cell Biol.* **157,** 851–864.
10. Torrente, Y., Tremblay, J. P., Pisati, F., et al. (2001) Intraarterial injection of muscle-derived CD34(+)Sca-1(+) stem cells restores dystrophin in mdx mice. *J. Cell Biol.* **152,** 335–348.
11. Gussoni, E., Soneoka, Y., Strickland, C. D., et al. (1999) Dystrophin expression in the mdx mouse restored by stem cell transplantation. *Nature* **401,** 390–394.
12. Jackson, K. A., Mi, T., and Goodell, M. A. (1999) Hematopoietic potential of stem cells isolated from murine skeletal muscle [see comments]. *Proc. Natl. Acad. Sci. USA* **96,** 14,482–14,486.
13. McKinney-Freeman, S. L., Jackson, K. A., Camargo, F. D., Ferrari, G., Mavilio, F., and Goodell, M. A. (2002) Muscle-derived hematopoietic stem cells are hematopoietic in origin. *Proc. Natl. Acad. Sci. USA* **99,** 1341–1346.
14. Asakura, A., Seale, P., Girgis-Gabardo, A., and Rudnicki, M. A. (2002) Myogenic specification of side population cells in skeletal muscle. *J. Cell Biol.* **159,** 123–134.

15. Walsh, F. S. and Ritter, M. A. (1981) Surface antigen differentiation during human myogenesis in culture. *Nature* **289,** 60–64.

16. Blau, H. M. and Webster, C. (1981) Isolation and characterization of human muscle cells. *Proc. Natl. Acad. Sci. USA* **78,** 5623–5627.

17. Ham, R. G., St. Clair, J. A., Webster, C., and Blau, H. M. (1988) Improved media for normal human muscle satellite cells: serum-free clonal growth and enhanced growth with low serum. *In Vitro Cell Dev. Biol.* **24,** 833–844.

18. Schubert, W., Zimmermann, K., Cramer, M., and Starzinski-Powitz, A. (1989) Lymphocyte antigen Leu-19 as a molecular marker of regeneration in human skeletal muscle. *Proc. Natl. Acad. Sci. USA* **86,** 307–311.

19. Illa, I., Leon-Monzon, M., and Dalakas, M. C. (1992) Regenerating and denervated human muscle fibers and satellite cells express neural cell adhesion molecule recognized by monoclonal antibodies to natural killer cells. *Ann. Neurol.* **31,** 46–52.

20. Michaelis, D., Goebels, N., and Hohlfeld, R. (1993) Constitutive and cytokine-induced expression of human leukocyte antigens and cell adhesion molecules by human myotubes. *Am. J. Pathol.* **143,** 1142–1149.

21. Furling, D., Coiffier, L., Mouly, V., et al. (2001) Defective satellite cells in congenital myotonic dystrophy. *Hum. Mol. Genet.* **10,** 2079–2087.

22. Zhou, S., Schuetz, J. D., Bunting, K. D., et al. (2001) The ABC transporter Bcrp1/ABCG2 is expressed in a wide variety of stem cells and is a molecular determinant of the side-population phenotype. *Nat. Med.* **7,** 1028–1034.

23. Zhou, S., Morris, J. J., Barnes, Y., Lan, L., Schuetz, J. D. and Sorrentino, B. P. (2002) Bcrp1 gene expression is required for normal numbers of side population stem cells in mice, and confers relative protection to mitoxantrone in hematopoietic cells in vivo. *Proc. Natl. Acad. Sci. USA* **99,** 12,339–12,344.

24. Goodell, M. A. (2002) Stem cell identification and sorting using the Hoechst 33342 side population (SP), in: *Current Protocols in Cytometry,* John Wiley, New York, pp. 9.18.11–19.18.11.

7

Cell Cultures of Autopsy-Derived Fibroblasts

Volker Meske,* Frank Albert,* and Thomas G. Ohm

1. Introduction

Until now, for most of the neurodegenerative diseases, such as Alzheimer's disease (AD), ideal animal systems do not exist. Hence, cell-biological experiments, which would help to elucidate the degenerative processes, cannot be performed with affected tissue. On the other hand, biopsy-derived human brain tissue from patients, as an alternative source for living cell material, is rare and, in autopsy-derived tissue, neuronal cells are generally already dead, before any experiments can be performed. There is evidence that peripheral tissue also expresses pathophysiological mechanisms relevant for brain dysfunction, and there are many reports dealing with disease-related abnormalities in the physiology of fibroblasts of AD patients *(1,2)*. The advantage of using autopsy-derived fibroblasts from deceased patients is that both the validity of the clinical diagnoses and the severity of neuropathological changes can be assessed reliably by subsequent histological investigations of the brain *(3)*.

Fibroblasts are robust cells that tolerate hypo-oxygen conditions *(4)*. For that reason, living cells can be isolated from autopsy-derived tissue specimen gained from individuals even with long postmortem delays (<48 h tested) *(5)*. We used connective tissue to isolate fibroblasts. There are several sources for connective tissue that can be used, for example, dermis of the skin, capsules and stroma of various organs, and mucous and serous membranes. We prefer skin specimens as a source, which are easy to obtain. The dissection of the tissue can be accomplished in several ways. The most noninvasive method is using explants (small parts of the tissue) and allow cells to migrate from the tissue samples.

*Frank Albert and Volker Meske contributed equally to this work.

From: *Methods in Molecular Medicine, vol. 107: Human Cell Culture Protocols, Second Edition*
Edited by: J. Picot © Humana Press Inc., Totowa, NJ

The second method is the mechanical disaggregation of the tissue, using the shear forces that occur during vigorous pipetting or pressing tissue into a mesh/sieve. The third method is based on the enzymatic digestion of the tissue using proteases (trypsin, collagenase, or elastase), which disrupt cell–cell and cell–matrix connections. We used both the explant method and the disaggregation method to obtain living fibroblasts from autopsy-derived skin specimens.

The explant method is easy to perform, but it takes a relatively long time before the first cells are visible (approx 1 wk). The disaggregation method is more labor intensive, but allows determination of the number of living cells gained from a given specimen before bringing them into culture. The disadvantage of this method is that cells are isolated from tissue by protease treatment. As the cells from postmortem tissue are already stressed, the treatment with proteases seems to have an additional negative impact on their viability. Trypsin, especially, was shown to cause cell-damaging effects under such conditions. We combined a proteolytical step using collagenase-D with a mechanical step. Nevertheless, the number of living cells gained from tissue sections with this method is small and varies, depending on the condition of the skin specimen.

We prefer the explant method, which is noninvasive and leads to stable healthy cell cultures. Nevertheless, we describe both methods in the following sections.

2. Materials

2.1. Explant Method

2.1.1. Tools

1. Dissecting scissors (Aesculap OC 415).
2. Scalpel (Bladeholder) (Aesculap, BB76, BB 80), disposable, sterile blades (Aesculap).
3. Forceps (Aesculap FB 401, FC 411, OC 22).
4. Petri dishes (Roth; diameter 100 mm T941.1, 200 mm T946.1).
5. Binocular microscope (e.g., Olympus SZX9).
6. Falcon tubes, 15 mL, sterile (Nunc 366060).
7. Cleaned and autoclaved cover slips, 24-mm diameter (Assisitent, cat. no. 1001) (*see* **Note 1**).
8. Six-well plates (Falcon, cat. no. 3046) (*see* **Note 2**).
9. Water bath, 37°C.
10. Autoclaveable bottles (Schott/Mainz, FRG) 100 mL, 250 mL.

2.1.2. Chemicals

1. Hank's balanced salt solution (HBSS) with Ca^{2+} and Mg^{2+} without phenol red (Gibco, cat. no. 14025-050).

2. Dulbecco's modified Eagle's medium (DMEM), high glucose (Gibco, cat. no. 41965-039).
3. Fetal calf serum (FCS), heat-inactivated (e.g., Gibco, cat. no. 10500-064).
4. Gentamycin, stock solution (50 mg/mL), 20 mL (Gibco, cat. no. 15750-037).
5. Fungizon, stock solution (250 µg/mL), 20 mL (Gibco, cat. no 15290-018).
6. L-Glutamine, stock solution (200 mM), 20 mL (Gibco, cat. no. 25030-032).
7. Silicon paste, cell-culture tested (Baysilon semi-viscose, Roth, cat. no. 080856.1).

2.1.3. Culture Media (see **Note 3**)

1. Culture-medium 1 (high antibiotic culture-medium): DMEM, containing 10% (v/v) FCS, gentamycin (200 µg/mL), fungizon (1.25 µg/mL), and a supplementation with 2 mM glutamine.
2. Culture medium 2 (standard growth medium): DMEM, containing 10% (v/v) FCS, gentamycin (50 µg/mL), and a supplementation with 2 mM glutamine.

2.2. Disaggregation Method

In addition to previous materials for explant method (*see* **Subheading 2.1.**).

1. A sterile 5-mL syringe with Luer-lock fitting.
2. Autoclaved glass Pasteur pipets (Roth, cat. no. E327).
3. Plastic filter holder (autoclaved) (Schleicher & Schuell, FP 025/1, cat. no. 461000).
4. Nylon-mesh filters, 100-µm pore size, sterile (Millipore, cat. no. NY1H02500).
5. Centrifuge with swinging-out rotor and buckets suitable for Falcon tubes (e.g., Eppendorf 5804 R).
6. Collagenase-D, activity >0.15 U/mg (Roche, cat. no. 1088858), dissolved in HBSS (with Ca^{2+} and Mg^{2+}) at a concentration of 2.5 mg/mL.

2.3. Splitting and Kryoconservation of Fibroblast Cultures

1. Falcon tubes (Nunc 366060).
2. Centrifuge with swinging-out rotor and buckets suitable for Falcon tubes (e.g., Eppendorf 5804 R).
3. HBSS without Ca^{2+}, Mg^{2+}, and phenol red (Gibco, cat. no. 14175-053).
4. Trypsin, stock-solution (2.5% [w/v]), 100 mL (Gibco, cat. no. 15090-046).
5. Dimethyl sulfoxide (DMSO), sterile, 5 × 5 mL (Sigma, cat. no. D 2650).
6. Isopentan (Merck).
7. Kryovials, sterile, 2 mL (Roth, cat. no. E309.1).
8. Culture medium 2 (*see* **Subheading 2.1.**).

2.4. Immunocytochemical Characterization of Cultured Fibroblasts

1. Bovine serum albumin (BSA) (Sigma, cat. no. A 2153).
2. Triton-X 100 (Sigma, cat. no. T 9284).
3. Antifibronectin, IgG made in rabbit (Sigma, cat. no. F 3648).
4. Antivimentin, IgM, monoclonal (Sigma, cat. no. V 5255).

5. Goat, anti-rabbit IgG, Alexa-546 conjugated (Molecular Probes cat. no. A 10010).
6. Biotinylated antibody: anti-mouse IgM (Vector, cat. no. BA 2020).
7. Avidin-D-FITC conjugate (Vector, cat. no. A 2001).
8. Mounting solution (Crystal/Mount, Biomeda, cat. no. MO2).
9. Phosphate-buffered saline (PBS), pH 7.4 and pH 8.5.
10. Paraformaldehyde (4%, w/v) in PBS (*see* **Note 4**).
11. Poly-D-lysine–coated cover slips (*see* **Note 5**).

3. Methods

Two different methods of obtaining fibroblasts are described below.

All work has to be performed under sterile conditions (laminar flow bench) and on a cooling block (4°C). All glassware and instruments have to be auto-claved before use. For sterilizing instruments during work, use 70% alcohol and a gas flame. Wear gloves. Chemicals listed are sterile and cell culture tested.

3.1. Explant Method

3.1.1. Skin Specimen

Skin specimens are taken from hairless areas of individuals at routine autopsy. Hairs are sources for a potential contamination of later cultures, shaving the area before dissecting the skin may reduce this risk. Another way to avoid a massive contamination of cell cultures is to clean the surface thoroughly with ethanol (70% v/v).

1. Lift the cleaned skin with forceps and cut off with a scalpel.
2. Place the tissue in a Falcon tube containing a sterile PBS/gentamycin (200 µg/mL) solution for the transfer to the laboratory.

3.1.2. Procedure

1. Wash skin sections three times with sterile HBSS (10 mL) in Petri dishes.
2. Turn the specimen with the skin-side to the glass bottom (*see* **Fig. 1B**) and remove fat tissue carefully with a scalpel (*see* **Fig. 1C**). The remaining fat cells are removed easily by scratching them off from the underlying connective tissue.
3. After scratching, rinse the tissue with fresh HBSS and transfer into a second Petri dish filled with HBSS with the skin-side on the bottom (*see* **Fig. 1D**).
4. Using forceps, lift the connective tissue of the dermis and cut off small parts of the tissue with dissecting scissors. Transfer these into a Petri dish containing

---

Fig. 1. *(see facing page)* **(A)–(I)**: A picture of the instruments used during the preparation and a sequence of pictures illustrating the main steps of tissue handling. For details, *see* **Subheading 3.1.3.**

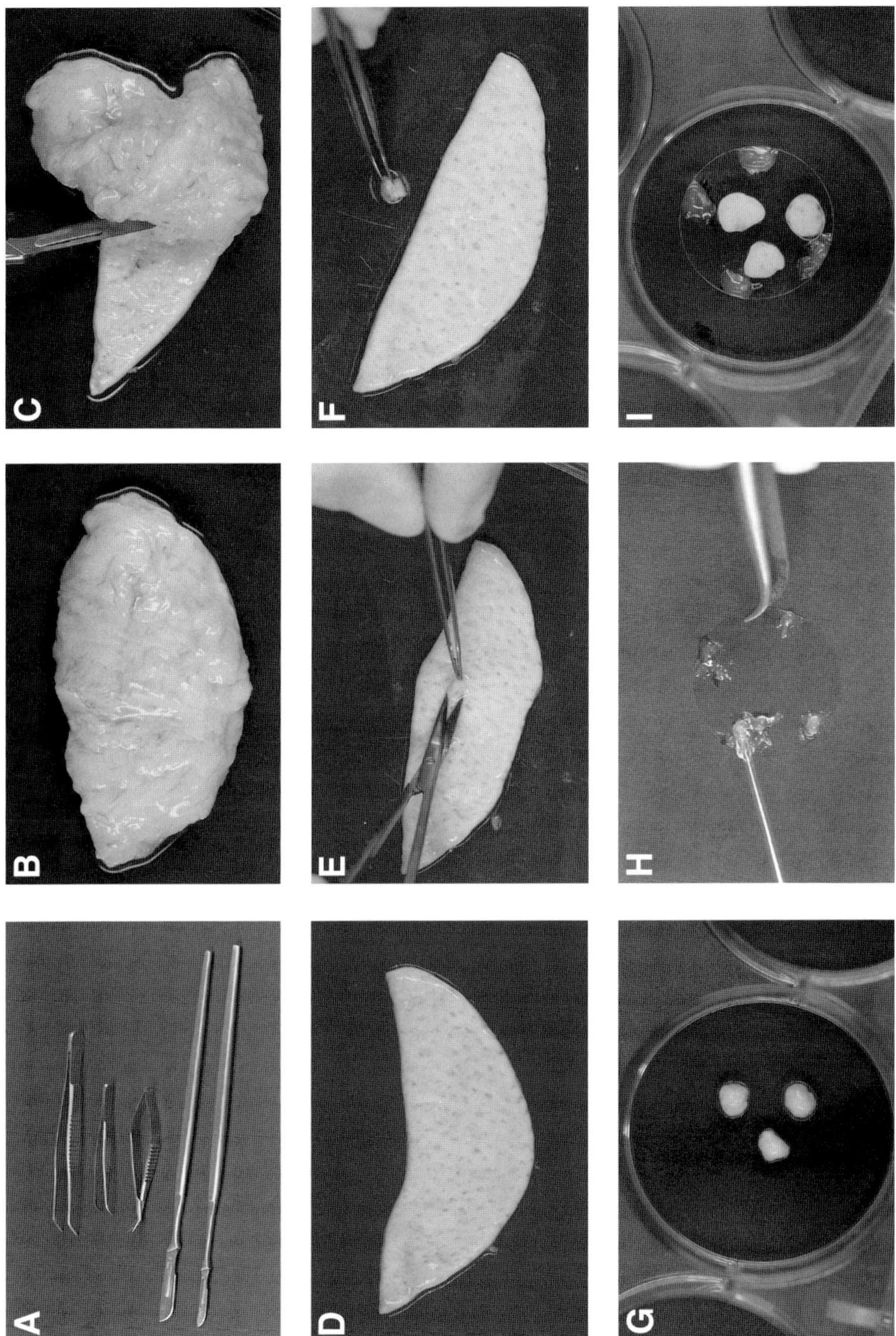

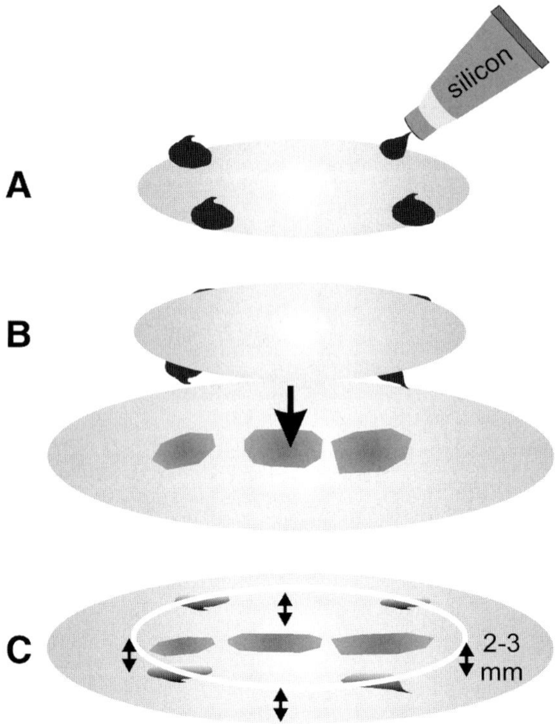

Fig. 2. Procedure to fix connective tissue at the plastic bottom of a six-well plate. For details, *see* **Subheading 3.1.3.**

HBSS (*see* **Fig. 1E,F**). Further dissect the connective tissue into small blocks with a size of approx 5 mm.

5. Transfer the tissue into a Falcon tube, and rinse the blocks three times with HBSS.
6. After rinsing place the tissue blocks at the bottom of six-well culture plate, three or four pieces per well (*see* **Fig. 1G**). A cover slip with four drops of silicon paste at the edge (*see* **Figs. 1H,2A**) is pressed upon the tissue blocks (*see* **Figs. 1I,2B**) in a way that a cleft of 2–3 mm between cover slip and plastic bottom remains (*see* **Fig. 2C**). This procedure ensures that the tissue is fixed at one point and will not flow away even during medium changes.
7. Fill the wells with prewarmed culture medium 1 (*see* **Note 6**).
8. Cells are cultured at 37°C in a humidified incubator in a 5% CO_2 atmosphere. It is necessary to check the filling of the wells every day. After 2 d of culturing, the medium is totally replaced by culture medium 2. Thereafter, regular medium changes (culture medium 2) are performed every 3 d. Half of the medium volume is carefully removed and replaced by fresh medium.
9. The growth of cells at the plastic bottom of the wells can be checked with an inverted microscope (**Fig. 3**) (*see* **Note 7**).

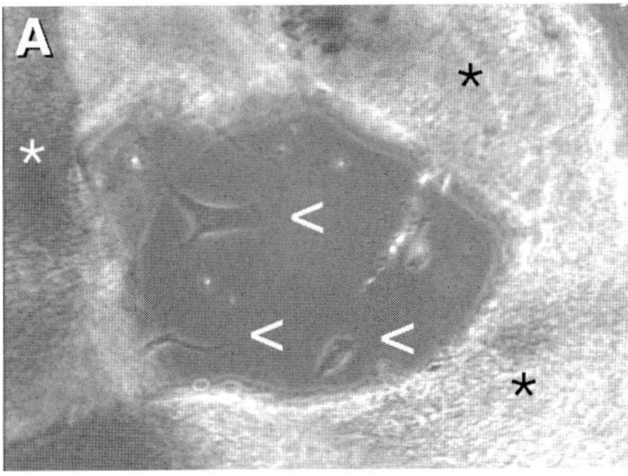
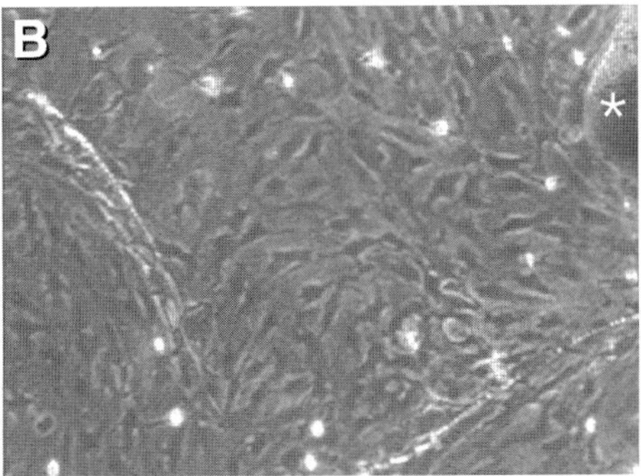

Fig. 3. Differential interference contrast (DIC) micrographs showing the development of a cell culture gained by the explant method. (**A**) Micrograph taken after 3 d in culture. Single cells (marked by open arrowhead) grow out from the connective tissue (marked by asterisks). (**B**) Confluent layer of cells after 4 wk of culturing. A tissue block is marked with an asterisk. Magnification: ×1000.

3.2. Disaggregation Method

3.2.1. Preparations

1. Prepare fresh medium the day you plan to do the fibroblast isolation and store it in sterile glass bottles at 4°C. Before use it has to be warmed up to 37°C.
2. Culture flasks and plates can be used without further treatment of the plastic surfaces.

3.2.2. Skin Specimen

Prepared as described in **Subheading 3.1.1.**

3.2.3. Procedure

The protocol follows the procedure of the explant method (*see* **Subheading 3.1.2.**) up to **step 5** in which connective tissue is cut into small pieces and washed three times with HBSS.

6. The next step is a digestion of the tissue with collagenase-D. Transfer the tissue blocks to a Falcon tube filled with 10 mL HBSS with Ca^{2+} and Mg^{2+} containing collagenase at a concentration of 2.5 mg/mL (*see* **Note 8**).
7. Incubate the tissue for 4 h at 37°C in a humidified incubator in a 5% CO_2 atmosphere.
8. Suck the resulting cell suspension repeatedly into a disposable sterile glass Pasteur pipet (*see* **Fig. 4B**) and then transfer into a sterile 5-mL syringe (with Luer-lock). Screw a sterile filter holder, containing a 100 µm, sterile nylon filter, on the opening of the syringe.

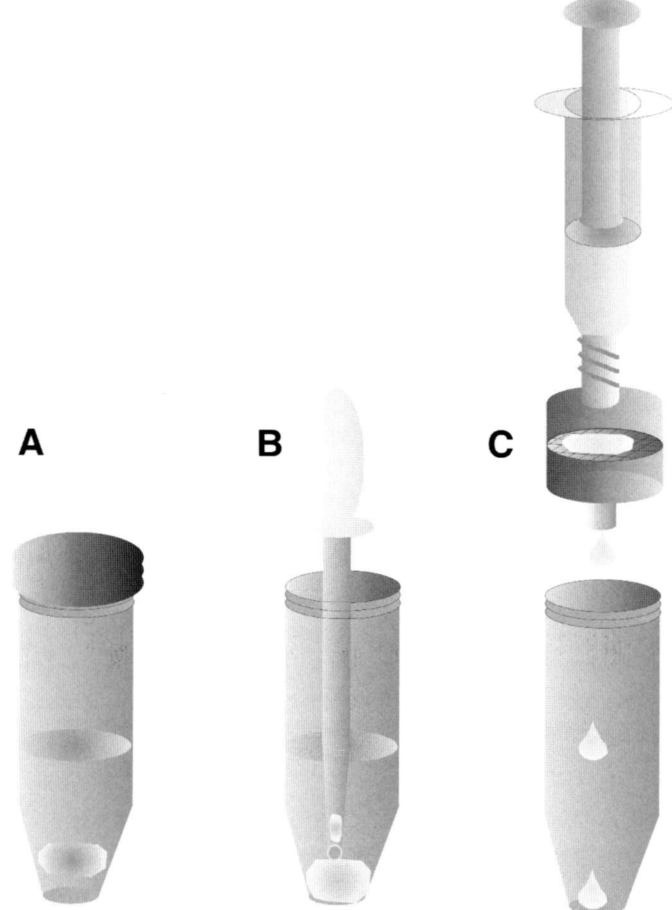

Fig. 4. Procedure to elute cells from a collagenase-digested connective tissue block. For details, *see* **Subheading 3.2.3.**

9. Press the suspension through the filter into a 15-mL Falcon tube (*see* **Fig. 4C**) and centrifuge eluted cells for 5 min, 200g at room temperature.
10. Resuspend pellets with growth medium (culture medium 1) and count living cells (*see* **Note 9**).
11. Plate the cells onto a plastic flask (25 cm^2 culture flasks, Nunc) with a density of 100–1000 viable cells/cm^2. Ensure that the bottoms of the culture vessels are evenly covered. Place the flasks for 1 h at 37°C in a humidified incubator with a 5% CO_2 atmosphere and allow the cells to adhere to the surface. Afterward, add prewarmed culture medium (culture medium 1) to a total volume of 4 mL/25 cm^2 flask and put the vessels back into the incubator.

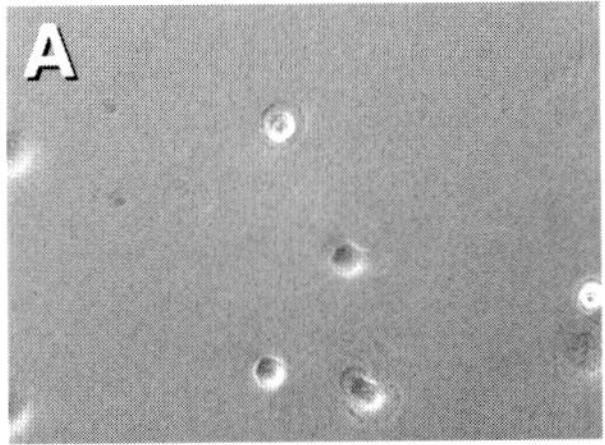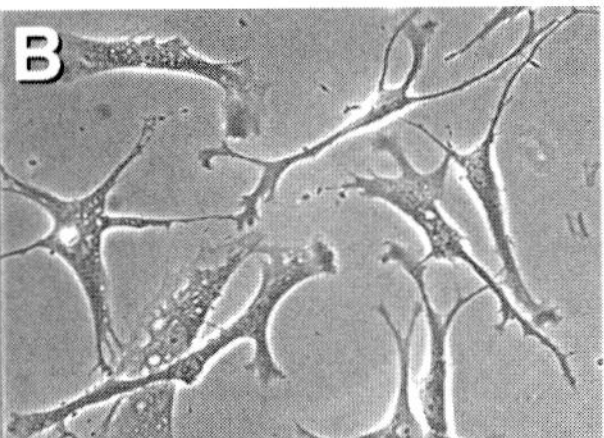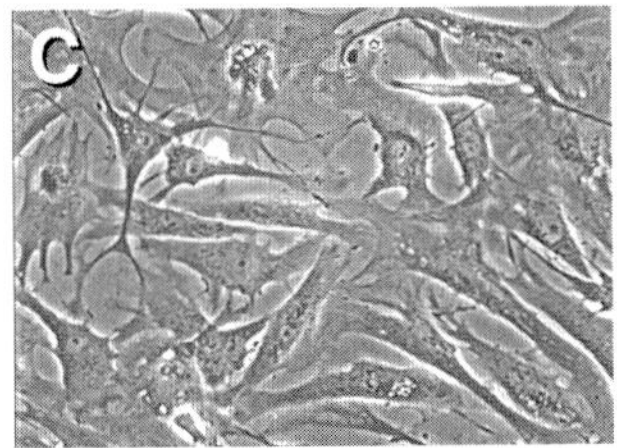

Fig. 5. DIC micrographs showing the development of a cell culture gained by the disaggregation method. The sequence shows the same culture after 3 h, 2 and 3 wk after the seeding of cells. Magnification: ×1000.

12. After 2 d culture medium 1 is replaced by medium 2. Regular medium changes are then performed every 3 d in which half of the culture medium (2–3 mL/flask) is replaced by fresh, prewarmed (37°C) medium (culture medium 2). Depending on the initial cell density, cultures reach confluency after 2–3 wk (*see* **Fig. 5**).

3.3. Contamination

1. Cell cultures should be treated with fungizone and high concentrations of gentamycin only for 2 d. After 2 d, the medium has to be replaced by a medium which contains only gentamycin at a moderate concentration (*see* **Note 10**).
2. If cases of contamination arise, discard the affected culture. Do not try to save the cells, because the risk that the germs infect other cultures within the incubator is high.

3.4. Splitting and Kryoconservation of Fibroblast Cultures

When cultures reach approx 90% of confluency, cultures can be split and partially kryoconserved for long-term storage.

1. Remove medium and wash cells three times with HBSS (w/o Ca^{2+}, Mg^{2+}).
2. Cover the cells with HBSS (w/o Ca^{2+}, Mg^{2+}) containing 0.25% (w/v) trypsin and place the culture plates at 37°C in a humidified incubator (5% CO_2 atmosphere) for approx 15 min.
3. The detachment of cells can be monitored by microscope. Gently tapping the culture plates/flasks on a solid surface during the trypsination may facilitate the detachment. Usually an incubation of 15 min is sufficient to detach most of the cells from the surface. Avoid extended exposures to the protease as cells will be irreversibly damaged.
4. Stop the trypsination by the application of equal volumes of culture medium to the protease solution.
5. Aspirate the cell suspension cautiously with a 5-mL pipet. Avoid strong shear forces during this procedure (switch to the slow mode if a Pipettus is used). Transfer the suspension into a 15-mL Falcon tube. Rinse the culture plate/flask with 2 mL culture medium and add it to the suspension.

6. Spin down the cells in a swinging-out rotor for 3 min at 200g, at room temperature.

To continue culturing the cells, proceed with step 7, to kryopreserve the cells, continue from step 10.

7. Discard the supernatant and resuspend the pellet with culture medium diluting the cell suspension to the desired volume (*see* **Note 11**).
8. Place the cultures in the incubator for 1 h, in order to allow cells to settle down and attach at the surface.
9. Subsequently, add warm culture medium to a total volume of 4 mL/culture vessel (25 cm² culture flask) and put them back into the incubator. Medium changes are performed after 6–7 d in regular intervals as described in **Subheading 3.2.3., step 12**.
10. For kryoconservation of cells (*see* **Note 12**), take the cell pellet resulting from the centrifugation after trypsination (**step 6**), wash this pellet (culture medium), and centrifuge twice more.
11. Resuspend the final cell pellet in 1.5–2-mL culture medium containing 20% (v/v) DMSO. Transfer the suspension to a sterile kryovial. The vials are placed in a vial holder and transferred in a thick-walled (2 cm) Styropor box filled with 200 mL isopentane.
12. Store the closed box in a deep freezer at –80°C overnight. This procedure ensures a slow freezing of the suspension. The vials can then be transferred into liquid nitrogen for a long-term storage.

3.5. Immunocytochemical Characterization of Cultured Fibroblasts

A double-positive staining for vimentin and fibronectin is described as an immunocytological marker for fibroblasts. We describe a double immunostain we routinely use for cell identification in our cell cultures. Antibodies from other manufactures may also be useful. Take care that the primary antibodies against vimentin and fibronectin are made in different species, so that the detection with the secondary antibodies gives independent and distinguishable signals for each protein.

1. Fix fibroblasts grown on poly-D-lysine–coated cover slips with 4% paraformaldehyde in PBS (w/v) for 15 min.
2. Wash three times with PBS, 5 min each, wash then incubate the cells for 15 min in PBS containing 0.1% (v/v) Triton-X 100.
3. Afterward, incubate the cells for 30 min in PBS containing 3% (w/v) BSA, in order to block unspecific binding of the following antibodies.
4. Wash a further three times with PBS, 5 min each, wash then incubate the cells for 1 h with markers of fibroblasts diluted in PBS containing 1.5% (w/v) BSA, e.g., rabbit anti-fibronectin (1:200) and monoclonal anti-vimentin IgM (1:250) antibodies.
5. After a triple wash with PBS, 5 min each, the secondary antibodies are added: (a) biotinylated anti-mouse IgM, diluted 1:100 in PBS/BSA (1.5% [w/v]) and

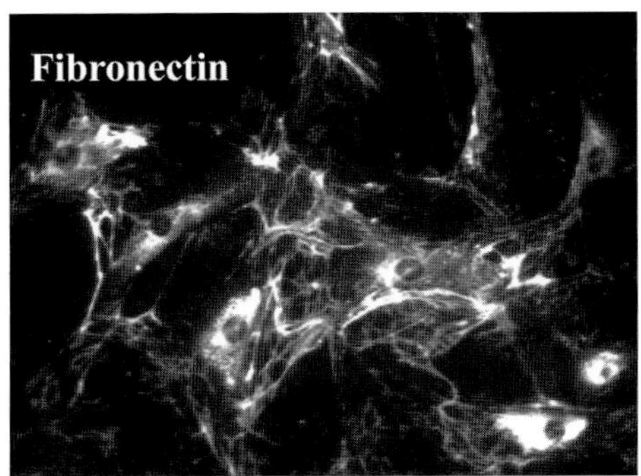

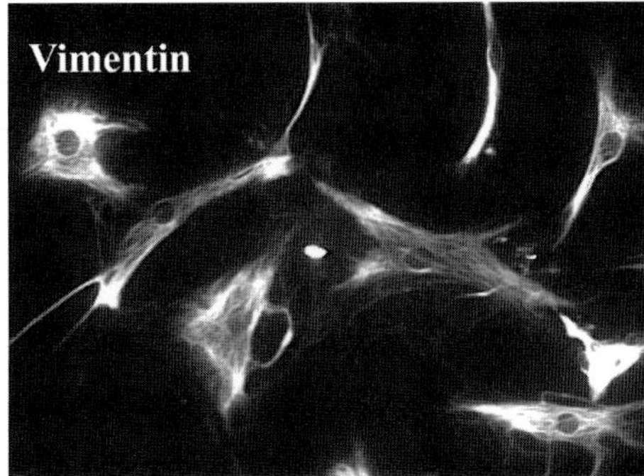

Fig. 6. Results from a double immunostain with fibronectin and vimentin for cell identification. For details of the staining-protocol, *see* **Subheading 2.4.** Note that cells show positive staining for vimentin and fibronectin, characteristic for fibroblasts. Magnification: ×1500.

(b) for fibronectin detection anti-rabbit IgG Alexa 546 conjugated (Biotrend), diluted 1:100 in the same buffer.The cells are incubated in secondary antibody for 1 h.

6. After three further washings with PBS incubate the cells for 1 h with an avidin-D-FITC–conjugate diluted 1:300 in PBS (pH 8.5).
7. After a final triple wash step with PBS (pH 8.5), mount the cells on a slide in a drop of mounting solution. Analyze the cells with a fluorescence microscope (Olympus IX 70) using filter settings for FITC or TRITC, respectively (*see* **Fig. 6**).

4. Notes

1. We used the following method to clean cover slips before use. The procedure ensures that organic substances and heavy metals are removed from the glass surface. Place the cover slips in a glass container with a lid and fill up with HNO_3 (concentrated, fuming, Merck cat. no. 1.00450). Caution: wear protective glasses and gloves. Close the container tightly and incubate at room temperature for 24 h. Decant the acid and wash the cover slips thoroughly with double-distilled water until the pH is neutral (8–10 times). Cover the glass plates with 100% ethanol and shake them for 10 min. Decant the ethanol and transfer the cover slips in an open glass Petri dish. Let the cover slips dry in an oven at 80°C. Close the Petri dish with a suitable glass lid and autoclave the glassware.
2. Culture flasks and plates can be used without further treatment of the plastic surfaces.
3. Prepare fresh medium the day you plan to do the fibroblast isolation and store it in sterile glass bottles at 4°C. Before use, it has to be warmed up to 37°C.
4. Preparation of fixative: Heat PBS up to 70°C and apply the paraformaldehyde while stirring the solution. If necessary, adjust pH with 0.1 *M* NaOH to 7.4. Filter

the solution through a Whatmann paper filter. Cool down the fixative to room temperature before use.

5. We use the following method to coat glass cover slips with poly-D-lysine. Cleaned glass cover slips are placed in wells of culture plates of suitable size. The surface of cover slips is covered with a 100 µg/mL Poly-D-lysine solution (Poly-D-Lysine, Sigma, MW: 70,000–100,000 P-6407 diluted with sterile double-distilled water). The cover slips are incubated in this solution for at least 2 h. Afterward, the poly-D-lysine is carefully sucked off and the cover slips are washed three times with sterile water. The remaining moisture at the surface is allowed to dry in the airstream of a sterile laminar flow bench. Cells can be seeded on pretreated cover slips as described in **Subheading 3.5.**

6. Make sure that the cleft is totally filled. The filling height should not transcend the upper surface of the glass plate more than 2–3 mm.

7. We use an IX-70 microscope (Olympus) with a ×20 DIC objective. First, cells are visible after 5–7 d and the culture reaches confluency after 2–3 weeks (*see* **Figs. 3A,B**). Usually, only few cells grow out at the glass surface of the cover slip, most cells are attached to the plastic bottom.

8. The screw lid of the tube should not be closed tightly (*see* **Fig. 4A**) to allow gas exchange.

9. Trypan blue staining: 30 µL Trypan blue solution (Trypan Blue Stain 0.4%, ready to use, Gibco cat. no. 12250-061) is mixed with 30-µL cell suspension and applied in a counting chamber.

10. Culture medium 1, initially used during the first 2 d of culturing, contains a cocktail of bactericide and fungicide with a broad action spectrum. The human cells tolerate this cocktail for a limited time. However, long exposure to especially antimycotics negatively influences the viability of human cells. For additional information about preventing and handling contaminations of primary cell cultures, *see* Vierck et al *(6)*.

11. We usually split the cells in a ratio 1:3 or 1:4. In detail: adjust the volume of the cell suspension gained from one flask (25 cm^2) to 6 (*8*) mL and distribute the volume to three (four) new flasks. Ensure that the bottoms are evenly covered. For immunostaining or cell biological experiments it is desirable to plate cells on glass cover slips. In cases more standardized, starting conditions are required, you have to count the number of living cells before seeding. Use a counting chamber and tryphan blue staining of cells for the vitality check (*see* **Note 9**). Dilute the cell suspension and add the desired number of cells in a volume of two ml into a culture flask (25 cm^2).

12. Usually we use the cells from one near confluent culture (80–90%) grown in a 25 cm^2 flask to establish one vial for kryoconservation.

References

1. Huang, H..M., Martins, R., Gandy, S., et al. (1994) Use of cultured fibroblasts in elucidating the pathophysiology and diagnosis of Alzheimer's disease. *Ann. N.Y. Acad. Sci.* **747,** 225–244.

2. Gibson, G., Martins, R., Blass, J., and Gandy, S. (1996) Altered oxidation and signal transduction systems in fibroblasts from Alzheimer patients. *Life Sci.* **59,** 477–489.

3. Braak, H. and Braak, E. (1991) Neuropathological stageing of Alzheimer-related changes. *Acta Neuropathol. Berl.* **82,** 239–259.

4. Taylor, W. G. and Camalier, R. F. (1982) Modulation of epithelial cell proliferation in culture by dissolved oxygen. *J. Cell. Physiol.* **111,** 21–27.

5. Meske, V., Albert, F., Wehser, R., and Ohm, T. G. (1999) Culture of autopsy-derived fibroblasts as a tool to study systemic alterations in human neurodegenerative disorders such as Alzheimer's disease—methodological investigations. *J. Neural Transm.* **106,** 537–548.

6. Vierck, J. L., Byrne, K., Mir, P. S., and Dodson, M. V. (2000) Ten commadments for preventing contamination of primary cell cultures. *Methods in Cell Sci.* **22,** 33–41.

8

Primary Culture and Differentiation of Human Adipocyte Precursor Cells

Vanessa van Harmelen, Thomas Skurk, and Hans Hauner

1. Introduction

Adipose or fat tissue is of mesenchymal origin and is comprised of adipocytes, adipose precursor cells, blood cells, endothelial cells, fibroblasts, and monocytes/macrophages. The adipocytes represent roughly two-thirds of the total cell number and, owing to their enormous cell size, make up more than 90% of the tissue volume. Adipocytes are major participants in the energy homeostasis of the human body by storing excess energy in the form of triglycerides and by releasing fatty acids to meet the energy needs of other organs depending on substrate balance and hormonal regulation *(1,2)*. In addition, adipocytes exert multiple auto-, para-, and endocrine functions; they secrete numerous signaling factors that are involved in the regulation of energy homeostasis and a variety of neuroendocrine, metabolic, and immune functions *(3–5)*. The cytokines leptin, tumor necrosis factor-α (TNF-α), interleukin 6 (IL-6), as well as angiotensinogen and plasminogen activator inhibitor-1 (PAI-1) are some of the products secreted by adipocytes *(3–5)*. When the balance between energy intake and expenditure is such that the intake exceeds the need of the body, adipose tissue expands and the individual becomes obese. During adipose tissue enlargement, most of the adipocyte factors are increasingly produced and it has been suggested that some of them might be involved in the pathogenesis of the complications associated with obesity, such as insulin resistance (TNF-α), hypertension (angiotensinogen), enhanced thrombogenesis (PAI-1), and premature atherosclerosis (IL-6) *(3)*.

To characterize the metabolic and endocrine function of adipocytes as well as adipocyte development under normal and pathophysiological conditions, a

From: *Methods in Molecular Medicine, vol. 107: Human Cell Culture Protocols, Second Edition*
Edited by: J. Picot © Humana Press Inc., Totowa, NJ

variety of culturing techniques can be used. A number of clonal preadipocyte cell lines of rodent origin have been available for more than 20 yr, however, adipocyte metabolism and endocrine function have been shown to differ substantially between humans and rodents *(6)*. Therefore, it is neccessary to use human cell culture models if specific questions concerning human adipocyte function are investigated. There are, in principle, three primary culturing methods used for studying human adipocyte function: (1) culturing adipose tissue explants *(7)*; (2) culturing freshly isolated mature adipocytes (after collagenase digestion) (7) and 3) culturing and in vitro differentiating of the stromal cell fraction of adipose tissue (which consists mainly of preadipocytes). In comparison to adipose tissue explants or freshly isolated adipocytes, in vitro differentiated preadipocytes offer some advantages. (1) These cells can be cultured under defined conditions (in serum-free medium) for a longer duration (up to 1 mo); (2) effects of external factors, i.e., from the circulation, can be excluded; and (3) the process of adipocyte development can be studied. Disadvantages include the long duration of adipocyte development (14–20 d) and variation in adipose differentiation from experiment to experiment.

This chapter describes the technique to isolate primary human adipocyte precursor cells and to differentiate them into adipocytes. This model offers a valuable tool for studying the mechanisms that lead to obesity and its complications.

2. Materials

2.1. Buffers and Media

1. Basal medium: Dulbecco's modified Eagle's medium (DMEM)/Ham's F12 containing phenol red (*see* **Note 1**). Both DMEM and nutrient mixture F12 are purchased in powder form from ICN Biomedicals (Cologne, Germany) and Invitrogen (Karlsruhe, Germany), respectively. Basal medium is prepared by dissolving both DMEM and F12 (50:50, v:v) supplemented with 15 mM HEPES, 14 mM NaHCO$_3$, 33 µM biotin, and 17 µM D-pantothenate. The pH is adjusted to 7.4 and the medium is sterilized by filtration. This medium can be kept for 1 mo at 4°C.
2. Phosphate-buffered saline (PBS): PBS is purchased from Sigma. It is delivered in tablet form and contains no calcium or magnesium. The tablets are dissolved according to the manufacturer's recommendations. The buffer (10 mM PBS) is sterilized by filtration and kept at 4°C for up to several months.
3. Collagenase solution: Collagenase (250 U/mL) is dissolved in PBS containing 2% bovine serum albumin (BSA) at pH 7.4. The collagenase is from Biochrom (Berlin, Germany, Worthington CLS type 1). The BSA (fraction V) is purchased from Sigma (Taufkirchen, Germany). The collagenase solution should be freshly prepared and sterilized by filtration for every preadipocyte isolation experiment to obtain the best yield of preadipocytes. The solution can also be stored at –20°C

and then be used after thawing gently at 37°C, but in this case the enzyme activity of the collagenase is reduced and the number of preadipocytes isolated per gram adipose tissue may be less as compared to when using freshly prepared collagenase solution (*see* **Note 2**).

4. Erythrocyte lysis buffer: This buffer contains 155 mM NH$_4$Cl, 5.7 mM K$_2$HPO$_4$, 0.1 mM EDTA at pH 7.3. When sterilized by filtration it can be kept at 4°C for several months.

5. Inoculation medium: This medium contsists of basal medium supplemented with gentamicin (50 µg/mL, from PAA, Cölbe, Germany) (*see* **Note 3**) and 10% fetal bovine serum (nonheat inactivated FBS, Biochrom, Berlin, Germany). FBS is kept in sterile aliquots at –20°C and never refrozen after thawing. The inoculation medium can be stored at 4°C for up to 2 wk.

6. Basal preadipocyte culture medium: Basal medium supplemented with human insulin (66 nM), triiodo-L-thyronine (1 nM), human transferrin (10 µg/mL), and gentamicin (50 µg/mL). The hormones and transferrin are from Sigma (Taufkirchen, Germany). For triiodo-L-thyronine a 1-mM solution is alkalized using 1 M NaOH and diluted to 2 µM stock solution in 50% EtOH. For insulin a stock solution of 22 µM in 10 mM HCl is prepared. The stock solution of transferrin is 1 mg/mL H$_2$O. The stock solutions are sterilized by filtration (except that in EtOH) and kept at –20°C for several months in aliquots. They are never refrozen after thawing. Once thawed, they can be kept at 4°C for 1 wk. Freshly prepared basal preadipocyte culture medium can be kept at 4°C and should be used within 1 wk.

7. Preadipocyte differentiation medium: This medium contains of basal preadipocyte culture medium supplemented with isobutyl-methylxanthine (IBMX from Serva, Heidelberg, Germany) at 0.5 mM, hydrocortisone (100 nM), and troglitazone (1 µg/mL from Sigma). The stock solution of IBMX consists of 20 mM alkalized with Na$_2$CO$_3$. This stock can be kept after sterile filtration at 4°C for several months. For hydrocortisone a stock solution of 0.1 mM in 50% EtOH is prepared. The stock solution of troglitazone consists of 1 mg/mL dissolved in dimethyl sulfoxide (DMSO). The stock solutions of both troglitazone and hydrocortisone are kept at –20°C for several months. After thawing they can be kept at 4°C for 1 wk. The troglitazone stock should be protected from light. Preadipocyte differentiation medium is always freshly prepared.

All media and buffers are kept in normally autoclaved glass bottles. Sterile filtration is done through 0.2-µm pore size filters [table top filters (Corning, NY)]. Stock solution aliquots of factors added to the medium are kept in regular sterile plastic tubes.

2.2. Additional Materials Required

1. Laminar air-flow hood.
2. Incubator with 5% CO$_2$ atmosphere.

3. Autoclaved sharp long-handled surgical scissors (15–18 cm), forceps, small scissors (10 cm).
4. Polystyrene 10-cm Petri dishes to prepare the fresh adipose tissue samples.
5. Sterile filters (polypropylene) with pore size 70- and 150-µm and tube (and funnel) to support the filters.
6. Sterile 50-mL plastic tubes with caps.
7. Centrifuge for 50-mL tubes (200g).
8. Inverse light microsope.
9. Neubauer chamber and trypan blue (0.4%, Sigma).
10. Shaking water bath at 37°C.
11. Sterile 1–50 mL pipets and pipet tips.
12. Sterile tissue-culture plastic ware (6-, 12-, and 96-well culture plates).

3. Methods

3.1. Tissue Collection

1. Adipose tissue specimens from the abdominal region [subcutaneous (sc) and visceral] are usually obtained from elective or laparoscopic abdominal surgery (hernia, gall stones, gynecology, adjustable gastric banding for obesity, and so on). During laparoscopic interventions the amount of tissue is usually rather limited (5 g). Abdominal (sc) adipose tissue can also be obtained using needle biopsies; however, the amount of tissue is even more limited (1–3 g) and there is a high contamination by blood cells. Adipose tissue from the mammary region is usually obtained from plastic surgery. In this case, it is often possible to obtain large amounts of material (100 g); however, this type of tissue contains much gland tissue. The number of stromal vascular cells that can be obtained per gram adipose tissue is dependent upon the site from which the tissue is removed. For mammary fat, the yield is around 300,000 cells per gram tissue *(8)*, whereas for the abdominal adipose tissues (visceral and sc) the yield is usually higher (around 700,000 cells per gram adipose tissue, unpublished observations). For ethical reasons, informed consent must be obtained from the patient/subject prior to tissue sampling.
2. When large pieces of tissue are obtained (>50 g, e.g., mammary tissue), the specimens are crudely prepared (under sterile conditions) in the surgical room, to remove skin and gland tissue and to isolate the fat lobes.
3. The adipose tissue samples are immediately transported to the laboratory in sterile glass bottles containing basal medium and arrive in the laboratory within 1.5-h after removing the tissue from the patient. Usually, the preadipocyte isolation experiment is started immediately; however, it is also possible to store the adipose tissue samples (in small pieces) in basal medium overnight at 4°C. We have observed that the adipose tissue differentiation capacity is similar in preadipocytes isolated immediately or after overnight storage at 4°C. However, the yield of preadipocytes per gram tissue is usually lower after an overnight storage at 4°C (unpublished observations).

3.2. Cell Isolation

1. The adipose tissue specimens are carefully liberated from remaining visible connective tissue and blood vessels in Petri dishes. Readily prepared pieces are kept in PBS in Petri dishes until the collagenase digestion procedure.
2. The adipose tissue pieces are put into 50-mL tubes, cut gently into very fine pieces using surgical scissors, and incubated in collagenase solution (3 mL solution/mL or gram adipose tissue) (*see* **Note 2**). For the incubation, the tubes are closed tightly and connected horizontally under water in a 37°C shaking water bath for 90 min. The tubes are never filled more than 45 mL in order to let the solution move freely in the tube and to allow the collagenase to reach all adipose tissue pieces.
3. After the digestion, the tubes are centrifuged at 200g for 10 min at room temperature; thereafter the supernatant containing mature adipocytes and collagenase solution is removed from the pellet that contains the preadipocytes (*see* **Fig. 1**). The pellet should not dry, so 1 mL solution should be kept at the bottom.
4. The pellet is resuspended (pellets from several tubes can now be pooled) in erythrocyte lysis buffer (1:9, v:v) (*see* **Note 4**) and incubated for not more than 10 min.
5. The suspension is filtered through a 150-μm filter into another 50-mL tube and centrifuged again as above (*see* **Note 5**). The supernatant is discarded and the pellet is resuspended in an appropriate volume of basal medium (10–20 mL, depending on the expected yield of preadipocytes).
6. The suspension is filtered through a 70-μm pore size filter into another 50-mL tube. An aliquot of the suspension is taken for cell counting (see below) and the rest of the suspension is centrifuged again as described above.

3.3. Determination of Cell Number

1. An aliquot of the cell suspension (25–50 μL) is diluted in trypan blue (50:50, v:v). In case a high cell number is expected, the aliquot can be diluted in basal medium first.
2. Two times, 10 μL is transferred to a Neubauer chamber and four squares are counted. Blue cells are considered to be dead, cells which appear light or colorless are considered to be alive.
3. From the mean value counted, the total cell number is calculated by: (mean counted cell number from the eight squares) × (10,000 (Neubauer chamber factor)) × (dilution factor of aliquot in trypan blue) × (total volume from which the aliquot is obtained).
4. After cell counting, the cell pellet is diluted in inoculation medium to a final concentration that is easy to use to seed the cells (e.g., 1 million cells/mL) (*see* **Notes 6** and **7**).

3.4. Cell Inoculation and Attachment

1. The advisable inoculation density is 30,000–50,000 cells/cm^2 to achieve optimal differentiation. A lower inoculation density may lead to suboptimal differentiation

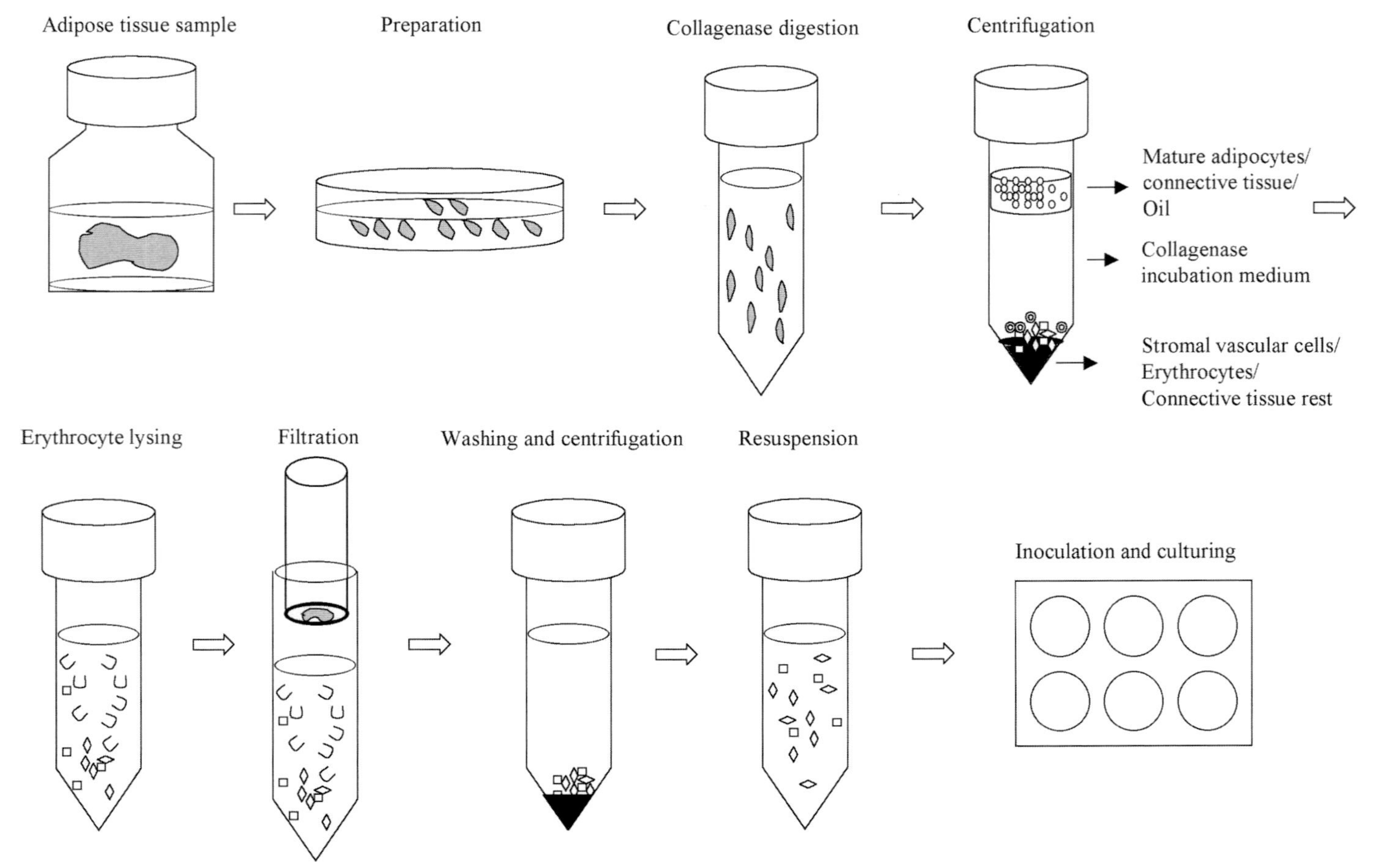

Adipose tissue sample
Preparation
Collagenase digestion
Centrifugation
Mature adipocytes/
connective tissue/
Oil
Collagenase
incubation medium
Stromal vascular cells/
Erythrocytes/
Connective tissue rest
Erythrocyte lysing
Filtration
Washing and centrifugation
Resuspension
Inoculation and culturing

(*see* **Note 8**). We mostly use 6- or 12-well plates, representing an area of 10 and 4.5 cm^2/well, respectively.

2. For optimal cell attachment, it is recommended that the cells be kept for 16–24 h in the inoculation medium (5% CO_2, 37°C). Although not all cells become attached during this time, the remaining cell number is sufficient to obtain a dense cell monolayer when the cells have spreaded after a few days in culture. Longer exposure of the cells than 2 d to inoculation medium, which contains of 10% fetal bovine serum (FBS), is associated with an increased mitogenic activity, i.e., the cell proliferate and they increase in number. However, in this case the differentiation capacity of the cells may be reduced.

3.5. Preadipocyte Differentiation

1. After cell adhesion, cells are washed two times with prewarmed PBS to remove nonattached cells (blood cells, dead preadipocytes) and serum.
2. Then differentiation of the preadipocytes is started by adding prewarmed preadipocyte differentiation medium to the cells (*see* **Note 9**).
3. The cells are cultured (5% CO_2, 37°C) for 3 d in the preadipocyte differentiation medium, whereafter the medium is changed to basal preadipocyte culture medium that is supplemented with hydrocortisone at 100 n*M* (*see* **Note 10**). This culture medium is renewed every 3–4 d (*see* **Note 11**). Visible lipid accumulation starts within 6–8 d under these conditions.
4. Within 16 d, the differentiated preadipocytes are completely filled with lipid droplets (*see* **Note 12**) and have changed their morphology to a spherical shape (*see* **Fig. 2b**) (*see* **Note 13**). The preadipocytes have now the functional characteristics of "mature" adipocytes although they still have a multilocular appearance (*see* **Note 14**). On average, 60% of the cells in the wells differentiate under these conditions.
5. The percentage of differentiated cells can be determined by counting differentiated and undifferentiated cells in five randomly selected areas (mm^2) under the microscope (*see* **Note 15**).

4. Notes

1. The cells are cultured in medium containing phenol red (pH indicator). It should be noted that in some assays (e.g., luminometric assay to measure glycerol release from the cells), phenol red interferes with the assay.
2. Poor attachment and spreading of cells might appear when the adipose tissue had been exposed to collagenase solution too long. Different batches of collagenase may differ substantially in quality and activity. Collagenase of the same batch is used throughout the whole series of experiments. When a new batch of collage-

---

Fig. 1. *(see opposite page)* Preparation and cultivation of stromal vascular cells from human adipose tissue.

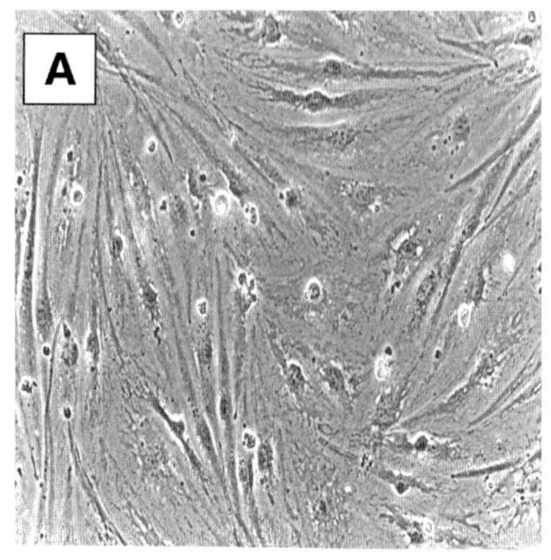

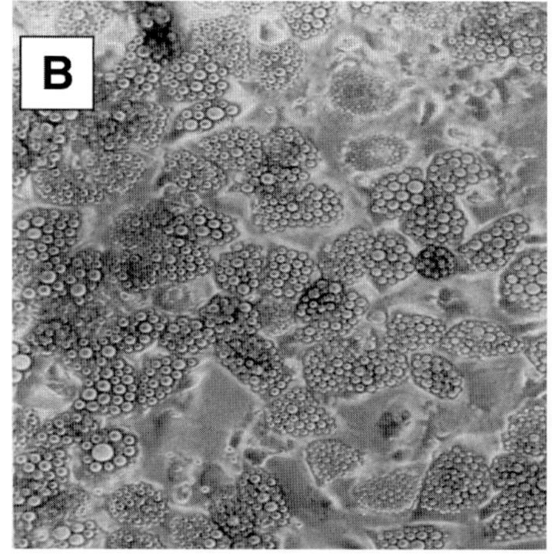

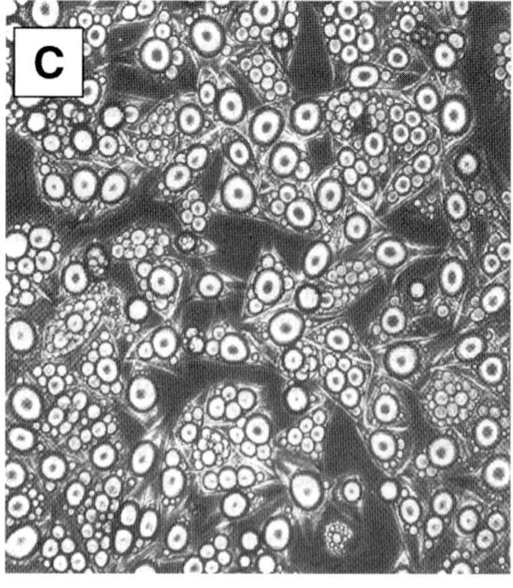

nase is needed, various batches are tested and compared to the old batch. The collagenase with the best activity is chosen and large quantities of the batch are reserved at the company.

3. To avoid contamination of the cultures by microorganisms, strictly sterile conditions are required and it is recommended that gentamicin be added to the culture media. Gentamicin works against Gram-negative and Gram-positive bacteria as well as against mycoplasms.

4. Erythrocytes disturb preadipocyte adherence and should therefore be eliminated using the described erythrocyte lysis buffer. This buffer is a hypotonic solution and makes the erythrocytes burst, because the plasma membrane of these cells is freely permeable to water in contrast to preadipocytes or other cell types *(9)*. This buffer does not interfere with preadipocyte attachment and differentiation, but should be washed off carefully.

5. Adipose tissue samples from the visceral region are densely vascularized. Therefore, it is very likely that there is a contamination by endothelial cells in the stromal cell fraction. The proportion of endothelial cells should be reduced to less than 5% of the total cell number in order to avoid major effects of the endothelial cells on preadipocyte functioning. Therefore, it is recommended, after the first filtration through 150 µm to do a second filtration immediately using a filter with pore size 30 µm. Endothelial cells reaggregate rapidly and are therefore retained by this filter *(10)*. However, some contamination by endothelial cells cannot be avoided. If precursor cells from different depots are compared, all samples must undergo the same filtration procedure. (The proceeding filtration step through 70 µm can be omitted in this case.)

6. It is possible to inoculate the cells directly in preadipocyte differentiation medium, i.e., cells are not exposed to inoculation medium first. However, the lack of FBS will cause a low cell adhesion (30–40%). The differentiation capacity of the preadipocytes on the other hand is higher in this case *(11)*.

7. One can also precoat the wells in the plates with a thin layer of fibronectin (0.02 mg/mL) overnight at 37°C and inoculate the cells in these wells in a serum-free medium (before seeding, the wells are rinsed with basal medium). Cell adhesion is high, but cannot fully replace the effect of serum from the inoculation medium. However, precoating with fibronectin does not influence negatively the differentiation capacity of the preadipocytes.

8. When the yield of preadipocytes is low, it is possible to let the cells proliferate first to confluency using fibroblast growth factor (FGF), whereafter the differentiation of the preadipocytes is started. FGF only moderately reduces preadipocyte

---

Fig. 2. *(see opposite page)* Micrographs of cultured preadipocytes during the differentiation process. (**A**) Human stromal vascular cells after 3 d in culture. (**B**) Differentiated preadipocytes from subcutaneous tissue on d 16. The cells are still multilocular at this stage. (**C**) Differentiated preadipocytes from sc tissue on d 30, the multiple lipid droplets have turned into a few large lipid droplets at this stage.

differentiation capacity *(12)*. Cells are inoculated at a density of 10,000/cm^2 in the usual inoculation medium (16–24 h). They are washed twice with PBS and then fed with basal preadipocyte culture medium with reduced hydrocortisone (10 n*M*), supplemented with recombinant hu-bFGF (1 n*M*, stock 1 µ*M* in PBS). Cells are grown until confluency (5–6 d). Then differentiation is started as usual.

9. As described above, hydrocortisone is present in the preadipocyte culture medium (100 n*M*) during the whole culturing period. This is because hydrocortisone, like insulin and triiodo-L-thyronine, has positive effects on the metabolic function and development of preadipocytes. However, hydrocortisone foremost has an adipogenic activity by controlling the expression of adipocyte specific genes *(13)*. Higher rates of preadipocyte differentiation may be reached by adding dexamethasone (1 µ*M*) instead of hydrocortisone to the medium. Dexamethasone is around 25 times more potent at the same concentration.

10. To maximize preadipocyte differentiation, the nonselective phosphosdiesterase inhibitor IBMX and/or the thiazolidinedione troglitazone are used in the differentiation medium for the first 3 d (an exposure longer than 3 d does not further improve differentiation). One can also use other thiazolidinediones such as pioglitazone, ciglitazone, or rosiglitazone, which are similarly adipogenic. They all stimulate preadipocyte differentiation by activating the transcription factor PPARγ (peroxisome proliferator activated receptor γ) that controls the expression of adipocyte specific genes *(14)*.

11. It is recommended to change the medium very gently and not more often than necessary. During lipid accumulation, the differentiating preadipocytes detach very easily from the well bottom and they become very vulnerable to vibration or other mechanical stress.

12. After d 16, when cells are completely filled with mutiple lipid droplets and express the adipocyte specific genes, cells can be cultured for another 10–14 d. However, they are becoming more difficult to handle because the multiple lipid droplets turn into a few large lipid droplets and cells are getting easily detached because of a decreasing production of matrix proteins (*see* **Fig. 2C**).

13. To facilitate lipogenesis from glucose, biotin and pathothenate are continuously present in the medium. Lipid accumulation can be accelerated by adding low concentrations of Intralipid (2% w:v) to the medium.

14. Some compounds are dissolved in EtOH or DMSO. A combination of such additives may lead to (sub)toxic concentrations and affect preadipocyte functioning considerably. It is important that a final concentration of 0.1% DMSO and 0.1% EtOH is not exceeded. When a test compound is dissolved in DMSO or EtOH, controls should be present in the experiment treated with the same concentration of DMSO or EtOH.

15. The differentiation capacity can vary considerably from experiment to experiment. Several technical aspects influencing differentiation are already mentioned in this chapter. However, this variation may also be due to inherent characteristics of the donors. For example, we have shown that mammary preadipocytes differentiate less well with increasing BMI *(8)*.

References

1. Carey, G. B. (1998) Mechanisms regulating adipocyte lipolysis. *Adv. Exp. Med. Biol.* **441,** 157–170.
2. Ramsay, T. G. (1996) Fat cells. *Endocrinol. Metab. Clin. North Amer.* **25,** 847–870.
3. Fried, S. K. and Russell, C. D. (1998) Diverse roles of adipose tissue in the regulation of systemic metabolism and energy balance, in *Handbook of Obesity* (Bray, G. A., Bouchard, C., and James, W. P. T., eds.), Marcel Dekker, New York, pp. 397–413.
4. Spiegelman, B. M. and Hotamisligil, G. S. (1993) Through thick and thin, wasting obesity and TNF-α. *Cell.* **73,** 625–627.
5. Lyon, C. J., Law, R. E., and Hsueh, W. A. (2003) Minireview: adiposity, inflammation, and atherogenesis. *Endocrinology* **144,** 2195–2200.
6. Ailhaud, G. and Hauner, H. (1998) Development of white adipose tissue, in *Handbook of Obesity* (Bray, G. A., Bouchard, C., and James, W. P. T., eds.), Marcel Dekker, New York, pp. 359–378.
7. Fried, S. K. and Moustaid-Moussa, N. (2001) Culture of adipose tissue and isolated adipocytes, in *Methods in Molecular Biology* (Ailhaud, G., ed.), Humana, Totowa, NJ, pp. 197–212.
8. van Harmelen, V., Skurk, T., Röhrig, K., et al. (2003) Effect of BMI and age on adipose tissue cellularity and differentiation capacity in women. *Int. J. Obes. Relat. Metab. Disord.* **27,** 889–895.
9. Alberts, B., Bray, D., Lewis, J., Raff, M., Roberts, K., and Watson, J. D. (1994) *Molecular Biology of the Cell* (Alberts, B., Bray, D., Lewis, J., Raff, M., Roberts, K., and Watson J. D., eds.), Garland, New York, p. 516.
10. Björntorp, P., Karlsson, M., Pertoft, H., Pettersson, P., Sjöström, L., and Smith, U. (1978) Isolation and characterization of cells from rat adipose tissue developing in adipocytes. *J. Lipid Res.* **19,** 316–324.
11. Entenmann, G. and Hauner, H. (1996) Relationship between replication and differentiation in cultured human adipocyte precursor cells. *Am. J. Physiol.* **270,** C1011–C1016.
12. Hauner, H., Röhrig, K., and Petruschke, T. (1995) Effects of epidermal growth factor (EGF), platelet-derived growth factor (PDGF) and fibroblast growth factor (FGF) on human adipocyte development and function. *Eur. J. Clin. Invest.* **25,** 90–96.
13. Hauner, H., Entenmann, G., Wabitsch, M., et al. (1989) Promoting effect of glucocorticoids on the differentiation of human adipocyte precursor cells cultured in a chemically defined medium. *J. Clin. Invest.* **84,** 1663–1670.
14. Spiegelman, B. M. (1998) PPARgamma: adipogenic regulator and thiazolidinedione receptor. *Diabetes* **47,** 507–514.

9

Human Mononuclear Phagocytes in Tissue Culture

Yona Keisari

1. Introduction

Peripheral blood human monocytes (HuMo) are the major source for human mononuclear phagocytes. Such monocytes when cultured, differentiate into monocyte-derived macrophages (HuMoDM), and undergo various structural, biochemical, and functional changes.

The most common method used for the separation of mononuclear cells from the blood, is the Ficoll-Hypaque density gradient centrifugation, essentially described by Boyum *(1)*. Ficoll-Hypaque at a density of 1.077 g/L is used to separate the denser granulocytes and erythrocytes from the lighter lymphocytes, monocytes, and thrombocytes. The mononuclear cells stay at the top of the Ficoll-Hypaque layer, whereas the denser cells sink to the bottom of the centrifuge tube.

Peripheral blood monocytes are purified from the mononuclear fraction by adherence to plastic. Adherence can be carried out either directly onto tissue-culture plates in which they will be further grown (24- or 96-well plates), or onto tissue-culture flasks from which they will be hence removed and recultured in the required plates or chambers *(2)*.

The adherent human MoDM bind firmly to plastic substrata and it is very difficult to remove the cells for quantitative measurements. There are several methods to enumerate or quantitate adherent monocytes/macrophages of which four are described below.

When an enriched monocyte cell suspension is required, MNC harvested on Ficoll-Hypaque gradients can be further separated on Percoll gradients into lymphocytes and monocytes. The method initially described by Ulmer and Flad *(3)*, was modified by Orlandi et al. *(4)*, which used only one Percoll concentration. After separation on Percoll, the monocytes that are less dense than

From: *Methods in Molecular Medicine, vol. 107: Human Cell Culture Protocols, Second Edition*
Edited by: J. Picot © Humana Press Inc., Totowa, NJ

lymphocytes stay on top of the Percoll layer, whereas the lymphocytes go through to the bottom of the tube.

Long-term incubation (7–21 d) of monocytes under tissue-culture conditions results in the differentiation of the cultured cells, and in the appearance of HuMoDM *(5)*. Yet, long-term incubation of the cells in culture results in a substantial loss of cells. In various studies, it was found that cytokines such as granulocyte-macrophage colony-stimulating factor (GM-CSF), IL-3 *(6–10)*, or IL-4 *(11,12)*, as well as PKC activators/tumor promoters *(13,14)* facilitate the survival and differentiation of human monocytes.

It was observed that exposure of monocytes to a combination of IL-4 and GM-CSF was more effective than each cytokine itself. Yet, the combination of GM-CSF (or IL-3) and IL-4 may give various results that depend on the order and timing of exposure to the cytokines, on the type of cells treated (e.g., monocytes or macrophages), and on the concentrations of the cytokines. Studies indicated that although GM-CSF is important for monocyte differentiation into macrophages, IL-4 provoked cell fusion and formation of giant cells *(15)*. Incubation of peripheral monocytes with GM-CSF and IL-4 for extended time periods may also lead to their differentiation into dendritic cells which express augmented antigen-presentation capabilities *(16)*. It is important to mention in this regard that, in order to obtain dendritic cells from monocytes, 10–100-fold higher concentrations of the cytokines are required, compared to those used to obtain differentiated macrophages.

In general, IL-4 is strongly implicated in the alternative macrophage activation pathway *(17)*, whereas IFN-γ is the key player in the classical macrophage activation pathway.

2. Materials

1. Earle's balanced salt solution (EBSS) and 10X EBSS.
2. Dulbecco's PBS without Ca^{2+} and Mg^{2+}.
3. Ficoll-Hypaque (density 1.077 g/L).
4. RPMI-1640, supplemented with 100 µg/mL streptomycin 100 U/mL penicillin, and 300 mg/mL (2 m*M*) L-glutamine.
5. Newborn bovine serum (NBS), or pooled human AB serum (HABS). Heat inactivated (56°C, 30 min).
6. Percoll (Pharmacia).
7. Human recombinant GM-CSF, IL-3, and IL-4.
8. Phorbol retinoyl acetate (PRA).
9. 12-*O*-Tetradecanoyl-13-phorbol acetate (TPA),
10. Mezerein (MEZ).
11. Bio-Rad Protein Assay (cat. no. 500-0006, Bio-Rad Laboratories, Munich, Germany). Dilute 1:5 in double-distilled H_2O, and filter before use. Prepare fresh for each assay.

12. Hemacolor color reagents (a product of Merck, Darmstadt, W. Germany), or Diff-Quik reagents (a product of Harleco, Gibbstown, NJ).
13. SDS 0.5% dissolved in double-distilled H_2O.
14. Methanol.
15. Isopropanol (2-propanol) for analysis.
16. MTT (3-(4,5-dimethyl-thiazol-2-yl)-2,5-diphenyl tetrazolium bromide) (5 mg/mL in PBS).
17. Limulus amebocyte lysate (LAL) reagent (Pyrogent, Whittaker, MA, Bioproducts Inc., Walkersville, MD) (*see* **Note 1**).

3. Methods

3.1. The Isolation of Mononuclear Cells (MNC) (see Note 2)

HuMos can be separated from heparinized blood samples or from the buffy coats of normal blood bank donors with sodium citrate as an anticoagulant. Blood bank buffy coats are obtained after centrifugation of blood bank bags (original volume of 400 mL), and removal of the erythrocytes and the plasma. The residual fraction (25–40 mL) contains more than 90% of the leukocytes of the blood donation, 10% of the erythrocytes, and 5% of the plasma.

1. Dilute buffy coats 1:4 (v/v) or heparinized blood samples 1:2 (v/v) with Dulbecco's PBS without Ca^{2+} and Mg^{2+}.
2. Add 40 mL of the diluted blood to 50-mL conical polypropylene tubes.
3. Add 10 mL of Ficoll-Hypaque to the bottom of each tube. To prevent mixing, insert a sterile Pasteur pipet into the diluted blood down to the bottom of the tube, and inject the Ficoll-Hypaque into the pipet by a 10-mL syringe (0.8 × 40-mm needle).
4. Centrifuge the tubes at 700*g* for 30 min at room temperature with brakes off.
5. Remove the cellular fraction that is on top of the Ficoll-Hypaque layer and transfer to a new 50-mL conical tube. For blood bank buffy coats from the same donor, pool cells from four tubes into one.
6. Wash three times with 50 mL cold Dulbecco's PBS without Ca^{2+} and Mg^{2+} at the following conditions:
   a. once at 350*g* for 10 min
   b. twice at 230*g* for 7 min
7. Resuspend the cells in supplemented RPMI-1640 containing 10% serum.
8. Place the tube for 16 h at 4°C to allow for separation of the mononuclear cells (cell pellet) from the thrombocytes (fluid suspension).
9. Discard the fluid gently by low rate aspiration and resuspend the cells in 10 mL cold supplemented RPMI-1640 + 2% serum.
10. For cell counts, dilute first 1:10 in supplemented RPMI-1640, and then 1:2 with 0.1% trypan blue solution, and count the cells with a hemocytometer.

The average yield is $540 \pm 130 \times 10^6$ MNC (range 400–770) per original 400 mL blood.

3.2. The Isolation of Mononuclear Phagocytes

3.2.1. Separation of Monocytes by Adherence in Microtiter Plates

1. Prepare mononuclear cells at 2–3×10^7 cells per mL in supplemented RPMI-1640 + 2% serum.
2. Add 0.1 mL of the cell suspension to each well (2–3×10^6 cells per well) of a 96- well tissue culture plate.
3. Incubate for 30 min at 37°C and 7.5% CO_2.
4. Remove nonadherent cells by washing the wells three times with warm (37°C) EBSS. For this purpose use a 5- or 10-mL syringe (without a needle).
5. Add 0.2 mL of supplemented RPMI-1640 + 10% serum.

The resulting monolayers contain 2.4 ± 0.2 to $3.0 \pm 0.4 \times 10^5$ cells/well (*see* **Note 3**).

3.2.2. Separation of Monocytes by Adherence in Tissue Culture Flasks

1. Resuspend mononuclear cells at 5×10^6 per mL in supplemented RPMI-1640 + 2% serum.
2. Add 5 mL of the cell suspension to 25-cm^2 or 20 mL to 75-cm^2 tissue culture flasks.
3. Incubate for 30 min at 37°C and 7.5% CO_2.
4. Remove nonadherent cells by washing the flasks extensively three times with warm (37°C) EBSS.
5. Add 5 or 25 mL, respectively, of supplemented RPMI-1640 + 10% serum.
6. Incubate for 16 h at 37°C and 7.5% CO_2.
7. Shake the flasks firmly and remove the medium that contains the detached monocytes. To remove still adherent cells, wash the flasks with cold EBSS using a 10-mL syringe and an 0.8×40-mm needle.
8. Centrifuge the cells at 350g for 10 min, and resuspend the cells in supplemented RPMI-1640 + 10% serum.
9. Culture 2–3×10^5 cells per well per 0.2 mL in 96-well plates, or 5×10^5 cells per well per 0.5 mL in 24-well plates.

3.2.3. Separation of Monocytes on a Isoosmotic Percoll Gradient

1. Prepare isoosmotic supplemented RPMI-1640+10% serum by adjusting the osmolality to 285 mosmol.
2. Prepare isoosmotic Percoll solution by mixing Percoll with EBSS (10X) 93:7 (v/v), and adjust to 285 mosmol.
3. Prepare a 46% solution of isoosmotic Percoll with isoosmotic RPMI-1640.
4. Resuspend the mononuclear fraction obtained after Ficoll-Hypaque separation in isoosmotic RPMI-1640 at 5–10×10^6 cells per mL.
5. Add 5 mL of the cell suspension to 10–12 mL tubes.
6. Add 5 mL 46% Percoll solution to each tube using a Pasteur pipet as described in **Subheading 3.1.** for Ficoll-Hypaque.

7. Centrifuge at 600*g* for 30 min at room temperature with brakes off.
8. Remove interface with a sterile pipet, and wash the cells twice with supplemented RPMI-1640.
9. Count cell number.
10. The recovery is 11.3 ± 3% (range 8–18%) of the MNC fraction, and 86 ± 6% of the cells are monocytes (range 77–95%).

3.3. Long-Term Cultures of Monocyte-Derived Macrophages (MoDM)

3.3.1. MoDM Cultured in the Presence of CSF

1. Prepare cultured monocytes in 96- or 24-well plates as described (*see* **Note 3**).
2. Add 50–250 U/mL (1–5 ng/mL) GM-CSF or IL-3 to the monocyte cultures (*see* **Note 4**).
3. Incubate without changing the medium for at least 10 d to obtain HuMoDM (*see* **Note 5**).
4. If extended incubation periods are required, add the indicated amount of CSF every 2 wk, by replacing half of the volume of the culture medium.

3.3.2. MoDM Cultured in the Presence of IL-4

1. Prepare cultured monocytes in 96- or 24-well plates as described (*see* **Note 3**).
2. Add 1–5 ng/mL IL-4 to the monocyte cultures (*see* **Note 6**).
3. Incubate without changing the medium for at least 10 d to obtain HuMoDM.

3.3.3. MoDM Cultured in the Presence of PKC Activators/Tumor Promoters (see **Note 7**)

1. Prepare cultured monocytes in 96- or 24-well plates as described (*see* **Note 3**).
2. Add TPA, MEZ, or PRA to monocyte cultures at a final concentration of 2–5 n*M* (*see* **Note 8**).
3. Incubate, without changing the medium, for at least 10 d to obtain HuMoDM.
4. If extended incubation periods are required, add the indicated amount of PKC activators every 3 wk by changing half of the volume of the culture medium (*see* **Note 9**).

3.4. General Methods for the Quantitation of Cultured Adherent Mononuclear Phagocytes

3.4.1. Cell Count

1. Remove the culture medium from the cultured monocytes.
2. Add ice-cold PBS (0.2 mL) without Ca^{2+} and Mg^{2+}.
3. Scrape gently the adherent cells and count with a hemocytometer.

3.4.2. Protein Determination (see **Note 10**)

1. Wash the cultured cells extensively with EBSS to remove medium and serum.
2. Lyse the cells with 200 µL of 0.1% Triton X for 30 min at 37°C.

3. Add 20-μL samples of the cell lysates to 200 μL of Bio-Rad reagent in 96-well plates.
4. Incubate the samples for 15 min and read at 600 nm in an automated spectrophotometer. Protein concentration is determined using a standard curve of BSA (5–1000 μg/mL).

3.4.3. Hemacolor Colorimetric Microtiter Assay (see **Note 11**)

1. Remove the supernatant from cell monolayers cultured in 96-well plates and dry the cells quickly in the air.
2. Fix the monolayers with methanol (50 μL per well) for 30 s (do not rinse the wells between **steps 2–4**).
3. Add 80–100 μL per well of Hemacolor Reagent 2 for 60 s.
4. Add 80–100 μL per well of Hemacolor Reagent 3 for another 60 s.
5. Rinse the plates three times with tap water.
6. Refill with water and decolorize for 5 min.
7. Remove the water and dry the plates extensively. (Following this step the stained cultures can be kept for several weeks in the dark.)
8. For stain extraction, add 0.2 mL per well of SDS (0.5%) dissolved in double-distilled H_2O for at least 90 min.
9. Measure OD at 600 or 630 nm with an automated microplate reader.

3.4.4. MTT Assay (see **Note 12**)

1. Remove the culture medium from cell monolayers cultured in 96-well plates.
2. Reconstituted each well with 0.2 mL supplemented RPMI-1640 containing 1 mg/mL MTT.
3. Incubate the cultures for 2–4 h at 37°C.
4. Remove the supernatants from the wells.
5. Add 0.2 mL per well of a lysing reagent containing 0.04 N HCl in isopropanol.
6. Mix the content of the wells thoroughly.
7. Read the plates in an automated microplate spectrophotometer at 570 and 630 nm as reference.

3.4.5. Alternative MTT Assay

1. If the cultured cells are not tightly adherent, and might be removed with the supernatant, it is recommended to remove only 0.1 mL of the culture medium from the cells cultured in 96-well plates.
2. Add to each well 0.025 mL of 5 mg/mL MTT in PBS.
3. Incubate the cultures for 2–4 h at 37°C.
4. Add 0.1 mL per well of a lysing reagent containing 0.04 N HCl in isopropanol.
5. Mix the content of the wells thoroughly.
6. Read the plates in an automated microplate spectrophotometer at 570 and 630 nm as reference.

4. Notes

1. All the media and buffers used should be assayed for the presence of bacterial endotoxin by the Gel-clot technique *(18)* using the Limulus Amebocyte Lysate (LAL) reagent. Reagents should be used only if no detectable LPS is found (sensitivity, 0.064 endotoxin units/mL).

2. **Steps 1–5** of this procedure should be carried out at room temperature (RT). All the reagents must be at RT, and the use of cold reagents should be avoided at this stage of the separation.

3. For extended incubation periods (more than 4 d) it is recommended not to culture cells in the wells at the periphery of the plates. These wells should be filled with sterile water to the top to reduce evaporation of liquid from the cultures.

4. Cytokines from different sources may differ in the optimal concentrations required for monocyte treatment. It is recommended to calibrate the cytokines when the method is implemented.

5. The optimal effect of CSF was achieved when added on the first day of culture. After 6 d in culture, the cells did not respond to the addition of CSF, and they behaved as nontreated cells. Microscopic observations of MoDM obtained in the presence of CSF revealed a homogenous population of large spread out cells, whereas nontreated cultures were more heterogeneous in their appearance and some small round cells were also apparent *(6,7)*.

6. The effect of IL-4 on monocyte viability was evident already at 0.1 ng/mL, but the best results were obtained at concentrations of 5–10 ng/mL. At concentrations above 1 ng/mL, IL-4 reduced the oxidative burst activity and IL-6 production by the macrophages *(12)*. IL-4 was also reported to reduce the production of chemotactic and cytostatic compounds by macrophages *(11)*.

7. Adherent monocytes cultured in 96-well plates showed a substantial loss (51%) of adherent cells in nontreated monocyte cultures after 2 wk of incubation. In comparison, HuMoDM cultures treated with various PKC activators/tumor promoters lost only 0–26% of the cells after incubation for 2 wk *(13,14)*.

8. Prepare stock solutions of PRA, TPA, and MEZ in DMSO at 10 μM and store in the dark at –20°C. When diluting the reagents in culture media before adding to the cells, the final concentration of DMSO should not exceed 0.1%

9. In our laboratory, we maintained MoDM cultures for 4 mo by adding TPA (2 nM) every 3 wk.

10. Determination of protein concentration of cells cultured in 96-well plates is according to the Bradford method *(19)*.

11. The Hemacolor colorimetric microtiter assay *(20)* uses reagents generally used to stain blood cells and cells in tissue cultures. The staining kit holds three solutions, solution 1 contains methanol for fixation, solution 2 contains a xanthene dye (orange color), and solution 3 is a thiazine solution containing a mixture of azure I dyes and methylene blue (blue reagent). A spectrophotometric analysis of a mixture of solutions 2 and 3 in SDS 0.5% revealed a peak of absorption at 517 nm caused by the xanthene dye, and a second peak at 634 nm caused by the thiazine

solution. Measurements of stained cells are carried out in an automated microplate reader using 630- or 600-nm filters. Diff-Quik reagents that serve a similar purpose can substitute Hemacolor reagents.

12. The MTT assay is based on the observation that tetrazolium salts are reduced to formazan by cellular respiratory enzymes. Only viable cells perform this activity, and thus the method may indicate the amount of viable cells present in culture. The method was initially described by Mosmann *(21)* for MTT, but other Tetrazolium reagents can also be used *(22)*.

References

1. Boyum, A. (1968) Isolation of mononuclear cells and granulocytes from human blood. *Scand. J. Clin. Lab. Invest.* **21 (Suppl. 97),** 77–89.
2. Treves, A. J., Yagoda, D., Haimovitz, A., Ramu, N., Rachmilewitz, D., and Fuks, Z. (1980) The isolation and purification of human peripheral blood monocytes in cell suspension. *J. Immunol. Methods* **39,** 71–80.
3. Ulmer, A. J. and Flad, H.-D. (1979) Discontinuous density gradient separation of human mononuclear leucocytes using Percoll as gradient medium. *J. Immunol. Methods* **30,** 1–10.
4. Orlandi, M., Bartolini, G., Chiricolo M., Minghetti, L., Franceschi, C., and Tomasi, V. (1985) Prostaglandin and thromboxane biosynthesis in isolated platelet-free human monocytes. I. A modified procedure for the characterization of the prostaglandin spectrum produced by resting and activated monocytes. *Prostaglandins Leukotrienes Med.* **18,** 205–216.
5. Zuckerman, S. H., Ackerman, S. K., and Douglas, S. D. (1979) Long-term human peripheral blood monocyte cultures: establishment, metabolism and morphology of primary human monocyte-macrophage cell cultures. *Immunology* **38,** 401–411.
6. Robin, G., Markovich, S., Athamna, A., and Keisari, Y. (1991) Human recombinant granulocyte-macrophage colony stimulating factor augments the viability and cytotoxic activities of human monocyte derived macrophages in long term cultures. *Lymphokine Cytokine Res.* **10,** 257–263.
7. Dimri, R., Nissimov, N., and Keisari, Y. (1994) Effect of human recombinant granulocyte-macrophage colony stimulating factor and IL-3 on the expression of surface markers of human monocyte derived macrophages in long term cultures. *Lymphokine Cytokine Res.* **14,** 237–243.
8. Elliot, M. J., Vadas, M. A., Eglinton, J. M., et al. (1989) Recombinant human interleukin-3 and granulocyte-macrophage colony-stimulating factor show common biological effects and binding characteristics on human monocytes. *Blood* **74,** 2349–2359.
9. Markowicz, S. and Engleman, E. G. (1990) Granulocyte-macrophage colony-stimulating factor promotes differentiation and survival of human peripheral blood dendritic cells in vitro. *J. Clin. Invest.* **85,** 955–961.
10. Eischen, A., Vincent, F., Bergerat, J. P., et al. (1991) Long term cultures of human monocytes in vitro. Impact of GM-CSF on survival and differentiation. *J. Immunol. Methods* **143,** 209–221.

11. te Velde, A. A., Klomp, J. P., Yard, B. A., de Vries, J. E., and Figdor, C. G. (1988) Modulation of phenotypic and functional properties of human peripheral blood monocytes by IL-4. *J. Immunol.* **140,** 1548–1554.
12. Keisari, Y., Robin, G., Nissimov, L., et al. (2000) Role of cytokines in the maturation and function of macrophages: Effect of GM-CSF and IL-4, in *The Biology and Pathology of Innate Immunity Mechanisms* (Keisari, Y. and Ofek, I., eds.), Kluwer Academic/Plenum, New York, pp. 73–89.
13. Keisari, Y., Bucana, C., Markovich, S., and Campbell, D. E. (1990) The interaction between human peripheral blood monocytes and tumor promoters: Effect on in vitro growth, differentiation and function. *J. Biol. Response Modif.* **9,** 401–410,
14. Markovich, S., Kosashvilli, D., Raanani, E., Athamna, A., O'Brian, C. A., and Keisari, Y. (1994) Tumor promoters/protein kinase C activators augment human peripheral blood monocyte maturation in vitro. *Scand. J. Immunol.* **39,** 39–44.
15. Dugast, C., Gaudin, A., and Toujas, L. (1997) Generation of multinucleated giant cells by culture of monocyte-derived macrophages with IL-4. *J. Leukoc. Biol.* **61,** 517–521.
16. Sallusto, F. and Lanzavecchia, A. (1994) Efficient presentation of soluble antigen by cultured human dendritic cells is maintained by granulocyte-macrophage colony-stimulating factor plus interleukin-4 and down regulated by tumor necrosis factor-alpha. *J. Exp. Med.* **179,** 1109–1118.
17. Gordon, S. (2003) Alternative activation of macrophages. *Nature Rev. Immunol.* **3,** 23–35.
18. Yin, E. T., Galanes, C., Kinsky, S., Bradshaw, R., Wessler, S., and Luderitz, O. (1972) Picogram-sensitive assay for endotoxin: Gelation of limulus polyuphemus blood cell lysate induced by purified lipopolysaccharide and lipid A from gram-negative bacteria. *Biochim. Biophys. Acta* **261,** 284–289.
19. Bradford, M. M. (1976) A rapid and sensitive method for the quantitation of microgram quantities of protein utilizing the principle of protein-dye binding. *Anal. Biochem.* **72,** 248–254.
20. Keisari, Y. (1992) A colorimetric microtiter assay for the quantitation of cytokine activity on adherent tissue culture cells. *J. Immunol. Methods* **146,** 155–161.
21. Mosmann, T. (1983) Rapid colorimetric assay for cellular growth and survival: Application to proliferation and cytotoxicity assays. *J. Immunol. Methods* **65,** 55–63.
22. Alley, M. C., Scudiero, D. A., Monks, A., et al. (1988) Feasibility of drug screening with panels of human tumor cell lines using a microculture tetrazolium assay. *Cancer Res.* **48,** 589–601.

10

Purification of Peripheral Blood Natural Killer Cells

Bice Perussia and Matthew J. Loza

1. Introduction

The ability to perform biological studies on natural killer (NK) cells requires effective methods for their isolation from hematopoietic cells of other lineages. NK cells are a discrete lymphocyte subset distinct from B and T cells based on both physical and phenotypic characteristics that can be exploited for their purification. Techniques based on differential cell buoyancy [centrifugation on discontinuous density gradients such as Percoll *(1)*] have been used to enrich NK cells from mixed lymphocyte populations but do not allow their purification to homogeneity. The mononuclear cell suspensions obtained, although enriched in NK cells, also contain variable proportions of other cell types (monocytes and/or activated T and B cells) *(2)*, and subsets of NK cells of higher density are lost in these preparations.

The most satisfactory purification techniques for NK cells, as well as for other leukocyte subsets, rely on their distinctive phenotype and make use of monoclonal antibodies (MAbs) directed to lineage-specific surface antigens (Ag). NK cell-specific surface markers have not been identified yet. Lack of surface expression of T-cell receptor/CD3 complex and surface immunoglobulins (Ig), concomitant with expression of CD16 (low-affinity receptor for the Fc portion of IgG, FcγRIIIA) *(3)*, CD56 (an N-CAM isoform) *(4)*, and CD161 (NKRP-1A) *(5)* (expressed, unlike CD56 and CD16, also on NK cells at a relatively developmentally immature functional stage) *(6,7)* identify NK cells within mononuclear cell populations. MAbs to these three differentiation Ag are available, and cells sensitized with them can be detected with a variety of secondary reagents to permit their identification and physical separation. Using

From: *Methods in Molecular Medicine, vol. 107: Human Cell Culture Protocols, Second Edition*
Edited by: J. Picot © Humana Press Inc., Totowa, NJ

MAbs, homogeneous NK cell preparations are isolated from mixed mononuclear cell populations following either of two schemes: (1) direct isolation using MAbs to surface Ag expressed on these cells (positive selection), or (2) depletion of all cells other than NK using a mixture of MAbs directed to Ag expressed on the former but absent from the latter population (negative selection). The advantage of positive selection is the rapid and specific isolation of NK cells. However, Ab binding to antigens capable of signal transduction, e.g., CD16, may lead to modulation of NK cells' biological functions *(8,9)*, making them unusable for selected applications. Negative selection techniques, instead, yield cells that are in their least altered state and are suitable for most functional studies. The choice between the two systems depends on the specific experimental requirements.

Here, we describe in detail the use of a dependable and relatively inexpensive method of NK cell isolation (indirect anti-Ig rosetting) that is suitable for both positive and negative selection, results in good yields, and allows easy and rapid manipulation of large numbers of cells. Indirect anti-Ig rosetting is based on the use of erythrocytes (E) coated with anti-mouse Ig Ab as a secondary reagent to detect cells that have bound at their surface murine MAbs recognizing lineage-specific Ag, leading to the formation of rosettes. The subsequent physical separation of Ag^+ (rosetted) and Ag^- (nonrosetted) cells is obtained following simple centrifugation on density gradients. Other reliable techniques exist that use secondary reagents coupled to different detection systems, but, unlike indirect anti-Ig rosetting, may not be practical for all investigators because of unavailability of specialized equipment, low yields, or prohibitive costs. For example, fluorochrome-labeled secondary reagents and fluorescence-activated cell sorting *(10)* is the most appropriate technique to purify NK cell populations with highest homogeneity. However, this requires availability of a flow cytometer with cell-sorting capabilities, is time-consuming and expensive, and has the disadvantage of allowing recovery of relatively low numbers of cells. Methods using magnetic beads *(11)*, although fast and efficient, are extremely expensive. Panning the Ab-sensitized cells on dishes coated with anti-mouse Ig *(10)* is efficient, fast, and economical, but may become impractical when large numbers of cells need to be processed. Complement (C)-dependent lysis *(10)*, practical and efficient, can be performed only with C-fixing MAb and may result in nonspecific toxicity and, consequently, the need for screening numerous batches of sera for optimal use. These methods (described accurately in the references provided) may, however, be used efficiently instead of indirect anti-Ig rosetting when specific needs make them appropriate. As an example, sorting $CD56^-CD3^-$ NK cells for cloning immature cells (*see* text) is safer to ensure depletion of contaminating immature T cells. The general approach to cell isolation discussed below can also be applied to a variety of additional needs, such as fractionation of NK cell subsets (e.g., CD8$^+$ and CD8$^-$ cells), substituting appropriate MAbs in the purification steps following the isolation of NK cells by negative selection.

Because the number of NK cells that can be obtained from peripheral blood is low and may not be sufficient for some studies, protocols are also provided to increase NK cell numbers in short-term cultures in vitro. Cells from these cultures can be used as a starting population to separate numbers of NK cells larger than those that would be obtained from equivalent volumes of fresh peripheral blood. NK cells prepared in this way, however, have some characteristics of activated NK cells *(11)* and it is advisable that results of studies using these cells are confirmed with primary resting NK cells. Additionally, immature CD3$^+$/CD56$^-$/CD161$^+$ NK cells are not present in appreciable percentages and numbers in these cultures. Thus, methods for generating large numbers of polyclonal and monoclonal populations of immature NK cells will also be discussed *(7,13)*.

2. Materials

1. Culture medium: RPMI-1640, supplemented with 10% heat-inactivated (45 min, 56°C) fetal bovine serum (FBS) (for the culture of the feeder B lymphoblastoid cell lines), or 5% human autologous plasma (for lymphocyte cultures), glutamine (2 mM) and, if desired, antibiotics (penicillin, 0.5 U/mL, streptomycin, 0.5 µg/mL) (complete medium). Autologous plasma is obtained, as a byproduct of the lymphocyte separation, at the top of the density gradient and the mononuclear cell interface (*see* **Subheading 3.1.1., step 3**).
2. Ficoll-Na Metrizoate density gradient (1.077 g/mL), such as Ficoll-Hypaque (F/H) (Pharmacia); it is stored at 4°C in the dark, and is commercially available.
3. Phosphate-buffered saline (PBS): 12 mM NaH$_2$PO$_4$, 12 mM Na$_2$HPO$_4$ buffer, pH 7.2, 0.15 M NaCl.
4. Saline: 0.15 M NaCl.
5. CrCl$_3$ solution: 0.1% CrCl$_3$•7H$_2$O in 0.15 M NaCl, pH 4.5 (stock solution). This must be prepared in advance and aged at least 1 mo before use. The stock solution must be stored at room temperature in a glass container protected from exposure to light; the shelf life for this solution is at least 1 yr. During the first week after preparing the solution, its pH needs to be checked every other day and adjusted to 4.5, if needed. Repeat the same once per week for the following 3 wk. This solution is used to couple anti-mouse antibodies to erythrocytes (E). CrCl$_3$ causes E agglutination by linking membrane proteins; thus, each new batch of CrCl$_3$ solution must be titrated to determine the optimal subagglutinating dilution to be used. For this, sheep E [25 µL of a 2% suspension in 0.15 M NaCl containing 0.1% bovine serum albumin (BSA)] are incubated (1:1, vol:vol) in round-bottom 96-well plates with serial 1:2 dilutions of the CrCl$_3$ stock solution in 0.15 M NaCl. The lowest dilution causing no E agglutination is determined after a 30-min incubation at room temperature.

6. Goat anti-mouse Ig (GaMIg) (1 mg/mL 0.15 *M* NaCl), adsorbed on human Ig and affinity-purified on mouse Ig. Affinity-purified GaMIg are commercially available but may contain human Ig-crossreactive Ab that, if present, may bind Ig-bearing cells (e.g., B cells or opsonized monocytes) and lead to contamination of the NK cell preparation with these cell types. Their depletion can easily be obtained adsorbing the GaMIg preparation over a human IgG-CNBr Sepharose 4B column. Phosphates need to be removed from the preparations by dialysis against 0.15 *M* NaCl (four changes are usually sufficient). After dialysis, the preparation is sterile-filtered (minimum concentration 1 mg/mL) and stored in 1–2 mL aliquots at 4°C for years without loss of titer. It is used to prepare the E detection system.

7. MAb reacting with leukocyte subsets: anti-T cells: CD3, CD4, CD5; anti-monocytes: CD14, CD32, CD64; anti-B cells: CD19; anti-NK cells: CD16, CD56, CD161. If needed (*see* **Notes 1** and **2**): anti-human E (anti-glycophorin A); anti-PMN (CD15). The murine B-cell hybrids producing these MAbs are all available from the American Type Culture Collection (ATCC); culture supernatants or ascites can be used. Alternatively, MAbs can be purchased from commercial sources.

8. Sheep erythrocytes (SE): these can be obtained from several commercial sources and are stored in Alsever's solution at 4°C for approx 1 mo. These cells are coated with the affinity-purified GaMIg using the procedure described in **Subheading 3.**

9. B-lymphoblastoid cell lines to be used as feeder cells: RPMI-8866, Daudi, or possibly other B-cell lines.

3. Methods

3.1. Preparation of the Starting Lymphocyte Populations

3.1.1. Peripheral Blood Lymphocytes (PBL)

1. Peripheral blood mononuclear cells (PBMC) are first prepared by density gradient centrifugation. The expected cell yield is approx 1×10^6 cells per mL of blood from healthy donors (range $0.5–2 \times 10^6$). Place 15 mL F/H in a 50-mL conical centrifuge tube and overlay 30 mL of blood (anticoagulated with heparin) slowly on top of this solution. For optimal recovery, the blood can be diluted (1:1, vol:vol, or more) with medium or PBS and care has to be taken not to disrupt the surface tension of the density gradient material.

2. Centrifuge at 800*g*, 20°C, for 35 min. Make sure that the centrifuge brake has been turned off to obtain a sharp PBMC band at the gradient's interface.

3. Collect the upper two-thirds of the plasma and centrifuge it (800*g*, 15 min) to spin down most platelets. Collect the supernatant (platelet-free autologous plasma) and store it at 4°C for use in the culture medium.

4. Carefully collect the mononuclear cell band at the interface using a 10-mL pipet and transfer it to a new 50-mL tube. Remove all cells in this band trying to take as little of the density gradient as possible. Mix the cell suspension with PBS (1:2, vol:vol) to dilute any carried over F/H.

5. Centrifuge the cells at 350*g*, 5 min, room temperature. Decant the cell-free supernatant (which is turbid due to the presence of platelets) and rap the tube against a solid surface to resuspend the pellet; if a large number of erythrocytes are present in the PBMC band, which depends on donor variability in erythrocyte density, it may be necessary to use vacuum aspiration to remove the cell-free supernatant and avoid cell loss (*see* **Notes 1–3**).

6. Resuspend the cells in PBS and centrifuge at 150*g*, 7 min. Repeat two additional washes (100*g* centrifugation), and, finally, resuspend the cells in complete medium for counting. Low-speed centrifugation is needed to reduce the platelet and erythrocyte contamination of the final cell suspension. The PBMC so obtained are used in the following steps and can also be used as a source of feeder cells in the cloning procedure described in **Subheading 3.3.**

7. The PBMC are depleted of most monocytes following an adherence step. For this, PBMC in complete medium are plated in tissue-culture-treated Petri dishes. The number of cells and volume of the cell suspension that will allow an even settling of cells depends on the size of the dish used. As an example, 50×10^6 cells in 5 mL medium form an evenly distributed monolayer when placed in 100-mm^2 dishes; proportionally lower numbers of cells (~approx 5×10^6 cells per 10 mm^2 surface) are placed in smaller dishes, but in this case a relatively larger volume of medium may be needed to cover the plate evenly. Be careful to avoid adding bubbles to the plates as they will prevent the cells from evenly contacting the bottom of the dishes.

8. Incubate the cells at 37°C for 30 min in a 5% CO_2 atmosphere.

9. Collect nonadherent cells without detaching the adherent monocytes: for this, add some PBS to each plate (~approx 6 mL/100 mm^2 dishes; not directly onto the cells but onto the wall of the Petri dish), and swirl/rock the solution back and forth to resuspend nonadherent cells. Transfer this supernatant to a tube.

10. Add PBS to one of the plates and repeat the washing step. Transfer the cell suspension to the next dish and continue until all plates have been washed. Repeat this step again, continuing until no significant number of nonadherent cells can be seen under an inverted microscope (usually four or five washes). All cells collected in the washes are pooled with those collected in **step 9.**

11. Centrifuge the cells, resuspend them in complete medium with 5% autologous plasma, and count. Viability is checked using erythrosin B or trypan blue to make the desired cell dilution for counting. Only live cells exclude the dye and appear translucent.

3.1.2. Short-Term PBL-B Lymphoblastoid Cell Lines Cocultures (12)

1. Grow the feeder B lymphoblastoid cell lines in culture medium containing 10% FBS, as needed; 2×10^5 feeder cells are needed for each 1×10^6 PBL that will be put into culture. The ability of B lymphoblastoid cell lines to act in vitro as feeders to sustain preferential proliferation of NK cells from PBL depends on the quality of the feeder cells before they are added to the cultures. Exponentially growing, viable cells are essential for successful cultures (*see* **Note 4**).

2. Irradiate these cells with 50 Gy. RPMI-8866 cells should be irradiated the day before they are needed and kept in a 37°C incubator (5×10^5 cells/mL) until use. Daudi cells can be irradiated and placed into culture on the same day.

3. Immediately before use, centrifuge the cells (150*g*, 5 min) and resuspend them in fresh complete medium (potentially inhibitory cytokines produced by these cells during the overnight incubation are removed in this way).

4. Mix the feeder cells with the PBL, prepared as in the previous subheading, at a 1:5 feeder cells:PBL ratio, and a final PBL concentration of 2.5×10^5/mL complete medium. This is the optimal ratio. However, it can be decreased to 1:10, in case the actual number of live irradiated feeder cells available is lower than expected.

5. Add 2 mL of the cell suspension to each well of a 24-well tissue-culture plate and place in an incubator (37°C, humidified 5% CO_2 atmosphere). Cultures can be set up in flasks (maximum size T25, 5 mL/flask, upright position). However, in this case, the yield of total cells, and especially NK cells, is lower.

6. On d 6 of culture, aspirate approx half of the medium from the wells and replace it with fresh medium (*see* **Note 5**).

7. On d 10, collect the cells from the cultures (*see* **Note 6**). The proportion of NK cells present can be determined by surface phenotyping (indirect immunofluorescence is the simplest method) the day before. On average, a fivefold increase in total cell number is achieved in the cultures at this time (e.g., approx 50×10^6 cells are recovered from cultures started from 10×10^6 PBL). Typically, NK (CD16$^+$/CD56$^+$) cells represent ~approx 70–80% and 50% of the cell population when RPMI-8866 and Daudi cells are used as feeders, respectively (this represents an approx 20-fold increase in the total number of NK cells compared to the starting PBL population) (*see* **Note 6**). The remainder cells are CD3$^+$ T cells. B cells and monocyte/macrophages are not detectable at the end of the culture. If active proliferation is observed before day 7, the cells can be collected earlier.

3.1.3. Immature NK Cell Cultures (7)

1. Eliminate the majority of mature NK and T cells by E-AET rosetting *(14)*, and the FcγR$^+$ cells from this population (residual mature NK cells, B cells, and monocytes) on EA monolayers *(14)*. This is important to ensure sufficient purity of the final preparation. From the remaining population, prepare a CD56$^-$ NK cell population as described in **Subheading 3.2.**, after sensitizing the PBL with MAbs to CD3, CD5, HLA-DR, CD56, and possibly other Ag expressed on mature NK cells, with the exception of CD161 MAbs.

2. Resuspend the CD56$^-$ cells (10^5 cells/mL) in complete medium with 5% autologous plasma, recombinant (r) human IL-2 (50 U/mL), neutralizing anti-IL-12 MAbs, at a concentration ensuring neutralization of 100 ng IL-12/mL. To obtain larger numbers after culture, feeder cells are added. These are constituted by a mixture of 50-Gy irradiated 5×10^5 PBMC (pool of three donors) and 10^5 B lymphoblastoid cells (Daudi preferred). Additionally, human rIL-4 (2 ng/mL) can be

added to cultures with or without feeder cells, in order to increase proliferation of the immature NK cells (*see* **Note 7**).

3. Place the cell suspension (1 mL per well in tissue-culture-treated 24-well plates, maximum 5×10^5 cells/well) and incubate in 8% CO_2, humidified atmosphere, 37°C.
4. Split as necessary, to maintain a viable cell culture. rIL-2, and fresh medium replaced 1–2 times/wk.

3.1.4. NK Cell Cloning (7)

NK cell clones can be generated from both total, mature CD161$^+$/CD56$^+$ and immature CD161$^+$/CD56$^-$ NK cells *(7)*. Their life-span, however, is inversely proportional to their developmental stage, with mature NK cell clones surviving shorter periods of time. Possibly, culture of immature NK cells, as described in the previous subheading, can be most efficiently achieved by combining several immature cell clones, produced as described here.

1. Prepare a total NK cell, or CD56$^-$ NK cell population (negative selection) as indicated in **Subheading 3.2.** The MAbs to be used are described in **Subheading 3.1.3., step 1**.
2. Resuspend the cells (20 cells/mL) in culture medium with 5% autologous plasma. For accuracy, prepare the final suspension through sequential 1:10 dilutions of a stock cell suspension (*see* **Note 7**).
3. In a separate tube, prepare an suspension of feeder cells in volume identical to that of the NK cell suspension. Feeder cells: mixture of PBMC ($1–2 \times 10^6$/mL, preferably a pool from three different individuals) and B lymphoblastoid cell line ($2–4 \times 10^5$/mL, Daudi preferred) both irradiated (50 Gy) on the same day and resuspended in the same culture medium.
4. Add to the suspension in **step 3**: PHA-L (1 µg/mL), human rIL-2 (100 U/mL), anti-IL-12 neutralizing MAb (2X concentrated), and 5% autologous plasma. When cloning from total NK cells, also add anti-TNF-α neutralizing MAb (*see* **Note 8**).
5. Mix equal volumes of the NK cell suspension and the feeder cell mixture. Plate 100 µL of the cell suspension in each well of a round-bottom 96-well microtiter plate (tissue-culture treated). Prepare at least five plates.
6. Incubate at 37°C, in a humidified 8% CO_2 atmosphere.
7. During the first 2 wk, add 50 µL complete medium, containing IL-2 on d 5 and every third day thereafter.
8. After about 2 wk, start checking the plates for clonal growth (there is no need to check at the inverted microscope until cell growth can be seen by eye).
9. Split cells from wells in which cell growth can be seen, as needed, but at the most 1:2 (vol:vol) (first in two wells of the 96-well plate, then in 24-well plates) (*see* **Note 9**).

3.2. Indirect Anti-Ig Rosetting for Positive and Negative Selection of NK Cells

3.2.1. Preparation of CrCl₃ SE Coated with GaMIg (15)

1. Wash SE three times with 0.15 *M* NaCl (never use PBS: phosphates inhibit the CrCl$_3$-dependent coupling of proteins to cell membranes). Spin the cells (800*g*, 10 min) and aspirate the cell-free supernatant (decanting the supernatant may result in loss of erythrocytes, if the cells were in a loose pellet).
2. Dilute the CrCl$_3$ stock solution in 0.15 *M* NaCl to the appropriate subagglutinating concentration previously determined for that batch (final optimal concentration is usually 0.01–0.005%). Filter-sterilize the solution using a 0.45-µm filter.
3. In a 50-mL conical centrifuge tube mix the following in the given order to prepare 50 mL of a 4% suspension of E-CrCl$_3$-GaMIg: 28 mL 0.15 *M* NaCl, 2 mL packed SE, 2 mL GaMIg (the Ig are dissolved in 0.15 *M* NaCl, usually at 1 mg/mL; depending on the batch, lower concentrations can be used), 8 mL CrCl$_3$ solution. Smaller volumes can be prepared, depending on the need, modifying the volumes of the different reagents but maintaining their relative proportions (*see* **Note 10**).
4. Incubate the suspension for 15 min at room temperature with occasional mixing/swirling.
5. Add PBS to stop the reaction and centrifuge (800*g*, 10 min).
6. Wash the E-CrCl$_3$-GaMIg twice with PBS as in **step 5**, and resuspend in 50 mL complete medium. The E suspension is stored at 4°C and can be used up to 1 mo. Each time before use, wash the cell suspension once with PBS to remove membranes of lysed E or free Ig which may have come off the cells and, if present, can compete with the intact SE for binding to the MAb-sensitized cells. Resuspend E at 4% in fresh complete medium.

3.2.2. Lymphocyte Sensitization with MAbs

1. Resuspend the PBL preparations from which NK cells are to be purified (20×10^6/mL complete medium) in an appropriately sized centrifuge tube.
2. Dilute the desired MAbs in PBS to the concentration previously determined optimal for rosette formation with cells known to express the Ag of interest, and mix (1:1, vol:vol) with the PBL. Cell-free culture supernatants, ascites, or purified Ig [or their F(ab′)$_2$ fragments] are appropriate for use. In general, culture supernatants work best at a 1:2–1:4 dilution, ascites at a 10^4–10^3 dilution, and purified Ig at 0.5–1 µg/mL. The optimal concentration to be used, however, has to be determined experimentally for each batch of Ab preparation (*see* **Note 11**). For negative selection of NK cells from PBL, use a mixture of anti-T (CD3, CD5), anti-B (anti-HLA-DR, CD19), and anti-monocyte (CD14) MAbs. If NK cells are to be purified from the PBL-B lymphoblastoid cells cocultures, avoid anti-HLA-DR; a mixture of anti-T and anti-monocyte MAbs. This is because the NK cells derived from these cultures are HLA-DR$^+$, and monocytes and B cells are not detectable. For positive selection, use in all cases a mixture of CD16 and CD56 anti-NK cell MAbs (add also CD161, unless depletion of immature NK cells is desired).

3. Incubate the cell suspension on ice for 30 min. Pre-chill the centrifuge and the tube carriers at this time (5–10°C).
4. Wash the excess unbound Ab with ice-cold PBS (5 min centrifugation, 200g, in the cold).
5. Decant the supernatant and wash twice more as in **step 4**.
6. Resuspend the cells in 10 mL ice-cold complete medium and place on ice. For optimal rosette formation, a maximum 200 × 10^6 cells can be placed in a 50-mL tube; up to 50 × 10^6 cells are instead placed in a 15-mL round-bottom culture tube. Never use conical tubes: the geometry of these tubes does not allow optimal rosette formation.

3.2.3. Rosette Formation with E-CrCl₃-GaMIg

1. Mix 2.5 mL of the 4% suspension of E-CrCl3-GaMIg, in ice-cold fresh complete medium, with 200 × 10^6 PBL presensitized, as above, with the desired combination of murine MAb. Volumes of E suspension are proportionally modified when using different cell numbers.
2. Centrifuge in the chilled carriers/centrifuge (200g, 7 min).
3. Incubate the pelleted cells on ice for 30 min.
4. Resuspend the cells with a Pasteur pipet until all clumps have been disaggregated (rosettes do not break apart). Place a drop of the cell suspension on a slide with cover slip to check for percent rosettes (optical microscopy, count at least 200 cells) (*see* **Note 12**).

3.2.4. Enrichment/Purification of NK Cells

1. After resuspending the pellet (*see* **Subheading 3.2.3., step 4**), underlay F/H carefully displacing the lymphocyte-SE mixture upward (13 or 5 mL F/H solution are underlaid in 50- and 15-mL tubes, respectively).
2. Being careful not to jar the tubes, centrifuge them at 800g for 15 min.
3. After centrifugation, the cells expressing the antigens recognized by the MAbs used are in the pellet (rosetted), and those not expressing them are at the interface of the gradient. If MAb-negative cells are to be obtained (negative selection), carefully transfer the nonrosetted cells from the interface of the gradient to a new tube and add 50% by volume sterile PBS (*see* **Notes 13–15**).
4. Wash the cells twice with PBS and resuspend them in complete culture medium.
5. To recover the Ab-positive cells (positive selection), aspirate the F/H, resuspend the pellet in a small volume of PBS, transfer the cells to a new tube (in order to avoid contaminating these cells with Ab-negative cells which may have adhered to the wall of the tube). Fill the tube with PBS and centrifuge. After decanting the PBS, resuspend the pellet loosely by rapping the tube against a solid surface and wash twice more.
6. In order to achieve the highest degree of purification (>98%), it is necessary to repeat the rosetting step, without adding new MAbs, on the cells obtained in **step 4** or **5** (be sure to keep the cells at 4°C). For this, the cells collected at the interface (negative selection, **step 4**) are pelleted with additional E-CrCl₃-GaMIg

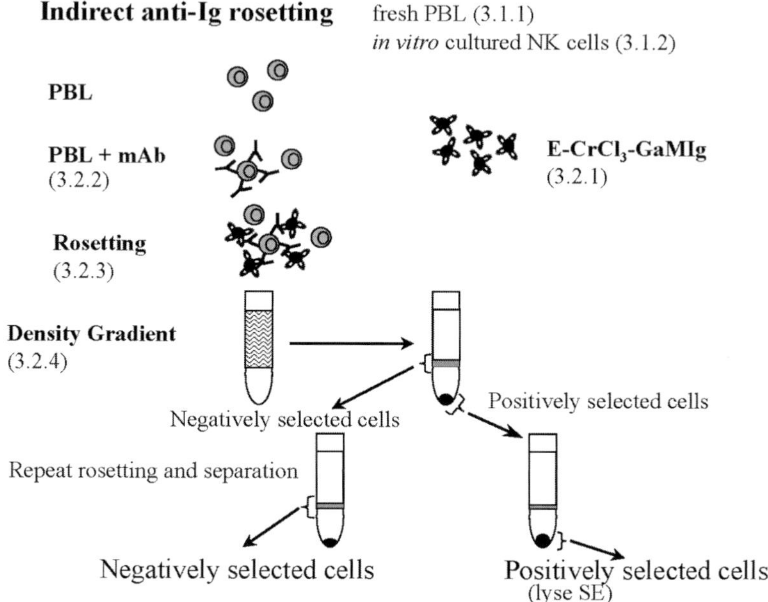

Fig. 1. Schematic outline of the indirect anti-Ig rosetting method of NK cell purification.

as before (**steps 1–3** in **Subheading 3.2.**); those collected from the pellet (positive selection, **step 5**) are resuspended in medium and pelleted again without adding more E. After a 30-min incubation on ice, the pellets are resuspended, F/H is underlayed, and the tubes are centrifuged as before (**steps 1** and **2**). NK cells are collected from the interface of the second F/H gradient performed with cells negatively selected (the pellet is discarded) or from the pellet of the second F/H gradient performed with positively selected cells (cells at the interface are discarded). In the latter case, SE can be lysed adding 0.5 mL H_2O and pipetting carefully for 1 min, after which the tube is quickly filled up with PBS and the cells are washed twice (*see* **Note 1**).

Following this protocol, homogeneous NK cell preparations are reproducibly obtained [>98% $CD16^+/CD56^+/CD3^-$ cells, as determined in indirect or direct immunofluorescence *(9)* using a panel of MAbs]. The actual yield is, in average, 60% (range 50–75%) of the theoretical one expected on the basis of the proportion of NK cells in the starting population: approx 8×10^6 and 30×10^6 NK cells can thus be obtained from 100×10^6 PBL or cultured lymphocytes, respectively, where NK cells are approx 15 and 80%. A scheme of the steps involved in the separation procedure is presented in **Fig. 1**.

4. Notes

4.1. Possible Problems in Preparing the Starting Lymphocyte Populations from Peripheral Blood

1. Occasionally, significant numbers of erythrocytes (E) may contaminate the PBL preparations. These cells have an unusual buoyancy, will be carried over in each step, and will actually be enriched during the purification procedure. Depending on the application, it may be required that the E be lysed. For this, use H_2O (*see* **Subheading 3.2.4., step 6**), or 0.15 *M* NH_4Cl, 0.01 *M* $KHCO_3$, 0.1 *M* EDTA, pH 7.2, buffer. To lyse E with this lysis buffer, the PBL suspension is incubated for 10 min on ice with the buffer (3 mL buffer each 100×10^6 cells), and washed once with PBS. Alternatively, antiglycophorin A MAbs can be added to the mixture of MAbs used for negative selection.

2. PMN usually are not found at the interface of the F/H when using peripheral blood from healthy donors. However, in rare donors and some patients, PMN with altered density are present that may not pellet through the density gradient, and variable proportions of them may contaminate the PBL preparation. They can be eliminated adding an anti-CD15 MAbs to the mixture used for negative selection.

3. Low PBL yield: This usually depends on cell loss following clumping during the isolation procedure. There are three primary causes for this:
   a. Insufficient or too-high speed washes of the PBMC: in this case a large a number of platelets may be carried over. These are subsequently activated (especially in the monocyte adherence step) and aggregate with other cell types.
   b. Cell clumps are easily formed when the cells are either spun at too-high speed, or resuspended inaccurately after washing.
   c. Release of DNA from dying cells which acts as a tenacious adhesive in vitro: addition of DNase I, 50 µg/mL, to these cells and incubation at 37°C for 2–3 min will clear the clumps (store aliquots of sterile DNase I, 5 µg/mL medium without FBS, for easy use). Continuing the purification without eliminating these clumps results in major cell loss.

4.2. Limited NK Cell Proliferation in the In Vitro PBL-B Lymphoblastoid Cell Cocultures

4. Individual laboratory clones of the RPMI-8866 and Daudi cell lines may vary in their ability to support NK cell proliferation from PBL; moreover, this property decreases after they are kept in culture for long periods of time (4–6 mo). Defrost a new batch of low-passage cells when the NK cell yields start declining (usually this happens gradually).

5. The precise feeding and harvesting schedule of cell cultures performed with for different feeder cell lines or clones of them may need to be optimized following the kinetics of NK cell proliferation. Maximum NK cell proliferation may actually occur one or more days earlier or later than that described above for the general case.

6. Although proliferation of NK cells in the cocultures is observed with PBL from most healthy donors, donor variability may account for occasional unsuccessful cultures.

4.3. Suboptimal Clonal Growth or Clonal Efficiency

This primarily depends on suboptimal conditions for cloning and maintaining cell survival. Although at times, a specific reason may not be easy to be found, several considerations allow increasing clone survival growth (and likely efficiency of the technique).

7. To ensure obtaining a reasonable numbers of clones, also set up plates containing a number of NK cells greater than 1/well (e.g., 10–20 cells/well). In case clonal growth is not observed in the more diluted plates, the clones obtained in the plates with higher cell concentration may be used, after determining their clonal origin *(16)*.
8. IL-12 is the primary monokine inducing NK cell development to their terminally differentiated stage *(7)*. It is produced endogenously in the cultures, by the accessory feeder cells (including some of the B-cell lines, e.g., the RPMI-8866). Adding a neutralizing anti-IL-12 MAb, especially if IL-4 is not added to the cultures, will help preventing clone differentiation. Addition of the neutralizing anti-TNF-α MAb will help prevent/retard death of the mature NK cell clones, although their life-span is in any case limited.
9. A too-low cell concentration may prevent NK cell clone growth. In general, a 1:2 split when the cells are packed but nice is appropriate. This may have to be repeated every day in the case of the immature NK cell clones. Usually, cells from four well-growing wells in 96-well plates can be combined in one well of a 24-well plate, and this further split. Splitting and transferring into flasks (maximum flask size T25, 5 mL/flask, upright position) can be done. However, the speed of growth and cell yield may decrease under these conditions. A maximum 1:2 split is recommended for the mature NK cell clones. This likely will need to be less frequent than that for the immature NK cell clones. When the clones are well growing, only IL-2 needs to be constantly provided to the cultures. Constant addition of TNF-α-neutralizing MAbs, and possibly MAb neutralizing Fas (commercially available) will help prevent cell death. The distinction between immature and mature NK cell clones is solely based on their phenotype *(7,13)*. For simplicity, this can be defined by immunofluorescence surface phenotyping only, using CD56 and CD161 MAbs. Immature NK cells are CD3$^-$/CD56$^{-/lo}$/CD161$^+$. Mature NK cells are CD3$^-$/CD161^{lo}/CD56$^+$. Immunofluorescence can be performed on cells from one of two wells of well-growing clones. The sooner it is performed, the better.

4.4. Lack of Rosette Formation

10. Insufficient amount of GaMIg coating the SE.
    a. Each new batch of GaMIg has to be titrated to determine the optimal concentration of Ig to be coupled to E. This is achieved testing for rosette formation

with SE suspensions prepared with serial dilutions of the GaMIg (from 2 to 0.25 mg/mL).

    b. Phosphates are present in one of the reagents used for $CrCl_3$ coupling. Because even trace amounts of phosphate inhibit the $CrCl_3$-dependent protein coupling to cell membranes, it is essential that E are washed with saline and that all reagents used are prepared and diluted in the same solution (avoid PBS and medium at any time).

    c. On rare occasion, coupling of GaMIg to E is unsuccessful for no apparent reason. Just start again. Each new batch of E-$CrCl_3$-GaMIg has to be prepared and tested in advance.

11. Incorrect dilutions of the MAbs used for PBL sensitization: To solve the problem, the working dilution for each MAb preparation used has to be determined, also for different batches of the same Ab preparation. For this, serial dilutions of the mAb are incubated with lymphocytes in round-bottom 96-well plates under the conditions described in **Subheadings 3.2.2.** and **3.2.3.** Rosette formation is assessed microscopically on an aliquot of the cells. Choose the dilution resulting in the maximum percentage of rosettes, corresponding to that of cells detected with the mAb by indirect immunofluorescence. Always use a positive control (e.g., anti-HLA-Class I mAb) for reference: this should give 100% rosettes. A negative control needs to be performed using an isotype-matched irrelevant MAb. This will exclude binding of NK cells to E independently from the sensitizing Ab.

12. Low numbers of rosettes compared to the proportion of Ab[+]-cells. If the reasons in **Notes 1** and **2** can be excluded, the most likely explanation for this is that the temperature was not kept sufficiently low during all steps. This may result in capping and down modulation of the antigen from the cell surface, thus preventing efficient rosette formation.

4.5. Contamination of the Final NK Cell Preparation with Other PBL Subsets

13. Perform all procedures in the cold and rosette twice.
14. In the case of negative selection, the most likely explanation is a short time of centrifugation of the F/H gradient: increase the time to 20 min and/or modify the centrifugation speed.
15. In the case of positive selection, inaccurate resuspension of the cell pellet after rosette formation will result in trapping rosette (Ab)-negative PBL in the clumps. These will sediment in the pellet of the F/H gradient, thus contaminating the rosette (Ab)-positive cells.

4.6. Low NK Cell Recovery

16. In the case of negative selection, this problem depends, for most part, on inaccurate resuspension of the cell pellets after rosette formation, before the separation on F/H. Rosette-negative PBL trapped in the clumps reach the pellet, resulting in loss of cells.

17. When the recovery of positively selected cells is low, loss of cells likely occurred during the E lysis step. Careful resuspension of the cells during the lysis and use of the NH$_4$Cl buffer (*see* **Note 1**) instead of H$_2$O should solve the problem.

References

1. Timonen, T., Ortaldo, J. R., and Herberman, R. B. (1981) Characteristics of human large granular lymphocytes and relationships to Natural Killer cells. *J. Exp. Med.* **153,** 569–582.
2. Perussia, B., Fanning, V., and Trinchieri, G. (1985) A leukocyte subset bearing HLA-DR antigens is responsible for in vitro alpha interferon production in response to viruses. *Nat. Immun. Cell. Growth Regul.* **4,** 120–137.
3. Perussia, B., Starr, S., Abraham, S., Fanning, V., and Trinchieri, G. (1983) Human natural killer cells analyzed by B73.1, a monoclonal antibody blocking Fc receptor functions. I. Characterization of the lymphocyte subset reactive with B73.1. *J. Immunol.* **130,** 2133–2141.
4. Lanier, L. L., Chang, C., Azuma, M., Ruitenberg, J. J., Hemperly, J. J., and Phillips, J. H. (1991) Molecular and functional analysis of human natural killer cell-associated neural cell adhesion molecule (N-CAM/CD56). *J. Immunol.* **91,** 4421–4426.
5. Lanier, L. L., Chang, C., and Phillips, J. H. (1994) Human NKR-P1A. A disulfide-linked homodimer of the C-type lectin superfamily expressed by a subset of NK and T lymphocytes. *J. Immunol.* **153,** 2417–2428.
6. Bennett, I. M., Zatsepina, O., Zamai, L., Azzoni, L., Mikheeva, T., and Perussia, B. (1996) Definition of a natural killer NKR-P1A+/CD56–/CD16– functionally immature human NK cell subset that differentiates in vitro in the presence of interleukin 12. *J. Exp. Med.* **184,** 1845–1856.
7. Loza, M. J. and Perussia, B. (2001) Final steps of natural killer cells maturation: a general model for type 1-type 2 differentiation? *Nature Immunol.* **2,** 917–924.
8. Perussia, B., Acuto, O., Terhorst, C., et al. (1983) Human natural killer cell analyzed by B73.1, a monoclonal antibody blocking FcR functions. II. Studies of B73.1 antibody-antigen interaction on the lymphocyte membrane. *J. Immunol.* **130,** 2142–2148.
9. Anegon, I., Cuturi, M. C., Trinchieri, G., and Perussia, B. (1988) Interaction of Fc receptor (CD16) with ligands induces transcription of interleukin 2 receptor (CD25) and lymphokine genes and expression of their products in human natural killer cells. *J. Exp. Med.* **167,** 452–472.
10. Chapter 3 In vitro assays for lymphocyte function and chapter 5 flow cytometry in *Current Protocols in Immunology* Vol. I (1991) (Coligan, J, Kruisbeek, A., Marguilees, D., Shevach, E., and Strober, W., eds.), Wiley, New York.
11. Naume, B., Nonstad, U., Stengker, B., Funderud, S., Smeland, E., and Espevic, E. (1991) Immunomagnetic isolation of NK and LAK cells. *J. Immunol.* **148,** 2429–2436.
12. Perussia, B., Ramoni, C., Anegon, I., Cuturi, M. C., Faust, J., and Trinchieri, G. (1987) Preferential proliferation of natural killer cells among peripheral blood

mononuclear cells co-cultured with B lymphoblastoid cell lines. *Nat. Immun. Cell Growth Regul.* **6,** 171–188.
13. Loza, M. J., Peters, S. P., Zangrilli, J. G., and Perussia, B. (2002) Distinction between IL-13$^+$ and IFN-γ^+ NK cells and regulation of their pool size by IL-4. *Eur. J. Immunol.* **32,** 413–423.
14. Perussia, B., Trinchieri, G., and Cerrottini, J. C. (1979) Functional studies of Fc receptor-bearing human lymphocytes: effect of treatment with proteolytic enzymes. *J. Immunol.* **123,** 681–687.
15. Goding, J. W. (1976) The chromium chloride method of coupling antigens to erythrocytes: definition of some important parameters. *J. Immunol. Methods* **10,** 61–66.
16. Taswell, C. (1981) Limiting dilution assays for the determination of immuno-competent cell frequencies. I. Data analysis. *J. Immunol.* **126,** 1614–1619.

11

Human Fetal Brain Cell Culture

Mark P. Mattson

1. Introduction

The human brain is a highly evolved and complex organ system, consisting of more than 10 billion nerve cells and at least three times as many glial cells. Because of its cellular complexity, it is important to develop technical approaches that allow isolation and study of nerve and glial cells under conditions in which their environment can be precisely manipulated. Such methods have proven valuable in discovering the cellular and molecular mechanisms of brain development, function, and disease in invertebrates and rodents *(1,2)*. However, several factors have contributed to the relative dearth of information concerning the mechanisms responsible for the development and proper function of the human brain compared to the knowledge base in lower species. A major impediment has been the ethical considerations surrounding the use of fetal human tissue obtained from elective abortions and the lack of an alternative source of viable normal human brain cells. Although research that employs human fetal tissue has been limited, what has been done has made a major impact in the development of prophylactics and therapeutics for several important human diseases. For example, the use of fetal tissue from elective abortions was key to development of the polio vaccine *(3)*.

In addition to ethical considerations, technical hurdles related to the ability to maintain long-term cultures and store cells in a cryopreserved state have greatly limited the type and number of experiments that can be performed. This chapter describes methods for procurement of human fetal brain tissue, and the preparation, maintenance, and long-term culture of neurons and glia from different brain regions. In addition, a protocol for cryopreserving fetal human brain cells is included, which provides a means to establish cell stocks that

From: *Methods in Molecular Medicine, vol. 107: Human Cell Culture Protocols, Second Edition*
Edited by: J. Picot © Humana Press Inc., Totowa, NJ

allow performance of experiments over an extended time period. Finally, examples of applications of such cultured cells to studies of human brain development and disease are briefly discussed.

2. Materials

1. Mg^{2+}- and Ca^{2+}-free Hank's balanced salt solution (HBSS) buffered with 10 mM HEPES (pH 7.2).
2. Serum-containing maintenance medium (SCMM): Eagle's minimum essential medium (EMEM) with Earle's salts (Gibco) supplemented with 20 mM KCl, 1 mM sodium pyruvate, 20 mM glucose, 1 mM L-glutamine, 2 mg/mL gentamicin sulfate, 10% fetal bovine serum (FBS) (v/v), and 27 mM sodium bicarbonate (pH 7.2) (all supplements from Sigma, St. Louis, MO).
3. Defined maintenance medium (DMM): Neurobasal medium with B27 supplements (Gibco, Grand Island, NY; *see* **Note 4**).
4. Cryopreservation medium: SCMM to which dimethyl sulfoxide is added to a final concentration of 8% (v/v).
5. Culture dishes: 35 mm (Coming, Oneonta, NY; cat. no. 25000) and 60 mm (Costar cat. no. 3060) dishes, glass-bottom 35-mm dishes from Mat-Tek Inc. (Ashland, MA, cat. no. P35G-14-0-C-gm), and 96-well plates (Coming cat. no. 25860).
6. Polyethyleneimine (50% solution; Sigma) is diluted 1:1000 in borate buffer, which consists of 3.1 g/L boric acid and 4.75 g/L borax in glass-distilled water, pH 8.4.

The nature and sources of other materials discussed can be found in the cited references.

3. Methods
3.1. Tissue Acquisition

Establishing a source of tissue requires correspondence with the directors of local women's surgical clinics, who are generally receptive to contributing to the scientific advancements that result from studies of human fetal tissue. Although surgery is most commonly performed during the first trimester of pregnancy, tissue from 12–15 wk fetuses has been utilized for several reasons, including that brain regions of interest can be identified more readily than at earlier gestational ages, and neuronal viability in cultures established from these gestational windows is much greater than in cultures established from older fetuses. Estimates of quantities of tissue required for particular experiments can be made based on numbers of viable cells obtained/fetus (*see* **Subheading 3.2.**).

1. Ethical considerations. Because of the delicate nature of the issue of use of human fetal tissue for biomedical research, guidelines have been established by various

governmental and institutional organizations in order to ensure that proper ethical principles are practiced *(4)*. The major source of human fetal brain tissue is elective abortions. The guidelines that were followed were developed in conjunction with the University of Kentucky Internal Review Board and include the following key points:

    a. Informed consent for tissue donation is obtained after the patient has consented, in writing, to the abortion.

    b. Individuals involved in the research are not involved in the abortion decision, and have no role in the timing of the abortion or the surgical procedures.

    c. Tissue is donated without any form of compensation to the patient.

    d. Confidentiality of the patient's identity and of the specific use of the tissue is maintained.

2. Handling and storage of brain tissue. Communication between the clinic staff and the investigator allows procurement of the tissue as soon as possible following the surgery. Stocks of sterile refrigerated HBSS are kept on hand at the clinic. A member of the clinic staff obtains the brain tissue and places it in the HBSS. From this point on, handling of tissue and cells is performed using sterile conditions under a laminar flow hood. Brain tissue can be held in cold HBSS for at least 6 h without reducing cell viability. The human tissue should be handled as biohazardous material, because it may harbor infectious agents, notably HIV and hepatitis. The investigator may choose to have a sample of tissue screened for such infectious agents, which would allow the opportunity to dispose of infected tissue. Alternatively, a sample of the patient's blood may be procured for HIV and hepatitis testing (with appropriate informed consent).

3.2. Cell Dissociation, Cryopreservation, and Culture

1. The most common method of elective abortion involves aspiration of the fetus. This method usually results in compression and fragmentation of the brain tissue, and so hinders identification of specific brain structures. The cerebral hemispheres remain relatively intact and can be recognized by their characteristic gyri. In many cases, other brain regions can also be identified, including the hippocampus and mesencephalon *(5,6)*. Once the brain region of interest is identified and isolated, the tissue is placed in cold HBSS and minced into pieces of approx 1 mm^3 using a scalpel (no. 10 blade).

2. Tissue pieces are transferred to a 15-mL conical-bottom culture tube and allowed to settle to the bottom. Typically, the amount of tissue per tube will be equivalent to approx one-half of a cerebral hemisphere or 2 hippocampi/tube. HBSS is removed, replaced with 4 mL of HBSS containing 2 mg/mL trypsin, and incubated for 20–30 min at room temperature. Tissue is rinsed twice with 5 mL HBSS, incubated for 5 min in HBSS containing 1 mg/mL soybean trypsin inhibitor, and rinsed twice with 5 mL HBSS (*see* **Note 1**). Ten milliliters of either SCMM or cryopreservation medium are then added to the tube, and cells are dissociated by trituration using a fire-polished Pasteur pipet. (The pipet is

fire-polished by holding vertically, tip down, in a Bunsen burner flame until the diameter of the tip is reduced by approx 50%). The resulting cell suspension typically contains 1–5 million viable cells/mL (*see* **Note 2**). Cells are then seeded into culture dishes in volumes calculated to achieve the desired plating cell density.

3. Cells dissociated in cryopreservation medium are aliquoted (0.5 mL) into 1-mL cryovials (with screwtops). Cells should be frozen relatively slowly (approx –1°C/min) to –80°C, which can be accomplished using a controlled-cooling freezer. Alternatively, cryovials can be placed in freezing containers (simply sandwich the vials between two, 3–4 cm thick, pieces of styrofoam) and are then placed in a –80°C freezer for 12–24 h. (It is important that the vials be completely insulated from the freezer air, so that they do not cool too rapidly.) Frozen vials are then transferred to liquid nitrogen for long-term storage. In order to establish cultures from the cryopreserved cell stocks, the vials are rapidly thawed by immersion in a water bath at 37°C. Cells have been stored for 3 yr using these methods with cell losses of only 20–40%.

4. We routinely use polyethyleneimine as a growth substrate, and apply it to the surface of the culture dish as follows: Polyethyleneimine is dissolved in borate buffer (1 : 1000 dilution [v/v]). The surface of the culture dish is covered with this solution and allowed to incubate at room temperature for 3–24 h.

5. The dishes are then washed in sterile deionized distilled water (5 × 3 mL), allowed to dry under a laminar flow hood, and then exposed for 10 min to UV light to ensure sterility. SCMM is then added to the dishes (1.5 mL/35-mm dish; 2.5 mL/60-mm dish; and 100 µL/well in 96-well plates), and they are placed in a humidified CO_2 incubator (6% CO_2/94% room air) at 37°C until the time of cell plating (typically 1–4 d following preparation of dishes).

6. Following cell plating, cultures are left in the incubator for 16–24 h to allow cells to attach to the growth substrate (*see* **Note 3**). The medium is then removed and replaced with a smaller volume (0.8 mL/35-mm dish; 1.5 mL/60-mm dish; 100 µL/well in 96-well plates), which improves long-term neuronal survival. Cultures can be maintained without medium change for up to 3 wk. To maintain cultures for longer time periods, 50% of the medium can be replaced on a weekly basis.

7. Cultures maintained as just described will contain both neurons and astrocytes (*see* **Fig. 1**); the astrocytes will progressively proliferate until they form a monolayer on the culture surface. Such cultures are valuable for a variety of studies. However, for many applications, it is desirable to utilize cultures that are essentially pure populations of either neurons or astrocytes. For example, in order to establish whether neurons are directly responsive to a neurotrophic factor, one must rule out the possibility of an indirect action mediated by glial cells (*7*). Nearly pure neuronal cultures can be obtained by maintaining the cells in DMM (serum-free medium) (*see* **Note 4**). Alternatively, cells can be maintained in SCMM to which the mitotic inhibitor cytosine arabinoside (10 ~µ*M*) is added on culture d 2. Essentially pure astrocyte cultures can be obtained by plating the cells on uncoated plastic dishes to which very few neurons will adhere. The death of

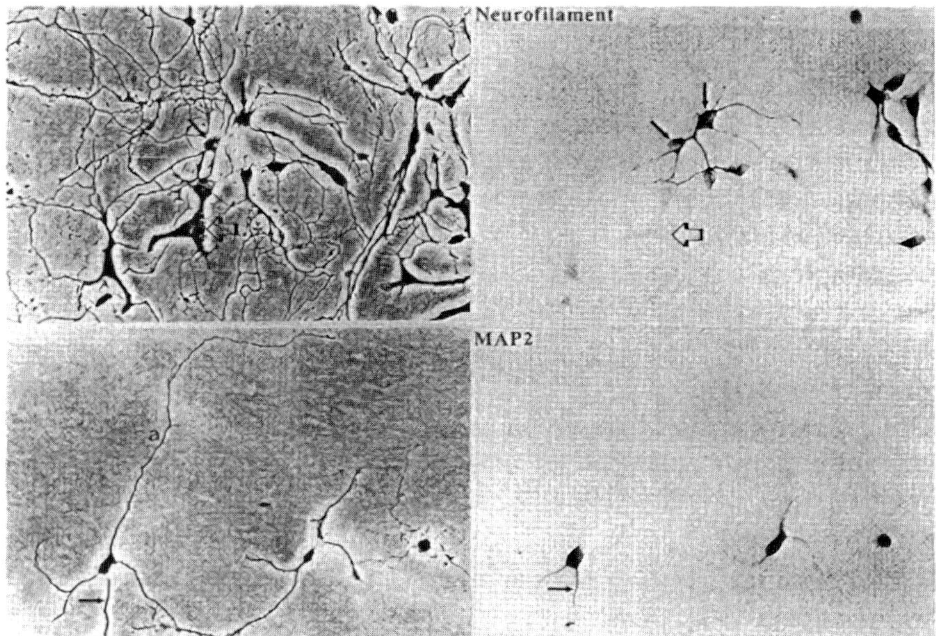

Fig. 1. Cell types present in human fetal brain cell cultures. (Upper panels) Phase-contrast (left) and bright-field (right) micrographs of neurons and (e.g., solid arrows) and astrocytes (e.g., open arrow) in a neocortical culture irnmunostained with antibody to the medium-sized (160-kDa) neurofilament protein. (Lower panels) Phase-contrast (left) and bright-field (right) micrographs of neurons tn a hippocampal cell culture irnmunostained with antibody to microtubule-associated protein-2 (MAP2). Note immunoreactivity of cell bodies and dendrites (e.g., arrow), but not axons (a).

any remaining neurons can be induced by doubling the medium volume and changing the medium twice per week.

3.3. Applications

1. Studying development: Fetal human brain cell cultures contain cells in various stages of development. Neurons in such cultures will readily elaborate neurites and form synaptic connections. Although relatively little work has been done to characterize the development of human fetal brain neurons in culture, the information that has been obtained has provided interesting insight into the similarities and differences between development of human neurons and their rodent counterparts. Human fetal cortical neurons grow more slowly and exhibit a longer "life-span" in culture than do fetal rat cortical neurons *(8)*. Rates of axon elongation were up to 10 times slower in cultured cortical neurons from fetal human brain compared to rat brain. The human fetal neurons also exhibited a protracted development of sensitivity to the excitatory neurotransmitter glutamate *(9)*. Stud-

ies of rat brain development had shown that expression of non-NMDA receptors (AMPA/kainate receptors) precedes the appearance of NMDA receptors; the expression of these receptors occurred over a period of approx 4–10 d in cell culture *(10)*. In addition, rat neocortical and hippocampal neurons became vulnerable to excitotoxicity within 1–2 d of the appearance of glutamate receptors. In contrast, human fetal neocortical neurons expressed non-NMDA and NMDA glutamate receptors over an extended 2–4 wk culture period. Moreover, the human neurons exhibited calcium responses to glutamate for several weeks prior to their being vulnerable to excitotoxicity *(9)*.

An important rationale for studying human fetal brain cells is that very little information is available concerning cellular and molecular mechanisms operative in human brain development compared to commonly studied laboratory animals. For example, a rapidly growing body of literature is available concerning the roles of neurotrophic factors in brain development in rodents. In contrast, parallel information in the human brain is lacking. Human fetal brain cell culture provides an opportunity to establish the similarities and differences in neurotrophic factor signaling in developing human and rodent brains. Basic fibroblast growth factor (bFGF) promotes neuronal survival in human neocortical cell cultures *(5)*. Immunostaining studies indicated that both neurons and astrocytes possess bFGF, which may be an endogenous source of trophic support. Regulation of neuronal growth cones, the motile distal tips of growing axons and dendrites, can be reliably studied in human fetal brain cell cultures using approaches developed in non-human systems *(11,12)*.

2. Researching neurodegenerative disorders: Both acute (e.g., stroke and traumatic brain injury) and chronic (e.g., Alzheimer's, Parkinson's, and Huntington's diseases) neurodegenerative disorders are major concerns in our society; taken together, they afflict tens of millions of Americans. An important impediment to conquering these disorders is that, in most cases, they can only be studied at the cellular and molecular levels postmortem. The ability to study living human brain cells under controlled conditions allows one to test hypotheses and establish cause–effect-type relationships. For example, in Alzheimer's disease, a 40–42-kDa peptide called amyloid beta-peptide (Abeta) forms insoluble deposits (plaques) in the brain (*see* **ref. *13*** for review). The Abeta arises from a much larger membrane-associated beta-amyloid precursor protein (betaAPP); mutations of betaAPP have causally been linked to a small percentage of cases of Alzheimer's disease. However, the roles of the betaAPP mutations and Abeta deposition in the pathogenesis of Alzheimer's disease are not well understood. Recent studies of cultured human fetal neocortical neurons showed that Abeta can cause neurons to be exquisitely sensitive to excitotoxic cell injury and death (*see* **Fig. 2**; ref. *14*). When added to human fetal neocortical cultures, Abeta accumulated on or in the plasma membrane (*see* **Fig. 2**). The mechanism of Abeta neurotoxicity involves elevation of intracellular calcium levels *(14)*, apparently resulting from free radical-mediated damage to plasma membrane proteins involved in the regulation of ion homeostasis *(15,16)*. The ability to study living

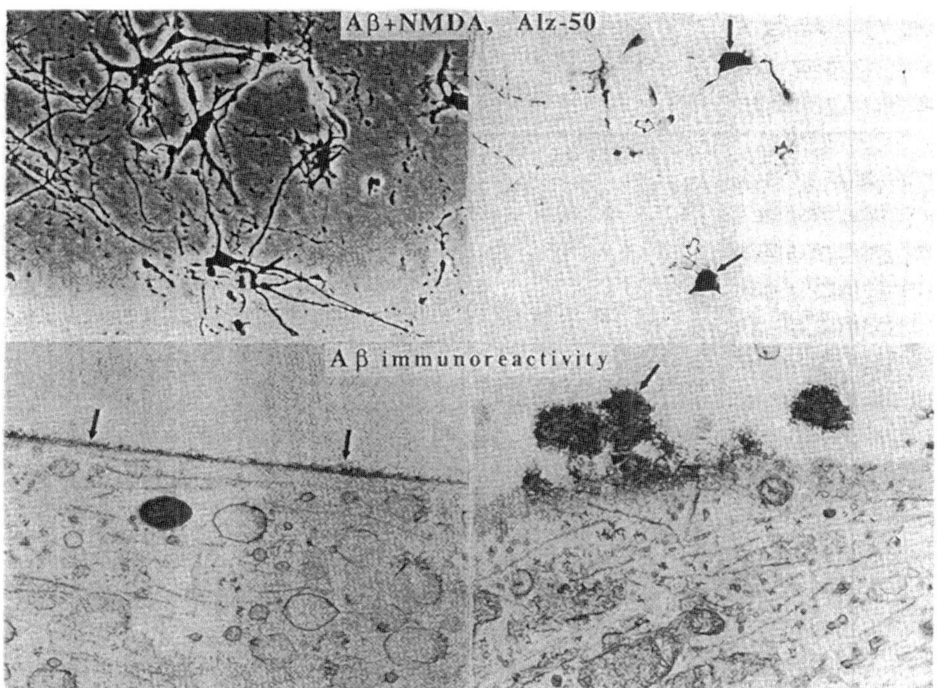

Fig. 2. Use of human fetal brain cell culture to study mechanisms of neuronal degeneration relevant to the pathogenesis of Alzheimer's disease. (Upper panels) Phase-contrast (left) and bright-field (right) micrographs of a field of cultured human fetal neocortical neurons (30 d in culture) shown 24 h following exposure to 40 μM amyloid beta-peptide (AP; amino acids 1–38) plus 50 μM NMDA. The cultures were immunostained with an antibody (Alz-50) that recognizes an altered form of the microtubule-associated protein "t" present in neurofibrillary tangles of Alzheimer's disease. Abeta increased neuronal vulnerability to NMDA excitotoxicity and induced the accumulation of Alz-50 immunoreactivity in the cell bodies of some neurons (solid arrows), but not others (open arrows). (Lower panels) Immunoelectron micrographs (peroxidase-DAB) showing the cell surface of neurons in human fetal neocortical cell cultures that had been exposed to Abeta for 24 h. Abeta immunoreactivity was localized primarily to the surface of the plasma membrane (e.g., arrows).

human neurons in culture has also provided insight into mechanisms of ischemic brain injury, a process relevant to hypoxic conditions in the developing brain (e.g., cerebral palsy) and stroke in the adult brain. Human fetal neocortical cultures have been employed to study the mechanisms whereby neurotrophic factors protect neurons against hypoglycemic and hypoxic insults. bFGF and nerve growth factor (NGF) each protected neocortical neurons against glucose deprivation-induced injury *(7)*. These neurotrophic factors prevented the delayed elevation of intracellular calcium levels otherwise induced by glucose deprivation. These findings suggest that endogenous neurotrophic factors may serve a neuroprotective function when the brain is subjected to ischemic conditions. Moreover, the data suggest that neurotrophic factors may prove useful in reducing neuronal injury in clinical settings, such as following stroke or traumatic brain injury *(17)*.

3. Actions of substances of abuse: A variety of substances of abuse (e.g., cocaine, alcohol, phencyclidine) are known to affect brain development adversely. Human fetal brain cell cultures can provide valuable information concerning the direct actions of substances of abuse on developing brain cells. Studies of cultured human fetal neocortical neurons have shown that phencyclidine, at concentrations believed to be reached *in utero*, can be neurotoxic *(18)*. The findings were consistent with a mechanism involving blockade of potassium channels by phencyclidine, resulting in membrane depolarization and elevated levels of intracellular calcium. Studies of the effects of ethanol and cocaine *(19)* on human fetal neurons remain to be performed. Further studies of cultured human fetal brain cells are needed and are likely to improve our understanding of human neurodegenerative disorders greatly, as well as mechanisms of brain damage resulting from substance abuse.

4. Notes

1. Bacterial and fungal contamination can be a problem. Its incidence can be reduced by washing the brain tissue several times in HBSS containing relatively high levels of antibacterial and antifungal agents (e.g., 1.0 mg/mL gentamicin and 5 mg/mL amphotericin B) prior to cell dissociation.
2. Deoxyribonuclease can be used to reduce cell clumping during the dissociation process. A concentration of 10 1.1g/mL deoxyribonuclease (Sigma) was used.
3. Fetal human neurons attach more slowly to the growth substrate than their rodent counterparts. In order to improve plating efficiency, it is therefore important not to change the culture medium for 16–24 h following plating.
4. DMM (Gibco "Neurobasal" plus B27 supplements) is an excellent serum-free medium for promoting long-term neuronal survival. However, this medium contains antioxidants, which act to reduce neuronal vulnerability to a variety of conditions, including exposure to excitotoxins, ischemia-like conditions, and amyloid toxicity.

References

1. Beadle, D. J., Lee, G., and Kater, S. B. (1988) *Cell Culture Approaches to Invertebrate Neuroscience*, Academic, London, U.K.
2. Mattson, M. P., Barger, S. W., Begley, J. G., and Mark, R. J. (1994) Calcium, free radicals, and excitotoxic neuronal death in primary cell culture. *Methods Cell Biol.* **46,** 187–216.
3. Hayflick, L. (1992) Fetal tissue banned . . . and used. *Science* **257,** 1027.
4. Vawter, D. E., Kearney, W., Gervais, K. G., Caplan, A. L., Garry, D., and Tauer, C. (1990) *The Use of Human Fetal Tissue: Scientific, Ethical and Policy Concerns.* Univ. Minnesota Press, MN.

5. Mattson, M. P. and Rychlik, B. (1990) Cell culture of cryopreserved human fetal cerebral cortical and hippocampal neurons: neuronal development and responses to trophic factors. *Brain Res.* **522,** 204–214.
6. Redmond, D. E., Naftolin, F., Collier, T. J., et al. (1988) Cryopreservation, culture, and transplantation of human fetal mesencephalic tissue into monkeys. *Science* **242,** 768–771.
7. Cheng, B. and Mattson, M. P. (1991) NGF and bFGF protect rat and human central neurons against hypoglycemic damage by stabilizing calcium homeostasis. *Neuron* **7,** 1031–1041.
8. Mattson, M. P. and Rychlik, B. (1991) Comparison of rates of neuronal development and survival in human and rat cerebral cortical cell cultures. *Mech. Aging Dev.* **60,** 171–187.
9. Mattson, M. P., Rychlik, Br., You, J.-S., and Sisken, J. E. (1991) Sensitivity of cultured human embryonic cerebral cortical neurons to excitatory amino acid-induced calcium influx and neurotoxicity. *Brain Res.* **542,** 97–106.
10. Mattson, M. P., Kumar, K., Cheng, B., Wang, H., and Michaelis, E. K. (1993) Basic FGF regulates the expression of a functional 71 kDa NMDA receptor protein that mediates calcium influx and neurotoxicity in cultured hippocampal neurons. *J. Neurosci.* **13,** 4575–4588.
11. Kater, S. B., Mattson, M. P., Cohan, C. S., and Connor, J. A. (1988) Calcium regulation of the neuronal growth cone. *Trends Neurosci.* **II,** 315–321.
12. Mattson, M. P. (1994) Secreted forms of β-amyloid precursor protein modulate dendrite outgrowth and calcium responses to glutamate in cultured embryonic hippocampal neurons. *J. Neurobiol.* **25,** 439–450.
13. Selkoe, D. J. (1993) Physiological production of the β-amyloid protein and the mechanism of Alzheimer's disease. *Trends Neurosci.* **16,** 403–409.
14. Mattson, M. P., Cheng, B., Davis, D., Bryant, K., Lieberburg, I., and Rydel, R. E. (1992) β-amyloid peptides destabilize calcium homeostasis and render human cortical neurons vulnerable to excitotoxicity. *J. Neurosci.* **12,** 376–389.
15. Goodman, Y. and Mattson, M. P. (1994) Secreted forms of β-amyloid precursor protein protect hippocampal neurons against amyloid β-peptide toxicity and oxidative injury. *Exp. Neurol.* **128,** 1–12.
16. Hensley, K., Carney, J. M., Mattson, M. P., et al. (1994) A model for β-amyloid aggregation and neurotoxicity based on free radical generation by the peptide: relevance to Alzheimer's disease. *Proc. Natl. Acad. Sci. USA* **91,** 3270–3274.
17. Mattson, M. P. and Scheff, S. W. (1994) Endogenous neuroprotection factors and traumatic brain injury: mechanisms of action and implications for therapies. *J. Neurotrauma* **11,** 3–33.
18. Mattson, M. P. and Rychlik, B. (1992) Degenerative and axon outgrowth-altering effects of phencyclidine in human fetal cerebral cortical cells. *Neurophannacology* **31,** 279–291.
19. Moroney, J. T. and Allen, M. H. (1994) Cocaine and alcohol use in pregnancy. *Adv. Neurol.* **64,** 231–242.

12

Culturing Human Schwann Cells

Victor J. Turnbull

1. Introduction

The method involved in the establishment of human adult Schwann cell cultures has steadily evolved over the last 20 yr. Unlike the more straightforward methods used with other species (e.g., rat), simple dissociation of human peripheral nerve tissue has been found to result in both very low cell yields and poor Schwann cell purity *(1,2)*. This is thought to be caused by the increased amount of connective tissue within human nerves and the mechanical damage sustained by the tightly wrapped Schwann cell processes. These problems were previously overcome by the pioneering work of Askanas et al. in 1980 who developed a method involving the continual re-explantation of small segments of peripheral nerve tissue (explants) *(3)*. The resulting cultures, derived from cells that had emigrated from the explants, were found to exhibit an increased Schwann cell purity upon successive re-explantations. The major drawback to this "re-explantation" method, however, was the length of time it took (months) to establish high-purity Schwann cell cultures.

Subsequently, it was found that a combination of both of these methods provided the optimal procedure for preparing human Schwann cell cultures. This was based on the observation that peripheral nerve explants could successfully be dissociated after only 1–2 wk in culture, resulting in a vast improvement in cell yield *(2,4,5)*. This initial short period of explant culture was suggested to provide several beneficial outcomes. It not only allowed Schwann cells to withdraw their tightly wrapped cell processes from axonal segments, but also took advantage of their intrisic ability to proliferate in response to axolemmal and myelin debris *(6,7)*. The depletion of connective tissue cells, the predominant

From: *Methods in Molecular Medicine, vol. 107: Human Cell Culture Protocols, Second Edition*
Edited by: J. Picot © Humana Press Inc., Totowa, NJ

cell type initially emigrating from the explants, may also improve the final Schwann cell purity.

The successful establishment of human Schwann cell cultures therefore requires a short period of peripheral nerve explant culture followed by explant dissociation. Important considerations include the source of peripheral nerve; the presence of contaminating peripheral nerve connective tissue cells; and the ultimate purpose/objective/requirements for the cultures.

2. Materials

1. Segments of normal human sciatic or posterior tibial nerve.
2. Dissecting equipment: forceps, scalpel/blades, dissecting microscope and two pairs of very fine point forceps (INOX no. 5 Dumont, Switzerland).
3. Dulbecco's modified Eagle's medium (DMEM): 4.5 g/L (w/v) D-glucose, 25 mM HEPES buffer, 1×10^5 units/L penicillin-G, 100 mg/L (w/v) streptomycin sulfate; store in the dark at 4°C.
4. Small sterile Petri dishes (60 mm diameter).
5. Heat-inactivated fetal calf serum (FCS); store at –20°C.
6. Cell-culture equipment: incubator, hemocytometer, inverted microscope, 37°C water bath.
7. Explant dissociation medium: DMEM containing 10% (v/v) FCS, 0.5 mg/mL (w/v) Collagenase A (Roche) and 1.25 U/mL Dispase I (Roche).
8. Poly-D-lysine (molecular weight [MW]: 30,000–70,000); store working solution (0.1 mg/mL [w/v]) at –20°C.
9. Small tissue-culture flasks (25 cm^2).
10. Sterile phosphate-buffered saline (PBS): 0.15 M NaCl, 0.1 M Na$_2$HPO$_4$, 15 mM KH$_2$PO$_4$, 25 mM KCl, pH 7.4; store at room temperature (RT).
11. Trypsin passaging solution: 0.1% trypsin in PBS (w/v); prepare from a 2.5% stock solution stored at –20°C.

3. Methods

The ensuing methods describe: (1) the dissection of human peripheral nerve and set up of the peripheral nerve explant culture, (2) the dissociation of the nerve explants resulting in the establishment of purified human Schwann cell cultures, and (3) a human Schwann cell passaging protocol.

3.1. Human Peripheral Nerve Explant Culture

Segments of normal human sciatic or posterior tibial nerves are obtained from lower limb amputation specimens (from patients ≥18 yr of age) within 8 h of surgical removal (*see* **Note 1**). Individual nerve fascicles are dissected from the peripheral nerve and sliced into small explants to give rise to the explant culture (*see* **Fig. 1**).

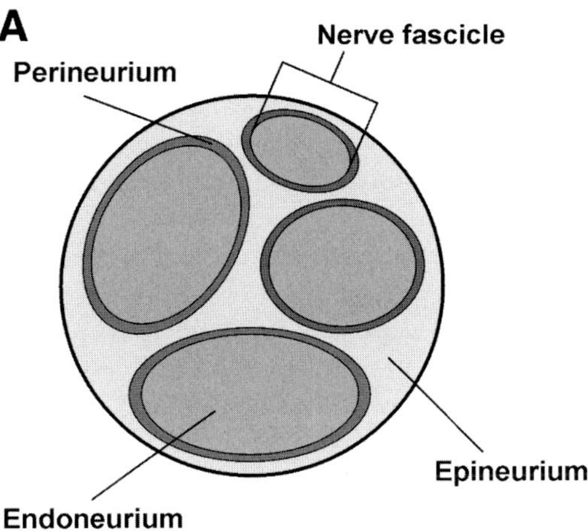

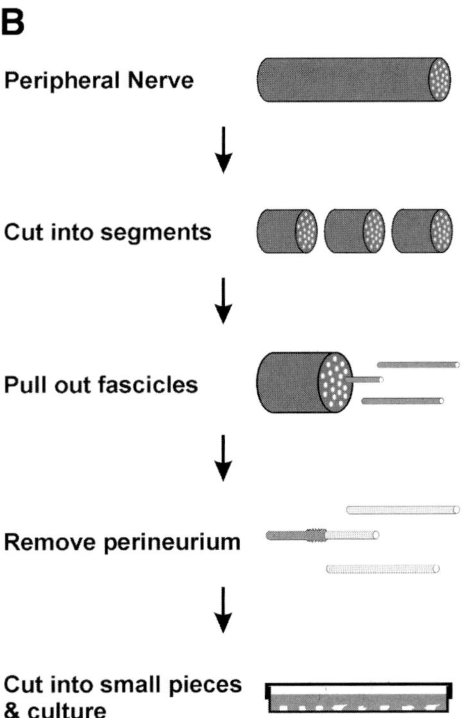

Fig. 1. **(A)** Schematic diagram showing a cross-sectional view of a peripheral nerve. **(B)** Schematic flowchart outlining the dissection of peripheral nerve for the establishment of peripheral nerve explant cultures.

3.1.1. Peripheral Nerve Dissection

1. Dissect the nerve from the limb using forceps and a scalpel blade.
2. Immediately place the nerve in a sterile specimen jar containing cold DMEM and place on ice (*see* **Note 2**).
3. On a sterile surface, transversely cut the nerve into 2-cm pieces with a scalpel blade and place in fresh DMEM on ice.
4. Each individual nerve segment is then sequentially dissected under a dissecting microscope in a Petri dish containing cold DMEM (*see* **Note 3**).
5. Firmly clasp one end of the nerve piece with fine forceps and slowly pull a single nerve fascicle from the opposing end with the other pair of fine forceps.
6. Remove the outer perineurial connective tissue sheath (*see* **Fig. 2**) by firmly clasping one end of the fascicle with fine forceps and sliding the other pair of forceps along the length of the fascicle (*see* **Note 4**). Then remove the remainder of the perineurial sheath trapped beneath the clasping forceps. It is important to remove as much of the sheath as possible as the remainder will act as a source of "contaminating" connective tissue cells in the resultant Schwann cell culture.
7. Place the nerve fascicle into a small Petri dish containing 5 mL of cold DMEM and keep on ice.

Repeat **steps 5–7** until all the nerve fascicles have been removed and then move on to the next nerve segment.

8. Cut the nerve fascicles into small explants, approx 2 mm in length, using a scalpel blade.
9. Replace the DMEM with prewarmed DMEM/FCS and incubate at 37°C in humidified air containing 7% CO_2.

3.1.2. Explant Culture

A period of explant culture is now required to allow time for Schwann cells to withdraw their processes. During this period, the majority of the explants will adhere to the culture surface and fibroblasts will emerge forming a monolayer of cells surrounding the explant. After 2 wk of culture the nerve pieces will be ready for enzymatic dissociation (*see* **Note 5**).

3.2. Dissociation of Peripheral Nerve Explants

1. Dissociate explants overnight at 37°C at a ratio of approx 50 explants per 3 mL dissociation media (*see* **Note 6**).
2. If the explants have not been totally dissociated, place them in fresh enzyme dissociation media and incubate for a further 2 h (*see* **Note 7**).
3. Transfer the resultant cell suspension into a 15-mL centrifuge tube and centrifuge for 5 min at 300*g*.
4. To ensure enzyme removal, wash the cells twice with 10 mL of DMEM (*see* **Note 8**).
5. Resuspend the final cell pellet in prewarmed DMEM/FCS and seed into poly-D-lysine coated tissue-culture vessels. An appropriate cell-seeding level is

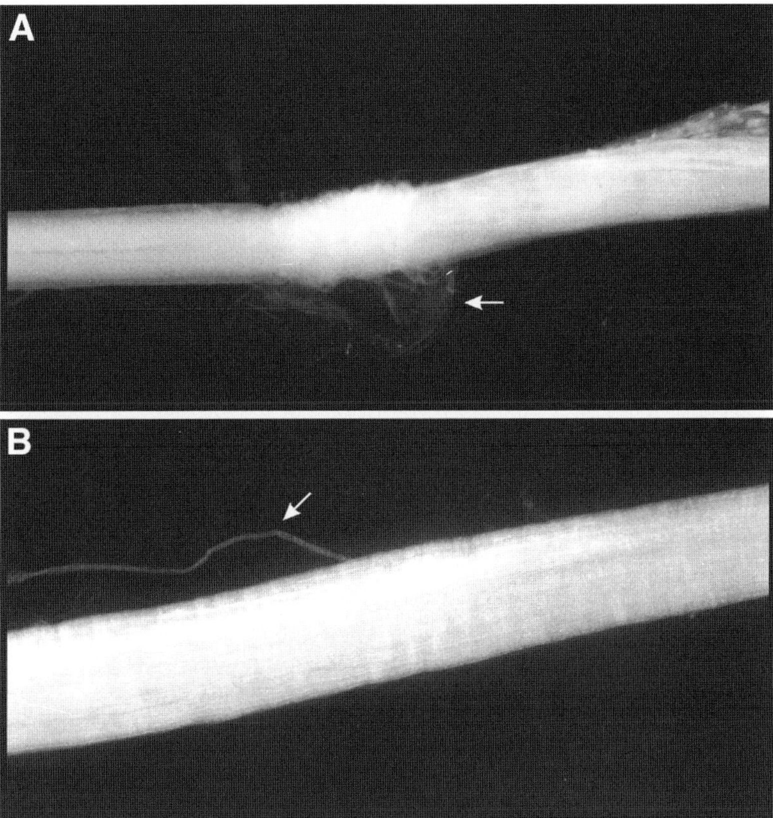

Fig. 2. Removal of the perineurial sheath during dissection of peripheral nerve fascicles. (**A**) The perineurial sheath (arrow) is pushed away from the fascicle using a pair of fine forceps. (**B**) Separation of individual nerve fibers (arrow) indicates that the perineurium has been stripped from the nerve fascicle (magnification ×50).

2×10^5 cells for small tissue-culture flasks (25 cm^2), or 1×10^4 cells for 13-mm diameter glass cover slips (*see* **Note 9**).

3.2.1. Mitogens

Schwann cell purity is an important feature of dissociated human Schwann cell cultures and should range between 90 and 99% using the above protocol (*see* **Fig. 3**). However, when cultured using standard media, the Schwann cell purity will rapidly decrease owing to the rapid growth of fibroblasts. Consequently, both heregulin-β1 (10 n*M*) and forskolin (2 µ*M*) should be added to the culture media (*see* **Note 10**). Heregulin-β1 is a potent mitogen for Schwann

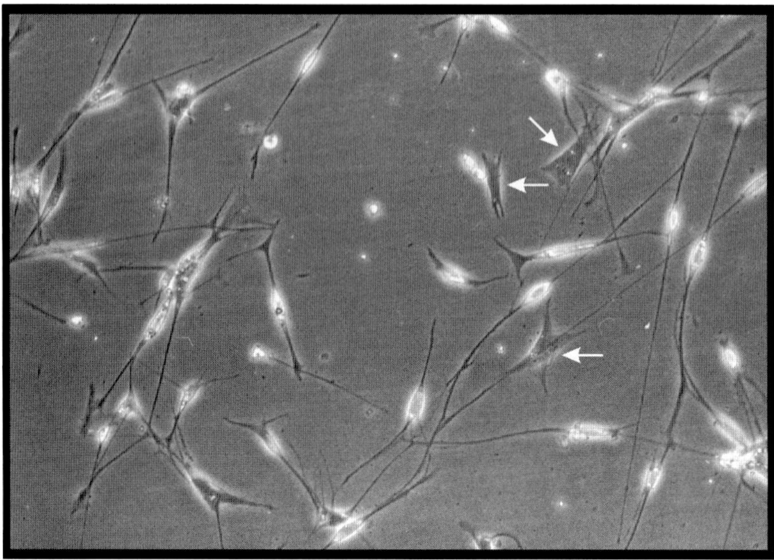

Fig. 3. Phase contrast appearance of a purified Schwann-cell culture 24 h after dissociation of peripheral nerve explants. Schwann cells exhibit a characteristic "phase-bright" bipolar/tripolar morphology. In contrast, contaminating fibroblasts (arrows) adopt a flat, polymorphic morphology (magnification ×200).

cells *(4,8)* whereas forskolin has a dual effect: it increases Schwann cell proliferation while inhibiting fibroblast growth *(2,9)*.

3.3. Passaging of Human Schwann Cells

Steps 1 and **2** of the following cell-passaging protocol have been optimized for Schwann cells cultured in small tissue-culture flasks. Passaging from different sized culture vessels will require the appropriate alterations in the volumes used.

1. Aspirate the culture media supernatant from the flask and wash the cells for 5 min with 5 mL of PBS (*see* **Note 11**).
2. Remove the PBS and incubate the cells at 37°C for 15 min with 5 mL of prewarmed trypsin passaging solution (*see* **Note 12**).
3. After transferring the resultant cell suspension into a 15-mL centrifuge tube, wash the flask with 5 mL DMEM/FCS and add to the centrifuge tube. Pellet the cells by spinning at 300*g* for 5 min.
4. After removing the supernatant, resuspend the cells in 10 mL of prewarmed DMEM/FCS and centrifuge as before.

5. Resuspend the cells into an appropriate volume of prewarmed DMEM/FCS and seed into a new poly-D-lysine-coated tissue-culture vessel.

4. Notes

1. One of the most critical factors involved in the establishment of human Schwann cell cultures is the source of peripheral nerve tissue. In this regard, it is highly recommended to set up collaboration with a neuropathologist based in the Anatomical Pathology Department of a hospital. Not only can they provide great assistance in sourcing the peripheral nerve tissue, they will often conduct the initial dissection thereby providing you directly with segments of peripheral nerve tissue. Lower limb amputation specimens can be used as a means of obtaining peripheral nerve tissue for the cultures as they are often stored at 4°C relatively soon after surgical removal. Their supply is usually very irregular so you need to ensure you are well prepared to set up the cultures on short notice. Human Schwann cell cultures have also been derived from sciatic nerves obtained postmortem from patients who have died from nonneurological disorders *(1,10)*. A postmortem interval of at least 20 h reportedly has no effect on cell viability *(10)*, but the possibility of a longer postmortem delay is a major concern and can often lead to very low cell yields. Peripheral nerves (phrenic, intercostal, or segments of lumbosacral plexus) have also been obtained from organ donors with the support of a transplant procurement team *(7,8)*. It is also feasible to derive cultures from peripheral nerve biopsy specimens, but this tissue is scarce, only available in small amounts, and is often compromised by the presence of pathological abnormalities.
2. A balanced saline solution [e.g., Hank's balanced saline solution (HBSS)] can be used in this step. As the dissection can take up to 2–3 h, and the procedure involves primary tissue, the use of DMEM, which contains antibiotics, is probably preferable.
3. Some blood and fatty debris from the nerve will often cloud the DMEM during the dissection procedure. Consequently, an occasional change of the DMEM and/or Petri dish is often required.
4. The best analogy to describe the perineurial sheath removal is to visualize straightening your arm, cupping the circumference of your sleeve with your hand, and sliding (or pushing) the arm of your shirt from your wrist to your elbow. With the nerve fascicle, one end is held with the forceps and another pair of forceps is used to slide (push) the sheath away from the fascicle. Obviously, some of the perineurial sheath will be trapped beneath the clasping forceps and will also have to be removed. Using this method ensures the whole sheath is removed from the fascicle in one simple and easy step and that you are not constantly picking and tearing off little pieces at a time. A close visual inspection of the fascicle will enable you to determine whether the perineurium has been removed, because, although the sheath is transparent, it has a slight brown tinge. As a result, the perineurium-free fascicle has a more "brighter white" appearance.
5. The media should be changed 24 h after initiation of the explant culture and thereafter approx twice a week. The precise timing of explant dissociation is not crit-

ically important. The explant culture period can range from a minimum of at least 1 wk *(5)* to at least 2 mo. The explants may be cultured for even longer periods, the maximum in our laboratory is 4 mo, but the final cell yield does tend to decrease. It is important, if the explants are to be cultured for any great length of time, to ensure the explants are regularly re-explanted (approx every 2 wk) into new dishes. This limits the buildup of surrounding fibroblasts that, in time, will form a network of cells that will be difficult to remove from the edges of the explants. If this occurs, the Schwann cell purity of the final dissociated cultures will be compromised. Depending on the ultimate requirement of the Schwann cell cultures, and the availability of human peripheral nerves, it may be desirable to keep a continuing supply of explant cultures. For example, the downregulation of cell-surface galactocerebroside (GalC) expression by human Schwann cells has been found to occur much more rapidly in dissociated culture than in explant culture *(11)*. In this case, dissociation of only a portion of the explants would leave a supply of explants for experimentation at a later date without severely affecting Schwann-cell GalC expression.

6. The presence of serum in the dissociation media is not a concern as neither of the enzymes is inhibited by serum. In general, each milligram of nerve explant tissue should yield greater than 2×10^4 cells *(1,2)*. In our laboratory, we perform the explant dissociation in six-well tissue-culture plates in the cell-culture incubator. The ability to view the plate under the microscope makes it easier to determine whether the explants have been adequately dissociated. An interesting variation to this dissociation method has been reported whereby complete explant dissociation was achieved in 3 h using higher enzyme concentrations *(12)*. This method yielded approx 1×10^6 cells for every 1 cm of the initial peripheral nerve segment. The suggestion that enzymatic injury to the plasma membrane of cells is lessened using this method (e.g., higher enzyme concentration for shorter time vs lower enzyme concentration for longer periods) is somewhat debatable.

7. Pipetting the dissociated explants up and down with a 1-mL pipet is an easy method for determining whether adequate dissociation has occurred. If a "gluggy" mass is formed that plugs up the opening of the pipet tip, a further incubation with fresh dissociation media is needed. However, total solubilization of the tissue is not necessary. Sufficient dissociation is obtained when the mix is comprised of clear, acellular connective tissue sheets surrounded by dissociated endoneurial cells. The insoluble matter is easily separated from the dissociation mix by pipetting it into a sterile tube (it sinks to the bottom) and removing the cell-containing supernatant. A couple of quick washes with DMEM will minimize cell loss.

8. Dissociated cells can be washed with PBS containing 0.04% ethylenediaminetetraacetic acid (EDTA) to inactivate the collagenase and dispase *(2)*; however, simple dilution of the enzymes appears to have no adverse effect on the resultant cultures.

9. Schwann-cell proliferation has been found to be strongly regulated by cell density *(13)*. In this study, cultures plated at a density of 3.6×10^3 cells/cm^2 exhibited a growth rate nearly five times greater than Schwann cells that were plated at

7.1 × 10³ cells/cm². The Schwann cells appeared to cease division at a density of approx 5.7 × 10⁴ cells/cm². There appears to be little difference between using poly-D-lysine or poly-L-lysine to prepare the culture surface. The manufacturers have suggested that using the D-isomer will avoid the production of biologically active L-amino acids that are produced by polymer degradation. Consequently, our laboratory has made the decision to err on the side of caution. Schwann cells also adhere well to other attachment factors (e.g., laminin), but these proteins are much more expensive. Positive identification of Schwann cells in culture can be performed using immunocytochemistry. All Schwann cells can be labeled with anti-S100 antibodies whereas anti-human fibronectin antibodies can be used to identify any contaminating connective tissue cells.

10. Because ethanol has been shown to inhibit the forskolin activation of adenylate cyclase *(14)*, the 10 m*M* stock solution of forskolin is prepared using 100% dimethyl sulfoxide (DMSO) and stored at –20°C. Forskolin is reportedly stable for at least 6 mo in solution *(15)*. Recombinant human heregulin can be stored for up to 2 yr at –20°C and can be diluted directly into culture media. It is important to note that poly-L-lysine is not an appropriate substrate when using heregulin and forskolin, as the expanding Schwann cells have been found to form clusters and detach from poly-L-lysine coated culture surfaces *(4)*. It was subsequently shown that laminin and ammoniated collagen were the most suitable substrates, with laminin providing the greatest enhancement to Schwann-cell growth rates.

11. This PBS incubation step is particularly important as the serum present within the culture media will inhibit the activity of the trypsin.

12. Passaging Schwann cells in a balanced saline solution comprising 0.05% trypsin and 0.02% EDTA for 5 to 10 min has also been described *(7,8)*.

Acknowledgment

The author wishes to sincerely thank Dr. M. Ayers for her help and advice regarding the preparation of this manuscript.

References

1. Morrissey, T. K., Kleitman, N., and Bunge, R. P. (1991) Isolation and functional characterization of Schwann cells derived from adult peripheral nerve. *J. Neurosci.* **11,** 2433–2442.
2. Rutkowski, J. L., Tennekoon, G. I., and McGillicuddy, J. E. (1992) Selective culture of mitotically active human Schwann cells from adult sural nerves. *Ann. Neurol.* **31,** 580–586.
3. Askanas, V., Engel, W. K., Dalakas, M. C., Lawrence, J. V., and Carter, L. S. (1980) Human Schwann cells in tissue culture: histochemical and ultrastructural studies. *Arch. Neurol.* **37,** 329–337.
4. Casella, G. T. B., Bunge, R. P., and Wood, P. M. (1996) Improved method for harvesting human Schwann cells from mature peripheral nerve and expansion in vitro. *Glia* **17,** 327–338.

5. Levi, A. D. O. (1996) Characterization of the technique involved in isolating Schwann cells from adult peripheral nerve. *J. Neurosci. Methods* **68,** 21–26.
6. Sobue, G., Brown, M. J., Kim, S. U., and Pleasure, D. (1984) Axolemma is a mitogen for human Schwann cells. *Ann. Neurol.* **15,** 449–452.
7. Morrissey, T. K., Levi, A. D. O., Nuijens, A., Sliwkowski, M. X., and Bunge, R. P. (1995) Axon-induced mitogenesis of human Schwann cells involves heregulin and p185^{erbB2}. *Proc. Natl. Acad. Sci. USA* **92,** 1431–1435.
8. Levi, A. D. O., Bunge, R. P., Lofgren, J. A., et al. (1995) The influence of heregulins on human Schwann cell proliferation. *J. Neurosci.* **15,** 1329–1340.
9. Rutkowski, J. L., Kirk, C. J., Lerner, M. A., and Tennekoon, G. I. (1995) Purification and expansion of human Schwann cells in vitro. *Nat. Med.* **1,** 80–83.
10. Boyer, P. J., Tuite, G. F., Dauser, R. C., Muraszko, K. M., Tennekoon, G. I., and Rutkowski, J. L. (1994) Sources of human Schwann cells and the influence of donor age. *Exp. Neurol.* **130,** 53–55.
11. Turnbull, V. J., Petratos, S., Papadopoulos, R., Gonzales, M. F., and Ayers, M. (2001) Variable galactocerebroside expression by human Schwann cells in dissociated and peripheral nerve explant cultures. *J. Neurosci. Res.* **65,** 318–321.
12. Calderón-Martínez, D., Garavito, Z., Spinel, C., and Hurtado, H. (2002) Schwann cell-enriched cultures from adult human peripheral nerve: a technique combining short enzymatic dissociation and treatment with cytosine arabinoside (Ara-C). *J. Neurosci. Methods* **114,** 1–8.
13. Casella, G. T. B., Wieser, R., Bunge, R. P., et al. (2000) Density dependent regulation of human Schwann cell proliferation. *Glia* **30,** 165–177.
14. Huang, R., Smith M. F., and Zahler, W. L. (1982) Inhibition of forskolin-activated adenylate cyclase by ethanol and other solvents. *J. Cyclic Nucleotide Res.* **8,** 385–394.
15. Daly, J. W., Padgett, W., and Seamon, K. B. (1982) Activation of cyclic AMP-generating systems in brain membranes and slices by the diterpene forskolin: augmentation of receptor-mediated responses. *J. Neurochem.* **38,** 532–544.

13

Well-Differentiated Human Airway Epithelial Cell Cultures

M. Leslie Fulcher, Sherif Gabriel, Kimberlie A. Burns, James R. Yankaskas, and Scott H. Randell

1. Introduction

The airway epithelium occupies a critical environmental interface, protecting the host from a wide variety of inhaled insults, including chemical and particulate pollutants and pathogens. The coordinated regulation of ion and water transport, mucous secretion, and cilia beating underlies mucociliary clearance. Physical trapping and removal of harmful substances, in combination with baseline or inducible secretion of antimicrobial factors, antioxidants, and protease inhibitors and recruitment of nonspecific inflammatory cells (neutrophils, monocytes), constitutes airway innate host defense.

Cystic fibrosis (CF) is a genetic disease in which impaired innate host defense results in repeated, severe airway infections. Airway epithelial cell cultures (AECCs) have been integral to our understanding of CF pathogenesis. Because CF is a monogenic, recessive, loss-of-function disorder, it is theoretically curable by gene therapy. However, the promise of gene therapy has not been fulfilled, mainly owing to vector inefficiency and safety concerns. AECCs will be an important tool for advancing gene therapy.

In addition to its key functional role in innate immunity, the airway epithelium modulates inflammation and adaptive immunity (dendritic cell function, specific T and B cells). Alterations in both innate and acquired immune function induced by the epithelium may contribute to the pathophysiology of asthma and chronic bronchitis. Many aspects of these profoundly important diseases remain poorly understood and AECCs will facilitate studies of both basic pathophysiology and development of novel therapies.

From: *Methods in Molecular Medicine, vol. 107: Human Cell Culture Protocols, Second Edition*
Edited by: J. Picot © Humana Press Inc., Totowa, NJ

The epithelium itself responds to injury and becomes modified during the progression of disease. The changes range from mild and transient cytopathology to epithelial hyperplasia or metaplasia, and ultimately, in some cases, malignant transformation. The bronchial epithelium is the source of the world's most prevalent lethal cancer, caused principally, but not exclusively, by exposure to tobacco products. AECCs enable analysis of epithelial growth and differentiation and may prove useful for studies relating to prevention, detection, monitoring, and treatment of lung cancer.

Finally, a significant and growing number of drugs are administered as aerosols. Transepithelial transport properties, as well as positive and negative effects on host cells, are key parameters requiring assessment. Thus, human AECCs are integral to the study of basic and applied aspects of airway biology, disease, and therapy.

1.1. Historical Perspective and Milestones

Epithelial cell cultures have been created from the human airway for more than 20 yr *(1)*. In the original method, finely minced airway tissue fragments were explanted and epithelial cells were harvested as outgrowths. Alternatively, protease dissociation creates suspensions of free epithelial cells *(2)*. The initial cell harvest usually contains some nonepithelial cells. Morphologic characteristics during passage in selective medium and immunostaining for cytokeratin have traditionally been used for cellular identification. Primary airway epithelial cells on plastic dishes can be repeatedly passaged. On plastic dishes, the cells assume a poorly differentiated, squamous phenotype. However, when freshly harvested or passaged primary airway epithelial cells are cultured under conditions enabling cellular polarization, a dramatic phenotypic conversion occurs, enabling the cells to more closely recapitulate their normal in vivo morphology. This was first recognized when animal or human airway epithelial cells were inoculated into devitalized tracheal or intestinal tubes and then implanted subcutaneously in compatible hosts *(3,4)*. A similar effect occurs in vitro when masses of cells assume a three-dimensional spheroidal shape *(5)* or if cells are grown on or within thick collagen gels *(6,7)*. However, the most widely utilized system enabling the cells to undergo mucociliary differentiation involves growing them on porous supports at an air–liquid interface, first shown by Whitcutt, Adler, and Wu *(8)*. These cultures demonstrate vectorial mucus transport *(9,10)*, high resistance to gene therapy vectors *(11)*, and cell-type-specific infection by viruses *(12)*, functions that cannot be studied using undifferentiated cells on plastic. The complex process of airway epithelial differentiation involves cell–matrix and cell–cell interactions, differentiation of mucous and goblet cells, and acquisition of characteristic epithelial ion transport properties. Numerous genes and proteins are induced during differentiation, including those characteristically present in secretory and ciliated cells *(13,14)*. Retinoic acid is essential to suppress or reverse squamous metaplasia in culture *(15)*. Much remains to be learned about the complex program regulating mucociliary differentiation and phenotypic modulation of the airway epithelium.

1.2. Summary and Purpose

Compared to undifferentiated cells on plastic, human airway epithelial cell cultures maintained at an air–liquid interface (ALI) represent a quantum leap toward the in vivo biology, and are an excellent model to probe airway epithelial function. Although they have been used for studies too numerous to cite here, the technical requirements, financial commitment, and experimental limitations inhibit their use in many laboratories. The approximate cost of passage 1 airway epithelial cells from a commercial supplier is $569 (U.S. dollars) per 0.5×10^6 cells. As a point of reference, expansion and subculture of this number of cells would typically generate 25 passage 2 ALI cultures 12 mm in diameter. An alternative is direct procurement of cells from human tissues, but this requires establishment of working relationships with surgeons and pathologists and compliance with appropriate regulations. Furthermore, the media is complex, with expensive individual components. The University of North Carolina established a Tissue Procurement and Cell Culture Core in 1984, under the auspices of the Cystic Fibrosis Foundation, to provide standardized cell cultures. From 1984 to 2003, the Core prepared cells from more than 6030 human tissue specimens, adopting new technologies to extend research capabilities. The purpose of this chapter is to share our detailed protocols and "tricks of the trade," thus, enabling others to overcome barriers toward using this relevant cell culture model.

For many years, the dogma was that only fresh primary cells seeded at high density could reliably form well-differentiated cultures, and that differentiation was dependent on a proprietary cell-culture supplement, Ultroser G (Biosepra SA, Cergy-Sainte-Christophe, France). U.S. importation of Ultroser G requires a permit from the Department of Agriculture. A breakthrough paper published by Gray et al. *(16)* showed successful differentiation of subcultured human airway epithelial cells with no proprietary reagents. These procedures significantly enhance the ability to study differentiation-dependent functions and have also increased the number and area of well-differentiated cultures produced from each tissue sample. The ability to store primary cells as frozen stocks stabilizes cell availability, enables the simultaneous production of cultures at different stages of maturity from the same patient sample, allows repeat experiments with the same specimen, and permits simultaneous performance of experiments with replicate cultures derived from multiple patients. The procedures detailed below represent an extension of the original methods given by Lechner and Laveck *(17)*, strongly influenced by the methodology of Gray et al. *(16)*, which evolved during years of practical experience in our laboratory.

2. Materials

2.1. Tissue Procurement

Airway epithelial cells can be extracted from nasal turbinate or polyp specimens, trachea, or bronchi procured locally through cooperation of surgeons and pathologists in accordance with relevant institutional, local, and national regulations. Surgical nasal specimens not requiring histopathologic examination or excess nonaffected portions of lung tissue after gross examination by a pathologist, such as bronchi after lobectomy or pneumonectomy for lung cancer, are common sources. These are transported to the laboratory in a specimen cup on wet ice containing a physiologic solution [sterile saline, phosphate-buffered saline (PBS), lactated Ringer's solution, or tissue-culture medium]. Lungs from potential organ donors are frequently unsuitable for transplantation as a result of age, smoking history, or acute injury such as aspiration, pulmonary edema, or pneumonia, but are useful for research. These can be obtained by development of appropriate protocols with federally designated organ procurement agencies that normally oversee collection and distribution of donated organs. In the U.S., nonprofit organizations such as the National Disease Research Interchange (www.ndriresource.org) facilitate provision of human biomaterials for research. When establishing protocols with organ suppliers, the laboratory must set criteria for organ acceptability (*see* **Note 1**). Lung tissues may be retrieved at time of autopsy but, in our experience, removal within several hours of time of death is necessary. Finally, one can circumvent the need for tissue procurement by purchasing human airway epithelial cells (www.cambrex.com).

2.2. Media

Two closely related media are used for culturing airway epithelial cells. Bronchial epithelial growth medium (BEGM) is used when the initial cell harvests are plated on collagen-coated plastic dishes or to expand passaged cells on plastic. ALI medium is used to support growth and differentiation on porous supports. BEGM composition is given in **Table 1** and the differences between BEGM and ALI are illustrated in **Table 2**. All base media and additives can be purchased commercially (*see* www.biosource.com for LHC basal media and additives and as specified below for others). Bovine pituitary extract (BPE) can be purchased commercially or made from mature bovine pituitaries. The decision whether to make BPE depends on the volume of media needed and, thus, savings realized.

Table 1
BEGM and ALI Composition

Additive		Final Concentration In Media	Company	Cat. no.
Bovine serum albumin		0.5 mg/mL	Sigma-Aldrich	A7638
Bovine pituitary extract, homemade[a]		0.8% (v/v)	Pel Freeze	57133-2
Bovine pituitary extract, commerical[a]		10 µg/mL	Sigma-Aldrich	P1476
Insulin		0.87 µM	Sigma-Aldrich	I6634
Transferrin		0.125 µM	Sigma-Aldrich	T0665
Hydrocortisone		0.1 µM	Sigma-Aldrich	H0396
Triiodothyronine		0.01 µM	Sigma-Aldrich	T6397
Epinephrine		2.7 µM	Sigma-Aldrich	E4642
Epidermal growth factor		25 ng/mL—BEGM 0.50 ng/mL—ALI	Atlanta Biological	C100
Retinoic acid		5×10^{-8} M	Sigma-Aldrich	R2625
Phosphorylethanolamine		0.5 µM	Sigma-Aldrich	P0503
Ethanolamine		0.5 µM	Sigma-Aldrich	E0135
Zinc sulfate		3.0 µM	Sigma-Aldrich	Z0251
Penicillin G sulfate		100 U/mL	Sigma-Aldrich	P3032
Streptomycin sulfate		100 µg/mL	Sigma-Aldrich	S9137
Gentamicin[b]		50 µg/mL	Sigma-Aldrich	G1397
Amphotericin[b]		0.25 µg/mL	Sigma-Aldrich	A2942
Stock 4	Ferrous sulfate	1.5×10^{-6} M	Sigma-Aldrich	F8048
	Magnesium chloride	6×10^{-4} M	J.T. Baker	2444
	Calcium chloride	1.1×10^{-4} M	Sigma-Aldrich	C3881
Trace Elements	Selenium	3.0 µM	Sigma-Aldrich	S5261
	Manganese	0.1 µM	Sigma-Aldrich	M5005
	Silicone	50 µM	Sigma-Aldrich	S5904
	Molybdenum	0.1 µM	Sigma-Aldrich	M1019
	Vanadium	0.5 µM	Sigma-Aldrich	A1183
	Nickel sulfate	0.1 mM	Sigma-Aldrich	N4882
	Tin	0.05 µM	Sigma-Aldrich	S9262

[a]*See* **Subheading 2.3.2.**
[b]Not in ALI.

Table 2
Differences Between ALI and BEGM Medium

	ALI	BEGM
Base media	LHC Basal:DMEM-H 50:50	LHC Basal 100%
Base Antibiotics	Pen/Strep (100 U/mL/100 µg/mL)	Pen/Strep (100 U/mL/100 µg/mL) Gentamicin 50 µg/mL Amphotericin 0.25 µg/mL
EGF	0.50 ng/mL	25 ng/mL
$CaCl_2$	1.0 mM	0.11 mM

2.3. Stock Additives for ALI and BEGM

Additives for media are filtered using 0.2-µM filters (unless product is sterile) and aliquots are stored at –20°C for up to 6 mo.

1. Bovine serum albumin (BSA) (300X 150 mg/mL): Add PBS directly to the BSA (Sigma-Aldrich, St. Louis, MO, cat. no. A7638) container to yield a concentration >150 mg/mL. Gently rock bottle at 4°C for 2–3 h until BSA is dissolved. Transfer to graduated cylinder and set volume to yield a final concentration of 150 mg/mL.

2. 100X BPE: Commercially prepared BPE is available from Sigma-Aldrich (cat. no. P1427) and is handled per manufacturer's instructions. It is used at a final concentration of 10 µg/mL. BPE can also be prepared from mature bovine whole pituitaries (Pel Freeze, Rogers, AR, cat. no. 57133-2). Thaw bovine pituitaries, drain, and rinse with chilled 4°C PBS. Add 2 mL of chilled PBS per gram of tissue. In a cold room, mince tissue in a Waring 2-speed commercial blender (Fisher Scientific, Pittsburgh, PA, cat. no. 14-509-17) at low speed for 1 min and then at high speed for 10 min. Aliquot suspension and centrifuge at 2500*g* for 10 min at 4°C. Collect supernatant and centrifuge again at 10,000*g* for 10 min. Harvest the final BPE supernatant. Homemade BPE is difficult to filter and needs to be filtered during media preparation as described in **Subheading 2.4.2.**

3. 1000X insulin (5 mg/mL): Dissolve insulin (Sigma-Aldrich, cat. no. I6634) in 0.9 *N* HCl.

4. 1000X transferrin (10 mg/mL): Reconstitute transferrin, human-holo, natural (Sigma-Aldrich, cat. no. T0665) in PBS.

5. 1000X hydrocortisone (0.072 mg/mL): Reconstitute hydrocortisone (Sigma-Aldrich, cat. no. H0396) in distilled water (dH$_2$O).

6. 1000X triiodothyronine (0.0067 mg/mL): Dissolve triiodothyronine (Sigma-Aldrich, cat. no. T6397) in 0.001 *M* NaOH.

7. 1000X epinephrine (0.6 mg/mL): Dissolve epinephrine (Sigma-Aldrich, cat. no. E4642) in 0.01 *N* HCl.

8. 1000X epidermal growth factor for BEGM, 50,000X for ALI (25 µg/mL): Dissolve human recombinant, culture-grade EGF (Atlanta Biological, Norcross, GA, cat. no. C100) in PBS.

9. Retinoic acid (concentrated stock = 1×10^{-3} M in absolute ethanol, 1000X stock = 5×10^{-5} M in PBS with 1% BSA): Retinoic acid (RA) is soluble in ethanol and is light sensitive. First, make a concentrated ethanol stock by dissolving 12.0 mg of RA (Sigma-Aldrich, cat. no. R2625) in 40 mL of 100% ethanol. Store in foil wrapped tubes at –20°C. To prepare the 1000X stock, first confirm the RA concentration of the ethanol stock by diluting it 1:100 in absolute ethanol. Read the absorbance at 350 nm using a spectrophotometer and a 1 cm light path quartz cuvet, blanked on 100% ethanol. The molar extinction coefficient of RA in ethanol equals 45,000 at 350 nm. Thus, the absorbance of the diluted stock should equal 0.45. RA with absorbance readings below 0.18 should be discarded. If the absorbance equals 0.45, add 3 mL of 1×10^{-3} M ethanol stock solution to 53 mL PBS and add 4.0 mL of BSA 150 mg/mL stock (*see* **Subheading 2.3., item 1**). For absorbance values less than 0.45, calculate the needed volume of ethanol stock as 1.35/absorbance and adjust the PBS volume appropriately.

10. 1000X phosphorylethanolamine (70 mg/mL): Dissolve phosphorylethanolamine (Sigma-Aldrich, cat. no. P0503) in PBS.

11. 1000X ethanolamine (30 µL/mL): Dilute ethanolamine (Sigma-Aldrich, cat. no. E0135) in PBS.

12. 1000X Stock 11 (0.863 mg/mL): Dissolve zinc sulfate (Sigma-Aldrich, cat. no. Z0251) in dH_2O. Store at room temperature.

13. 1000X Penicillin–streptomycin (100,000 U/mL and 100 mg/mL): Dissolve penicillin-G sodium (Sigma-Aldrich, cat. no. P3032) and streptomycin sulfate (Sigma-Aldrich, cat. no. S9137) in dH_2O for a final concentration of (100,000 U/mL and 100 mg/mL, respectively).

14. 1000X gentamicin (50 mg/mL): Sigma-Aldrich, cat. no. G1397. Store at 4°C. Used for BEGM only.

15. 1000X amphotericin B (250 µg/mL): Sigma-Aldrich, cat. no. A2942. Used for BEGM only.

16. 1000X Stock 4: Combine 0.42 g ferrous sulfate (Sigma-Aldrich, cat. no. F8048), 122.0 g magnesium chloride (J.T. Baker, Phillipsburg, NJ, cat. no. 2444), 16.17 g calcium chloride-dihydrate (Sigma-Aldrich, cat. no. C3881), and 5.0 mL hydrochloric acid (HCl) to 800 mL of dH_2O in a volumetric flask. Stir and bring total volume up to 1 L. Store at room temperature.

17. 1000X trace elements: Prepare seven separate 100 mL stock solutions (*see* **Table 3**). Using a volumetric 1-L flask, fill to the 1-L mark with dH_2O. Remove 8 mL of dH_2O. Add 1.0 mL of each stock solution and 1.0 mL of HCl (conc.). Store at room temperature.

2.4. Making LHC Basal Medium, BEGM, and ALI Medium

The overall approach to making media depends on the culture scale of the individual laboratory. For example, purchase of pre-made base media and addi-

Table 3
Stock Solutions for Trace Elements

Component	Sigma-Aldrich Cat. no.	Amount/ 100 mL	Molarity
Selenium (NaSeO$_3$) highly toxic	S5261	520 mg	30.0 mM
Manganese (MnCl$_2$ • 4H$_2$O) harmful	M5005	20.0 mg	1.0 mM
Silicone (Na$_2$SiO$_3$ • 9H$_2$O) corrosive	S5904	14.2 g	500 mM
Molybdenum [(NH$_4$)$_6$Mo$_7$O$_{24}$ 4H$_2$O]	M1019	124.0 mg	1.0 mM
Vanadium (NH$_4$VO$_3$) highly toxic	A1183	59.0 mg	5.0 mM
Nickle (NiSO$_4$ • 6H$_2$O) toxic	N4882	26.0 mg	1.0 mM
Tin (SnCl$_2$ • 2H$_2$O) corrosive	S9262	11.0 mg	500 µM

tives may represent a logical choice for small-scale efforts. However, laboratories making large quantities of media may choose to make base media and additive stocks in house. Small batches, i.e., 500 mL or 1 L of BEGM or ALI medium, are easily assembled within the reservoir of a bottle top filter, whereas 6-L quantities are made in a volumetric flask and are sterilized by pumping through a cartridge filter. Both scales of media preparation are illustrated below.

1. *LHC Basal Medium:* For small-scale production of BEGM or ALI, it is recommended to purchase the premade LHC basal medium (Biosource, Camarillo, CA, cat. no. P118-500). For large-scale production, LHC basal medium powder can be specially ordered from Sigma-Aldrich (*see* **Note 2**). In a 5-L volumetric flask, dissolve the 5 L prepackaged mixture in 4 L of dH$_2$O. Add 5 g NaHCO$_3$, 150 mL of 200 mM L-glutamine (Sigma-Aldrich, cat. no. G7513), stir, and adjust pH to 7.2–7.4. Bring total volume up to 5 L. Filter into sterile 500 mL bottles using 0.2-µm Vacucap (VWR, West Chester, PA, cat. no. 28143-315). Store at 4°C.
2. *BEGM Medium:* BEGM medium is prepared using 100% LHC basal medium. For small-scale production, thawed additives are dispensed into media in the top of a bottle top filter unit. Note that some additives are not 1000X stock solutions. For media made with homemade BPE that is difficult to filter, use a 0.4-µm filter unit. For commercial BPE, a 0.2-µm filter is acceptable. To add homemade BPE to media, thawed BPE aliquots are first centrifuged at 1500*g* for 10 min to remove debris and cryoprecipitate, prefiltered through a 0.8-µm syringe filter, and added to the media just as the last few milliliters of media are being filter-sterilized. Large-scale media production requires a peristaltic pump system, such as a Mas-

terflex pump (Cole-Parmer Instruments, Vernon Hills, IL, cat. no. EW77910-20). Additives are dispensed into base media in a large flask. Masterflex tubing (Cole-Parmer, cat. no. 96400-17) is rinsed with 70% ETOH followed by dH_2O; and appropriate connections are made to filter-sterilize the media through a Gelman 0.45-μm filter cartridge (Fisher Scientific, cat. no. 28-146-179) into sterile 500-mL bottles. Store media at 4°C.

3. *ALI Medium:* ALI medium uses a 50:50 mixture of DMEM-H (Gibco, Carlsbad, CA, cat. no. 11995-065) and LHC basal medium as its base. Additives are thawed and dispensed into base media at the proper concentrations. ALI medium is filtered according to small- or large-scale production methods given in **Subheading 2.4.2.** Note that some additives are not 1000X stock solutions and that base ALI medium omits gentamicin and amphotericin. To prepare low endotoxin medium, use LHC basal medium (Biosource) DMEM (Fisher Scientific, cat. no. BW12-604F), BPE (Sigma-Aldrich), and a low endotoxin grade of BSA (Sigma-Aldrich, cat. no. A2058).

2.5. Antibiotics

It is assumed that many primary human tissues contain yeasts, fungi, or bacteria. Media for primary cultures can be supplemented with gentamicin (50 μg/mL) and amphotericin (0.25 μg/mL). In our experience, fewer episodes of contamination will result from using amphotericin (Sigma-Aldrich, 1.25 μg/mL), ceftazidime (Fortaz®, GlaxoSmithKline, RTP, NC, 100 μg/mL), tobramycin (Nebcin®, Eli Lilly & Co., Indianapolis, IN, 80 μg/mL), and vancomycin (Vancocin®, Eli Lilly & Co., 100 μg/mL). When processing tissues that are chronically infected from CF patients, additional antibiotics are used for at least the first 3 d of culture. Supplemental antibiotics are chosen based on microbiology reports as described in a prior publication *(18)*. In the event of repeated fungus or yeast contamination, nystatin (Sigma-Aldrich, final concentration of 100 U/mL, cat. no. N1638,) and Diflucan® (fluconazole for injection, Pfizer, NY, final concentration of 25 μg/mL) can be added for the first 3 d of primary cell culture. When antibiotics are obtained from the hospital pharmacy instead of suppliers of tissue-culture reagents, sterile liquids for injection may be added directly to media, whereas powders are weighed, dissolved in medium, and filter-sterilized. Antibiotics received from the pharmacy as powders contain a given amount of antibiotic and unknown quantities of salts and buffers. The purity of the antibiotic is determined by comparing the weight of the powder in the vial to the designated antibiotic content listed by the manufacturer. Once reconstituted, antibiotics from powders are stored at 4°C, and used within 1 d *(18)*.

2.6. Cell-Culture Medias, Reagents, and Solutions

All solutions are filter-sterilized and stored at –20°C unless otherwise noted.

1. F-12 nutrient mixture (Ham) powder with 1 mM L-glutamine: To make 5 L of Ham's F-12 from powder (Gibco, cat. no. 21700075) add 4 L of dH$_2$O to a volumetric flask. Add 5 × 1 L packs to flask and supplement with 50 mL of 1.5 M HEPES (Sigma-Aldrich, cat. no. H3375), 100 mL of 0.714 M NaHCO$_3$, 4.0 mL gentamicin (Sigma-Aldrich, cat. no. G1397), and 5 mL of 1000X pen/strep (*see* **Subheading 2.3., item 13**). Adjust pH to 7.2. Bring total volume up to 5 L and store at 4°C.
2. Cell freezing solution: Combine 2 mL of 1.5 M HEPES, 10 mL of fetal bovine serum (Gibco, cat. no. 200-6140AJ), and 78 mL Ham's F-12. Gradually add 10 mL DMSO (Sigma-Aldrich, cat. no. D2650).
3. 1% Protease XIV with 0.01% DNase (10X stock): Dissolve Protease XIV (Sigma-Aldrich, cat. no. P5147) and DNase (Sigma-Aldrich, cat. no. DN-25) in desired volume of PBS and stir. A 1:9 dilution in minimum essential medium (MEM) (*see* **Subheading 2.6., item 6**) is used for cell dissociation.
4. Soybean trypsin inhibitor (1 mg/mL): Dissolve soybean trypsin inhibitor (Sigma-Aldrich, cat. no. T9128) in Ham's F-12, store at 4°C.
5. 0.1% trypsin with 1 mM ethylene diamine tetraacetic acid (EDTA) in PBS: Dissolve Trypsin Type III powder (Sigma-Aldrich, cat. no. T4799) in PBS. Add EDTA from concentrated stock for a final concentration of 0.1% trypsin with 1 mM EDTA. pH solution to 7.2–7.4.
6. MEM: Supplement 500 mL Joklik's MEM (Sigma-Aldrich, cat. no. M8028) with 5 mL L-glutamine (Sigma-Aldrich), 0.40 mL gentamicin (Sigma-Aldrich), and 0.50 mL 1000X pen/strep (*see* **Subheading 2.3., item 13**). Store at 4°C.

3. Methods

3.1. Overview

AECs can be obtained from nasal or lung tissue specimens and can be seeded directly onto porous supports for primary ALI cultures or can be first grown on plastic for subculturing passage 1 or passage 2 cells to porous supports. An overview of the process is given in **Fig. 1**.

3.2. Type I and III Collagen Coating of Plastic Dishes

Primary and thawed cryopreserved cells are plated onto collagen-coated plastic dishes, whereas cells passaged without freezing do not require coated dishes. Add 2.5 mL of a 1:75 dilution of Vitrogen 100 (Cohesion, Palo Alto, CA) in dH$_2$O per 100-mm dish. Incubate for 2 h at 37°C. Aspirate remaining liquid and expose open dishes to UV in a laminar flow hood for 30 min. Plates can be stored for up to 6 wk at 4°C.

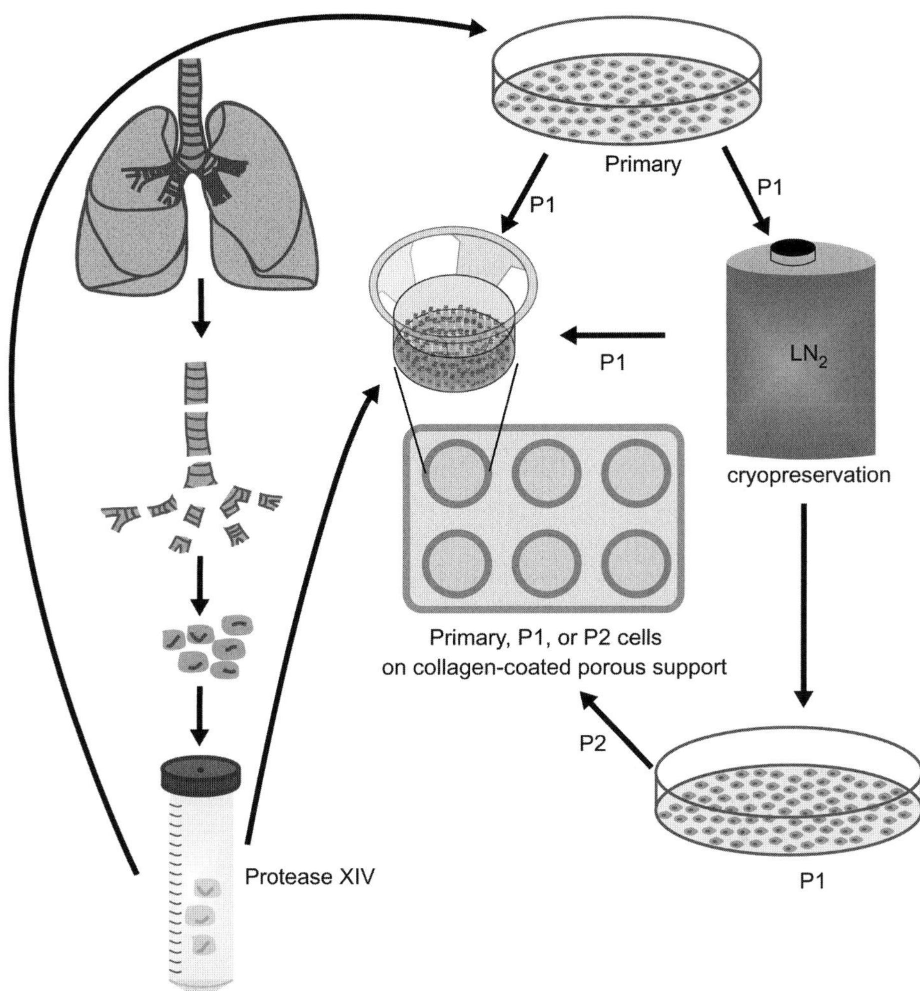

Fig. 1. Overview of the human AECC process. Primary cells may be plated directly on porous supports or on plastic dishes for subsequent cryopreservation and subpassage.

3.3. Type IV Collagen Coating of Porous Supports

A variety of porous supports are suitable for AECCs. Transwell®-COL PTFE membrane inserts, 12- or 24-mm diameter (Corning, Inc., Acton, MA, cat. nos. 3493 and 3491, respectively) are provided collagen precoated by the manufacturer. Transwell®-Clear (Corning, Inc., cat. nos. 3460 and 3450), Snapwell™ (Corning, Inc., cat. no. 3801), and Millicell-CM membranes (Millipore, Billerica, MA, cat. nos. PICM01250 and 03050) must be coated with type IV collagen for successful long-term cultures. For unknown reasons, 0.4 μM, rather than the 3.0 μM, pore-size membranes consistently make superior cultures. To coat, first resuspend 10 mg collagen type IV (Sigma-Aldrich, cat. no. C7521) in 20 mL dH_2O and add 40 μL of concentrated acetic acid. Incubate for 15–30 min at 37°C until fully dissolved. Syringe filter (0.2 μm) and store aliquots at –20°C. Thaw frozen stock and dilute 1:10 with dH_2O. Add 150 μL per 12-mm Transwell Insert, Costar Snapwell Insert, or 12-mm Millicell-CM membrane or 400 μL per 24-mm Transwell Insert, Costar Snapwell Insert, or 30-mm Millicell-CM membrane. Allow to dry at room temperature in a laminar flowhood overnight. Expose to UV in a laminar flowhood for 30 min.

3.4. Isolating Primary AECs

Primary AECs originate from nasal turbinates, from nasal polyps, and from normal and diseased lungs. When handling human tissues, always follow standard safety precautions to prevent exposure to potential bloodborne pathogens, including gloves, lab coat, and eye protection. Tissue is transported to the laboratory in sterile containers containing sterile chilled lactated Ringer's (LR) solution (Abbot Laboratory, North Chicago, IL, cat. no. 7953), MEM, or another physiologic solution. Nasal tissue samples are usually processed without further dissection but whole lungs require significant dissection as described below.

1. Assemble the following on a clean countertop or in a laminar flowhood.
    a. Absorbent bench covering (3M Health Care, St. Paul, MN, cat. no. 1072).
    b. Large plastic sterile drape (3M Health Care, cat. no. 1010).
    c. Ice bucket containing sterile specimen cups (Tyco Health Care Group LP, Mansfield, MA, cat. no. V2200) filled with LR solution.
    d. Use instrument sterilizer (Fine Science Tools, Foster City, CA, cat. no. 18000-45) or preautoclaved instruments. Suggested tools include curve-tipped scissors, delicate 4.5″ (Fisher Scientific, cat. no. 08-951-10); heavy scissors, straight, sharp, 11.5 cm (Fine Science Tools, cat. no. 14058-11); forceps, blunt-pointed, straight, 15 cm (Fine Science Tools, cat. no. 11008-15); rat-tooth forceps 1 × 2, 15.5 cm (Fine Science Tools, cat. no. 11021-15); scalpels, #10 (Bard-Parker™, Becton Dickinson and Co., Hancock, NY, cat. no. 371610); sterile covered sponges, 4″ × 4″ (Tyco Health Care Group LP, cat. no. 2913).
2. Dissect airways by removing all excess connective tissue and cutting into 5–10 cm segments. Clean tissue segments, removing any additional connective tissue and lymph nodes and rinsing in LR solution. Slit segments longitudinally and cut into 1 × 2 cm portions with a scalpel. Transfer to specimen cup containing chilled LR solution. *See* **Fig. 2A–D** for dissection process.
3. Because human tissue samples are likely to contain yeasts, bacteria, or fungi, we begin antibiotic exposure as soon as possible. Prepare 250 mL of J-MEM plus desired antibiotics (*see* **Subheading 2.5.**), named Wash Media. Aspirate LR solu-

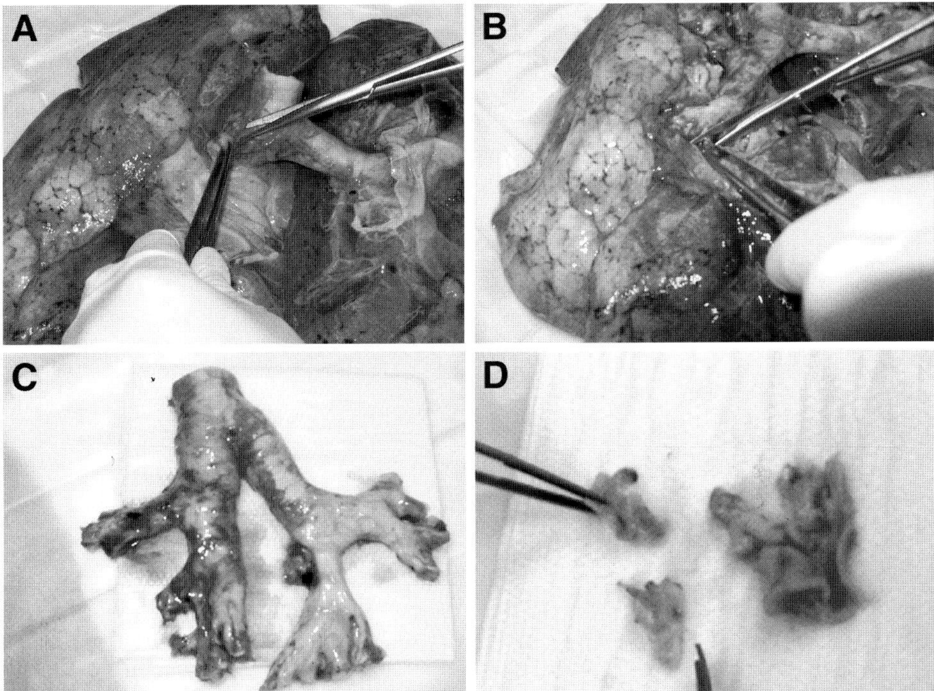

Fig. 2. Dissection process for human lungs. Adherent structures are removed from proximal airways (**A**) and distal airways are freed from lung tissue (**B**). The airway tree is excised (**C**), cut into portions and adherent structures are removed. Airways are slit open and cut into segments (**D**). Segments are placed in specimen cups (not shown) and processed as described in the text.

tion from tissue and add Wash Media, swirl, and replace wash media three times. Transfer washed tissue segments into 50-mL conical tubes containing 30 mL Wash Media plus 4 mL Protease/DNase solution. (Approximate tissue to fluid ratio of 1:10, final volume = 40 mL.) Place tubes on a rocking platform in a cold room at 4°C, selecting 50–60 cycles/min.

4. Tissues from chronically infected patients or any specimens containing abundant secretions are treated to remove mucus and other debris. The tissues are soaked in a solution containing supplemental antibiotics (*see* **Subheading 2.5.**), dithiothreitol (DTT) (Sigma-Aldrich, cat. no. D0632), and DNase (Sigma-Aldrich, cat. no. DN-25). To prepare this Soak Solution, add 65 mg DTT and 1.25 mg DNase to 125 mL of Wash Media and filter sterilize. The final concentration of DTT and DNase are 0.5 mg/mL and 10 µg/mL, respectively. Aspirate LR solution from tissue and add 60 mL Soak Solution, swirl, and soak 5 min. Repeat Soak Solution step. Next, rinse tissue three times in Wash Media to remove DTT/DNase. Transfer tissue to 50-mL tubes containing 30 mL Wash Media plus 4 mL protease/

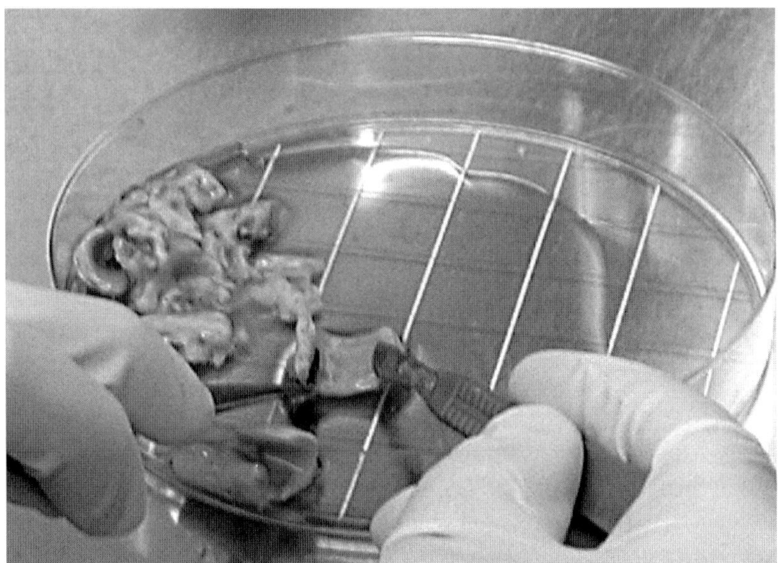

Fig. 3. Epithelial cell removal after protease dissociation. The concave, luminal airway surface is gently scraped with a convex scalpel blade to remove adherent cells.

DNase (final volume = 40 mL) and place tissue on platform rocker at 4°C, 50–60 cycles/min for 48 h.
5. Nasal turbinates, polyps, and small bronchial specimens undergo the same procedure, except that these tissues can be dissociated in 24 h in 15-mL tubes containing 8 mL Wash Media plus 1 mL protease solution.

3.5. Harvesting Cells

Follow standard sterile tissue-culture techniques under a laminar flow hood.

1. End dissociation of tissue by pouring contents of 50-mL tubes into a 150-mm tissue-culture dish; add FBS (Gibco) to a final concentration of 10% (v/v) to neutralize protease.
2. Scrape epithelial surface with a convex surgical scalpel blade #10 as illustrated in **Fig. 3**. Rinse tissue surfaces and collection dish with PBS and pool solutions containing dissociated cells into 50-mL conical tubes.
3. Centrifuge at 500g for 5 min at 4°C. Wash cells once in media, resuspend in a volume calculated to be approx 5 × 10^6 cells/mL, and count using a hemocytometer.

3.6. Plating Cells

Primary AECs may be cultured directly on porous supports in ALI medium at a density of 0.1–0.25 × 10^6-cells per cm^2, which is equivalent to 0.8–2.0 × 10^5 cells per 12 mm support or 0.7–1.75 × 10^6 cells per 24–30 mm support (*see* **Note 3**). Alternatively, to generate P1 or P2 cells for subculture to porous supports, primary cells can be plated in BEGM on collagen-coated plastic dishes at a density of $2–6 \times 10^6$ per 100-mm dish. Primary cell media should be supplemented with additional antibiotics (*see* **Subheading 2.5.**) for the first 3 d after plating, and should be changed every 2–3 d or as needed to prevent acidification (*see* **Subheading 3.7.3.**).

3.7. Cell Culture Maintenance

3.7.1. Primary Cells on Plastic

Assess attachment to plastic dishes 24 h after plating primary cells. If the cells attached well and the dish contains few clumps of floating epithelial cells, wash the cells with PBS and feed with BEGM plus antibiotics (*see* **Subheading 2.5.**). Large floating clumps of cells can be "rescued" to increase cell yield. Harvest the media in 50-mL conical tubes. Gently wash dishes with PBS and add to harvested clumps. Pellet cells at 500*g* for 5 min. Aspirate the supernatant and add 10–15 mL of freshly prepared declumping solution containing 2 m*M* EDTA, 0.5 mg/mL DTT, 0.25 mg/mL collagenase (Sigma-Aldrich, cat. no. C6885), and 10 µg/mL DNase in PBS. Incubate 15 min to 1 h at 37°C while visually monitoring clump dissociation. Add FBS to 10% (v/v), centrifuge at 500*g* for 5 min, remove supernatant, and resuspend pellet in BEGM for counting. Plate at a density of $2–6 \times 10^6$ per 100-mm collagen coated dish. Medium is changed every 2–3 d.

3.7.2. Passaging Primary Cells on Plastic

When primary cultures reach 70–90% confluence, they are ready for passage. We believe it is important to harvest hard-to-detach cells while minimizing trypsin exposure of cells that release quickly. Thus, we use a double-trypsinization process. Rinse cells with PBS, add 2 mL of trypsin/EDTA per 100-mm dish and incubate 5–10 min at 37°C. Gently tap dish to detach cells. Rinse cells with PBS and harvest into 50-mL conical tube containing 20 mL STI solution on ice. Add another 2 mL of trypsin/EDTA to dishes and repeat, visually monitoring detachment. Pool harvested cells and centrifuge at 500*g* for 5 min. Aspirate supernatant and resuspend cells in desired volume of media for counting.

3.7.3. Media Change in ALI Cultures

Primary, passage 1, or passage 2 AECs may be grown on collagen-coated porous supports. Remove media on the top with a Pasteur pipet attached to a vacuum, and rinse the apical surface with PBS. Prior to confluence, replace media in the apical and basolateral compartment but after confluence do not add media to the apical compartment. The volume of media added to the apical and basolateral chambers depends on the specific porous support. Transwell insert and Costar Snapwell inserts hang in 12- or 6-well plates. Millicell CM membranes stand on legs and can be kept in a variety of dishes. During periods of rapid cell growth, cells on Transwell insert in the standard configuration will yellow the media rapidly and will require daily media changes. We have devised Teflon adapters to enable 12-mm Transwell inserts to be kept in six-well plates with a larger, 2.5-mL, basolateral reservoir, which decreases the media change frequency; 24-mm Transwell Insert may be kept in Deep Well Plates (Collaborative Biomedical Products, Bedford, MA, cat. no. 01-05467) with 12.5 mL media. Typically 6×12 mm or 2×30 mm Millicell CM inserts are kept in 10 mL of media in a 100-mm dish.

3.8. Cryopreservation of Cells

1. Primary AECs are trypsinized from plastic dishes (now P1 cells) and cryopreserved for long-term storage in liquid nitrogen. Cells are resuspended in Ham's F-12 media at a concentration of $2–6 \times 10^6$ cells/mL.
2. Keep cells on ice and slowly add an equal amount of freezing media (*see* **Subheading 2.6., item 2**) to the cell suspension.
3. Place cryovials in Nalgene Cryo Freezing container (VWR, cat. no. 5100-0001) and place in –80°C freezer for 4–24 h. An insulated box can be used as a substitute.
4. Transfer vial(s) from the –80°C freezer to liquid N_2 (–196°C) for long-term storage.

3.9. Thawing Cells

1. Warm Ham's F-12 and plating media to 37°C. Note: Warm media must be added gradually so that the DMSO concentration gradient is not too steep.
2. Thaw the cryovial in a beaker of 37°C water. Remove cryovial and wipe outside with 70% ethanol. Transfer cells to a 15-mL conical tube.
3. Dilute the cell suspension by slowly filling the tube with warm Ham's F-12. Centrifuge at $600g$ for 5 min at 4°C.
4. Gently resuspend cells in the appropriate plating media, count cells, and assess viability.

3.10. Histological Methods

Histologic assessment of ALI cultures is a useful experimental tool but the thin, pliable membrane poses unique challenges. To maintain the integrity of cells grown on membranes for morphologic analysis, cultures are generally processed without removing the membrane from the support, and then re-embedded to enable production of cross sections. This applies to cultures processed for frozen, paraffin, or plastic sectioning, and for transmission electron microscopy. Representative examples of frozen, paraffin, and plastic sections are shown in **Fig. 4A–C.**

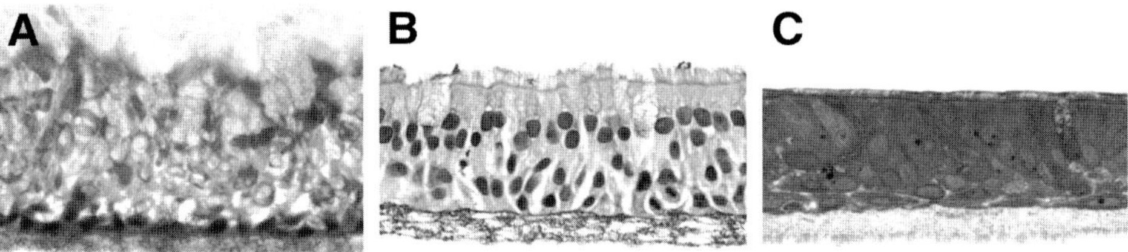

Fig. 4. Histologic assessment of AECCs. Frozen **(A)**, paraffin **(B)**, and plastic **(C)** sections of well-differentiated cultures. Original magnification = 500X, A and B = H&E stain, C = Richardsons stain.

3.10.1. Frozen Sections

To obtain frozen sections of cells grown on membranes, media is removed and, if desired, the cells are rinsed with PBS. Excess fluid is blotted and the culture is sandwiched between two layers of embedding media (Fisher Scientific, cat. no. NC9418069), using a weigh boat to support the bottom layer. The sandwiched membrane is then frozen and removed from the support with a scalpel and cut into slices within a chilled cryostat chamber. The slices are placed on edge in a disposable embedding mold that is then filled with embedding media and frozen to create a tissue block for sectioning on a cryostat *(19)*. When performed carefully, the tissue remains frozen throughout the double-embedding process and cross sections of the epithelium are produced when the final block is sectioned.

3.10.2. Paraffin and Plastic Sections

Paraffin sections of cells grown on membranes are obtained by fixing, dehydrating to 100% ethanol, and clearing in Slide Brite (Sasco, Albany, GA) before infiltration with paraffin embedding media *(20)*. For plastic embedding, the sample is transferred from 100% ethanol to 50:50 solutions of plastic embedding media before infiltration with 100% plastic. For both paraffin and plastic, the membrane is sandwiched between two layers of embedding media, then hardened as usual and cut from the support and into slices. The slices are placed on edge in an embedding mold, covered with embedding media, and again hardened as usual to create a tissue block resulting in cross sections.

3.10.3. Transmission Electron Microscopy Sections

Cultures for transmission electron microscopy are treated similarly as those processed for paraffin sections except using glutaraldehyde fixation and osmium tetroxide post fixation. The cultures are dehydrated into 100% ethanol and then infiltrated with graded mixtures of resin and ethanol and finally pure resin *(21)*, avoiding propylene oxide which will dissolve the membrane support. A flat wafer of the culture is then polymerized, cut into slices, and re-embedded.

3.10.4. Scanning Electron Microscopy

Samples are processed for scanning electron microscopy by fixing in glutaraldehyde and postfixing in osmium tetroxide followed by dehydration to 100% ethanol. While still in the support, the culture is critical point dried and mounted using a carbon conductive tab (Ted Pella, Redding, CA, cat. no. 16084-1). The membrane is removed from the support with a scalpel and is coated with gold for viewing in a scanning electron microscope *(22)*.

3.11. Electrophysiologic Assessment of AECCs

Cystic fibrosis is the most common fatal genetic disorder of the Caucasian population *(23)*. The cloning of the CF gene *(CFTR)* marked a new era in our understanding of the pathophysiology of CF *(24–26)*. Heterologous expression and bilayer reconstitution studies showed *CFTR* to be a cAMP-regulated Cl^- channel *(27,28)*. Mutations in *CFTR* also result in defective regulation of the epithelial Na^+ channel, ENaC *(29)*, and alter the function of an epithelial Ca^+-activated Cl^- channel, CaCC *(30,31)*. Thus, the CF epithelium is characterized by the absence of cAMP-mediated Cl^- conductance, and hyperactiviation of ENaC and CaCC. Human AECCs, mounted in Ussing chambers, have been used to characterize ion transport properties of CF and normal tissues and for testing of potential pharmacologic or genetic therapies.

For study in Ussing chambers, CF and normal AECCs are plated onto Costar Snapwell (Corning, Inc., cat. no. 3801) tissue-culture inserts precoated with collagen type IV (*see* **Subheading 3.2.**). Cells are visually evaluated for confluence, development of cilia, and maintenance of an ALI. Transepithelial resistance (R_T) and potential difference (PD) are measured using an EVOM device (World Precision Instruments, Sarasota, FL), as per manufacturer's instructions. Monolayers generating at least a 1-mV PD and a 150 Ω • cm^2 R_T are used for Ussing chamber studies, which typically occurs 10–14 d after plating. Transepithelial voltage (V_T), R_T, and short-circuit current (I_{SC}), are measured using Ussing chambers specifically designed for Snapwell inserts (Physiologic Instruments, La Jolla, CA, cat. no. P2300). Cells are typically bathed in Krebs Bicarbonate Ringer's solution (KBR) on the basolateral side and a modified KBR, high K^+, low Cl^- solution (HKLC), on the apical side, to enable focusing on apical Cl^- secretion *(30–32)*. All bathing solutions are bubbled with 95% O_2, 5% CO_2 and maintained at 37°C. Voltage is clamped to zero, and pulsed to 10 mV for 0.5-s duration every minute. Electrometer output is digitized online

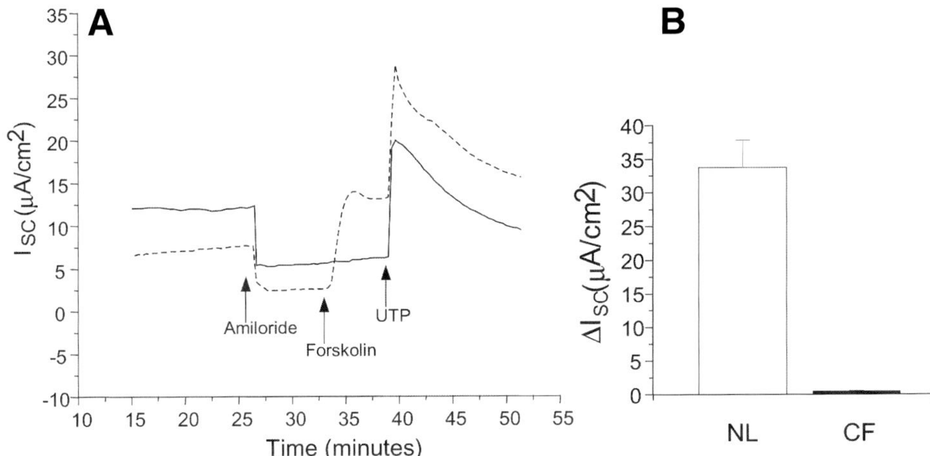

Fig. 5. Electrophysiologic assessment of AECCs. Representative I_{SC} traces of normal (dashed) and CF (solid) human airway epithelial cultures studied in Ussing chambers as described in the text. Amiloride (10^{-4} M), forskolin (10^{-5} M) and uridine triphosphate (UTP) (10^{-4} M) are added to determine the presence of ENaC, CFTR, and CaCC, respectively (**A**). Summary I_{SC} values from normal (open bars, $n = 18$) and CF (filled bar, $n = 19$) human bronchial epithelial cultures in response to bilateral addition of forskolin (10^{-5} M) (**B**).

and I_{SC}, R_T, and calculated V_T are displayed on a video monitor and stored on a computer hard drive. Drugs are added from concentrated stock solutions to either lumenal and/or serosal sides of the tissue.

Representative tracings of both normal and CF airway epithelial cultures are shown in **Fig. 5A**. The most reliable and reproducible difference between CF and normal epithelial cultures is the absence of a cAMP-mediated I_{SC} response in CF cultures (*see* **Fig. 5B**). Other ion transport properties used to distinguish CF from normal cultures include greater percentage inhibition of the basal current by amiloride (10^{-4} M) in CF cultures, and an elevated response of the calcium-activated Cl$^-$ conductance as measured by the I_{SC} response to purinergic receptor activation by uridine triphosphate.

4. Notes

1. To protect the safety of research personnel, we do not accept specimens from individuals with known infection with human immunodeficiency virus, hepatitis B, hepatitis C, or tuberculosis. Samples from individuals on immunosuppressive therapy, especially long-term, may pose increased risk. All human tissue samples must be treated as potentially biohazardous and handled using standard precau-

tions. It is a research team decision, related to the scientific goals, whether to accept specimens from individuals with an extensive smoking history. AECs can be procured successfully from lungs with acute lung injury or pneumonia. The latter can usually be cultured successfully by selecting appropriate antibiotics (*see* **Subheading 2.5.**). A range of clinical data (laboratory values including blood gases, X-ray, or bronchoscopy findings) can be used to help determine lung acceptability. Owing to the lack of systematic studies, there are no hard and fast rules guiding the relationship between physiologic function and successful cell culture, but an arterial P_{O2} of greater than 150 mmHg on 100% inspired oxygen is a reasonable lower limit.

2. If anticipated usage of LHC basal medium exceeds 550 L per year, powdered stock can be custom ordered from Sigma-Aldrich. The composition is given in **Table 4**.

3. Seeding densities. Primary and passaged primary human airway epithelial cells are mortal and their growth characteristics depend on a sufficient seeding density. Furthermore, attachment and growth of cells from different individuals and preparations may vary. Thus, generous seeding densities of primary cells on porous supports are required to consistently obtain confluent cultures that differentiate and maintain a long-lasting, patent ALI. Although it is tempting to expand primary cells on plastic to geometrically increase cell number, growth capacity of mortal cells is finite, and "overexpansion" should be avoided. The seeding guidelines herein will generally enable successful differentiated cultures persisting for at least 45 d. Primary cells to be first grown on plastic dishes should be seeded at no less than 1×10^6, and preferably 2–6×10^6, cells per 100-mm collagen coated dish. Seeding densities for smaller or larger dishes should be calculated mathematically based on surface area. Under these conditions the cells should grow to >70% confluence within 7–10 d. If longer periods are required to reach >70% confluence, subsequent growth may be impaired. Cells at >70% confluence, but not >95% confluence, should be trypsinized for cryopreservation or subpassaged to a porous support. Alternatively, the cells can be can be expanded one more round at a seeding density of 1×10^6 cells per uncoated 100-mm tissue-culture dish for expansion to passage 2. Seeding densities for primary, passage 1, and passage 2 cells on porous supports should be in the range of 1.5×10^5 cells/cm². Thus, 12-mm Millicell CM or 12-mm Transwell membranes are typically seeded with approx 125,000 cells each, whereas 30-mm Millicell CM or 24-mm Transwell membranes are seeded with approx 1×10^6 cells. This seeding density will result in confluence, or near confluence, within 1–3 d after seeding, at which point an ALI should be established. Lower seeding densities may be fully successful with some specimens, which can be determined empirically using aliquots of passage 1 or 2 cells. Unfortunately, prescreening is not possible when plating primary cells, and greater variability is anticipated between different patient preparations.

Acknowledgment

The authors wish to acknowledge Lisa Brown for editing and graphic design.

Table 4
LHC Basal Media Formula

Component	Sigma-Aldrich Cat. no.	Formula g/L	g/550L Batch
L-Arginine HCl	A6969	0.421	231.55
L-Alanine	A7469	0.009	4.95
L-Asparagine•H_2O	A7094	0.015	8.25
L-Aspartic acid	A7219	0.004	2.2
L-Cysteine HCl monohydrate	C6852	0.042	23.1
L-Glutamic acid	G8415	0.0148	8.14
Glycine free base	G8790	0.0076	4.18
L-Histidine HCl	H5659	0.0336	18.48
L-Isoleucine	I7403	0.004	2.2
L-Lysine monohydrochloride	L8662	0.0366	20.13
L-Leucine	L8912	0.1312	72.16
L-Methionine	M5308	0.009	4.95
L-Proline	P5607	0.0346	19.03
L-Phenylalanine	P5482	0.01	5.5
L-Serine	S4311	0.1262	69.41
L-Tryptophan	T8941	0.0062	3.41
L-Tyrosine disodium	T2269	0.00756	4.158
L-Threonine	T8441	0.0238	13.09
L-Valine	V0513	0.0702	38.61
Adenine HCl	A8751	0.03088	16.984
D-Biotin	B4501	0.00002	0.011
Choline chloride	C1879	0.028	15.4
Folic acid	F7876	0.00079	0.4345
D-(+)-glucose	G8270	0.5405	297.275
Myoinositol	I5125	0.018	9.9
Niacinamide	N3376	0.00004	0.022
D-Pantothenic acid hemicalcium	P2250	0.00026	0.143
Putrescine•$_2$HCl	P7505	0.00016	0.088
Pyridoxine hydrochloride	P9755	0.00006	0.033
Riboflavin	R4500	0.00004	0.022
Thiamine hydrochloride	T4625	0.00034	0.187
Thymidine	T9441	0.00073	0.4015
Vitamin B_{12}	V2876	0.00041	0.2255
Cupric sulfate pentahydrate	C7631	0.000002	0.0011
Potassium chloride	P4504	0.112	61.6
Sodium phosphate dibasic	S0876	0.284088	156.2484
Sodium chloride	S9625	6.084	3346.2
D-(+)-glucose	G8270	0.27025	148.6375

(continued)

Table 4 *(continued)*

Component	Sigma-Aldrich Cat. no.	Formula g/L	g/550L Batch
Pyruvic acid sodium	P2256	0.055	30.25
Phenol red sodium	P4758	0.00124	0.682
Sodium acetate anhydrous	S8750	0.301	165.55
D-(+)-glucose	G8270	0.27025	148.6375
Dl-6,8-Thioctic acid	T5625	0.00021	0.1155
HEPES free acid	H3375	5.4	2970

References

1. Lechner, J. F., Haugen, A., McLendon, I. A., and Pettis, E. W. (1982) Clonal growth of normal adult human bronchial epithelial cells in a serum-free medium. *In Vitro* **18,** 633–642.
2. Gruenert, D. C., Finkbeiner, W. E., and Widdicombe, J. H. (1995) Culture and transformation of human airway epithelial cells. *Am. J. Physiol.* **268,** L347–L360.
3. Terzaghi, M., Nettesheim, P., and Williams, M. L. (1978) Repopulation of denuded tracheal grafts with normal, preneoplastic, and neoplastic epithelial cell populations. *Cancer Res.* **38,** 4546–4553.
4. Yankaskas, J. R., Knowles, M. R., Gatzy, J. T., and Boucher, R. C. (1985) Persistence of abnormal chloride ion permeability in cystic fibrosis nasal epithelial cells in heterologous culture. *Lancet* **1,** 954–956.
5. Jorissen, M., Van Der Schueren, B., van den Berghe, H., and Cassiman, J. J. (1989) The preservation and regeneration of cilia on human nasal epithelial cells cultured *in vitro. Arch. Otorhinolaryngol.* **246,** 308–314.
6. Wu, R., Yankaskas, J., Cheng, E., Knowles, M. R., and Boucher, R. (1985) Growth and differentiation of human nasal epithelial cells in culture. Serum-free, hormone-supplemented medium and proteoglycan synthesis. *Am. Rev. Respir. Dis.* **132,** 311–320.
7. Benali, R., Tournier, J. M., Chevillard, M., et al. (1993) Tubule formation by human surface respiratory epithelial cells cultured in a three-dimensional collagen lattice. *Am. J. Physiol.* **264,** L183–L192.
8. Whitcutt, M. J., Adler, K., and Wu, R. (1988) A biphasic chamber system for maintaining polarity of differentiation of cultured respiratory tract epithelial cells. *In Vitro Cell. Dev. Biol.* **24,** 420–428.
9. Matsui, H., Grubb, B. R., Tarran, R., et al. (1998) Evidence for periciliary liquid layer depletion, not abnormal ion composition, in the pathogenesis of cystic fibrosis airways disease. *Cell* **95,** 1005–1015.
10. Matsui, H., Randell, S. H., Peretti, S. W., Davis, C. W., and Boucher, R. C. (1998) Coordinated clearance of periciliary liquid and mucus from airway surfaces. *J. Clin. Invest.* **102,** 1125–1131.

11. Pickles, R. J., McCarty, D., Matsui, H., Hart, P. J., Randell, S. H., and Boucher, R. C. (1998) Limited entry of adenoviral vectors into well differentiated airway epithelium is responsible for inefficient gene transfer. *J. Virol.* **72,** 6014–6023.

12. Zhang, L., Peeples, M. E., Boucher, R. C., Collins, P. L., and Pickles, R. J. (2002) Respiratory syncytial virus infection of human airway epithelial cells is polarized, specific to ciliated cells, and without obvious cytopathology. *J. Virol.* **76,** 5654–5666.

13. Bernacki, S. H., Nelson, A. L., Abdullah, L., et al. (1999) Mucin gene expression during differentiation of human airway epithelia *in vitro*. Muc4 and muc5b are strongly induced. *Am. J. Respir. Cell Mol. Biol.* **20,** 595–604.

14. Zhang, Y. J., O'Neal, W. K., Randell, S. H., et al. (2002) Identification of dynein heavy chain 7 as an inner arm component of human cilia that is synthesized but not assembled in a case of primary ciliary dyskinesia. *J. Biol. Chem.* **277,** 17,906–17,915.

15. Yoon, J. H., Koo, J. S., Norford, D., Guzman, K., Gray, T., and Nettesheim, P. (1999) Lysozyme expression during metaplastic squamous differentiation of retinoic acid-deficient human tracheobronchial epithelial cells. *Am. J. Respir. Cell Mol. Biol.* **20,** 573–581.

16. Gray, T. E., Guzman, K., Davis, C. W., Abdullah, L. H., and Nettesheim, P. (1996) Mucociliary differentiation of serially passaged normal human tracheobronchial epithelial cells. *Am. J. Respir. Cell Mol. Biol.* **14,** 104–112.

17. Lechner, J. F. and LaVeck, M. A. (1985) A serum-free method for culturing normal human bronchial epithelial cells at clonal density. *J. Tiss. Cult. Meth.* **9,** 43–48.

18. Randell, S. H., Walstad, L., Schwab, U. E., Grubb, B. R., and Yankaskas, J. R. (2001) Isolation and culture of airway epithelial cells from chronically infected human lungs. *In Vitro Cell Dev. Biol. Anim.* **37,** 480–489.

19. Bancroft, J. and Stevens, A. (1996) *Theory and Practice of Histological Techniques*, Battle Press, Columbus, OH, pp. 69–80.

20. Sheehan, D. and Hrapchak, B. (1980) *Theory and Practice of Histotechnology*, Battle Press, Columbus, OH, Vol. **2,** 59–66.

21. Hayat, M. A. (1989) *Principles and Techniques of Electron Microscopy* **3,** 79–92.

22. Goldstein, J. (1984) *Scanning Electron Microscopy and X-ray Microanalysis*, pp. 495–540.

23. Boat, T. F., Welsh, M. J., and Beaudet, A. L. (1989) Cystic fibrosis, in *The Metabolic Basis of Inherited Disease* (Scriver, C. R., Beaudet, A. L., Sly, W. S., Valle, D., Stansbury, J. B., Wyngaarden, J. B., and Frederickson, D. S., eds.), 2649–2680.

24. Rommens, J. M., Iannuzzi, M. C., Kerem, B.-T., et al. (1989) Identification of the cystic fibrosis gene: Chromosome walking and jumping. *Science* **245,** 1059–1065.

25. Riordan, J. R., Rommens, J. M., Kerem, B.-T., et al. (1989) Identification of the cystic fibrosis gene: Cloning and characterization of complementary DNA. *Science* **245,** 1066–1073.

26. Kerem, B., Rommens, J. M., Buchanan, J. A., et al. (1989) Identification of the cystic fibrosis gene: Genetic analysis. *Science* **245,** 1073–1080.

27. Anderson, M. P., Rich, D. P., Gregory, R. J., Smith, A. E., and Welsh, M. J. (1991) Generation of cAMP-activated chloride currents by expression of CFTR. *Science* **251,** 679–682.
28. Kartner, N., Hanrahan, J. W., Jensen, T. J., et al. (1991) Expression of the cystic fibrosis gene in non-epithelial invertebrate cells produces a regulated anion conductance. *Cell* **64,** 681–691.
29. Stutts, M. J., Canessa, C. M., Olsen, J. C., et al. (1995) CFTR as a cAMP-dependent regulator of sodium channels. *Science* **269,** 847–850.
30. Gabriel, S. E., Makhlina, M., Martsen, E., Thomas, E. J., Lethem, M. I., and Boucher, R. C. (2000) Permeabilization via the P2X7 purinoceptor reveals the presence of a Ca^{2+}-activated Cl^- conductance in the apical membrane of murine tracheal epithelial cells. *J. Biol. Chem.* **275,** 35,028–35,033.
31. Tarran, R., Loewen, M. E., Paradiso, A. M., et al. (2002) Regulation of murine airway surface liquid volume by CFTR and Ca^{2+}-activated Cl^- conductances. *J. Gen. Physiol.* **120,** 407–418.
32. Donaldson, S. H., Hirsh, A., Li, D. C., et al. (2002) Regulation of the epithelial sodium channel by serine proteases in human airways. *J. Biol. Chem.* **277,** 8338–8345.

14

Isolation and Culture of Human Alveolar Epithelial Cells

Carsten Ehrhardt, Kwang-Jin Kim, and Claus-Michael Lehr

1. Introduction

1.1. Background

The human lung comprises more than 40 different cell types. The morphology and function of constituent cells of the proximal, conducting airway epithelium differ drastically from those of the more distal, alveolar epithelium. This chapter will concentrate on the isolation and culture of human alveolar epithelial cells that line the peripheral gas exchange region of the lung.

Whereas immortalized cell lines emanating mostly from the different cells of the tracheal/bronchial epithelium of human and other animals' lungs are available *(1)*, no cell lines that possess significant functional properties of alveolar epithelial cells (AEC) are reported to date *(2)*. Primary culture of AEC is, therefore, used for most in vitro studies of alveolar epithelial function (e.g., transport and various metabolic pathways). The primary culture of human AEC involves isolation, purification, and culture of alveolar epithelial type II (ATII) cells from human tissue obtained after lung resections. These ATII cells, when plated on permeable supports or plasticware, acquire the type 1 cell-like phenotype and morphology under appropriate culture conditions *(3)*. Owing to the lack of availability of human tissue and some ethical issues pertaining to use of human tissues in certain countries, most studies were based on isolation and culture of cells from the lungs of small laboratory animals including mouse, rat, and rabbit *(4–8)*. However, not much information on species differences in this specific area of cellular research has been systematically studied yet.

From: *Methods in Molecular Medicine, vol. 107: Human Cell Culture Protocols, Second Edition*
Edited by: J. Picot © Humana Press Inc., Totowa, NJ

1.2. AECs

ATII cells constitute about 60% of AECs and about 15% of all lung parenchymal cells, although they cover less than 5% of the alveolar air spaces of adult human lungs *(9)*. It is well known that ATII cells synthesize, secrete, and recycle some components of the surfactant that regulates alveolar surface tension in the distal airspaces of mammalian lungs. ATII cells govern extracellular surfactant transformation by regulating, for example, pH and Ca^{2+} of the hypophase, and play various roles in alveolar fluid balance, coagulation/fibrinolysis, and contribute to host defense. ATII cells proliferate, differentiate into type I (ATI) cells, and remove apoptotic ATII cells by phagocytosis *(10)*, thus, contributing to epithelial repair following lung injury. ATII cells are thought to be progenitor cells for type I cells, especially in injury/repair of epithelial tract lining the distal airspaces of the lung. For a summary of the latest state in type II cell research, the reader is encouraged to read Fehrenbach's excellent review that appeared recently *(10)*.

Compared with its neighbor, the ATII cell, the ATI cell has received little attention. The general functions of ATI cells are relatively unexplored, because specific marker molecules that could be used for definitive identification of ATI cells were identified only very recently. Several ATI cell-specific gene products described below (for a review, *see 11*) were used to address many problems in alveolar cell biology/molecular biology. Nonetheless, the identification of several proteins expressed by this cell and their presumed activities suggest more-sophisticated cell functions than mere gas exchange. The putative functions of type I cells include control of proliferation of peripheral lung cells, metabolism and/or degradation of peptides and peptide growth factors, generation of cyto/chemokines, regulation of alveolar fluid balance, and transcellular ion and water transport.

1.3. Differentiation Markers for AECs

The study of differentiation of type II cells into type I cells crucially depends on the possibility to distinguish both cell types. Beside the pure morphological characterization (e.g., presence of lamellar bodies, cuboidal shape), a number of alternative approaches to distinguish ATII from other cell types have been developed, such as modified Papanicolaou staining, cell-type-specific lectins, and immunohistochemical/immunocytochemical markers *(12)*. The expression of markers, however, may be altered because of the specific culture conditions and the situation is further complicated by the transient appearance of an intermediate phenotype during differentiation, in that both ATI and ATII cell-like features may coexist in such cells. It has to be taken into account that so far there is no clear evidence showing that the differentiation of ATII cells definitively yield terminally differentiated ATI cells, necessitating the use of a term, "type I cell-like phenotype" to reflect this fact in the literature *(13)*.

It has been reported that specific lectins label apical membranes of either type I or type II cells *(14)*. The lectins *Ricinus communis 1, Bauhinia purpurea*, and *Lycopersicon esculentum* bind to ATI cells, but not ATII cells, while *Maclura pomifera* binds to ATII, but not ATI cells. These findings strongly suggested that ATI cells express membrane glycoproteins (and/or glycolipids) that are distinctly different from those expressed on the apical cell membranes of ATII cells.

A number of molecular markers, specific for type I cells, have been described in the recent past. The first one was T1α, a 36-kDa glycoprotein found in rodent lungs. Antibodies have been developed against lung proteins that specifically label type I cells in human lung with patterns that match those of rodent T1α *(15)*. However, the human data are unclear, because antibodies to rodent T1α do not recognize ATI cell antigens in normal adult human lung. This likely indicates that O-glycosylation of the human protein(s), differs significantly from that of the rodent proteins.

Aquaporin-5 (AQP-5), a second ATI cell marker, is a member of the large family of aquaporin proteins, most of which are water channels. AQP-5 is a transmembrane protein of approx 27–34-kDa that resides in the ATI cell apical plasma membrane. Immunohistochemical and immunocytochemical studies by several investigators with various antibodies, as well as Northern and Western analyses of isolated cells, have shown that AQP-5 is uniquely expressed by ATI cells in the peripheral regions of the lung.

It was shown many years ago that ATI cells contain numerous, small, flask-shaped membrane invaginations, or caveolae, that open to the alveolar lumen or interstitial space. The presence of such caveolar structure in ATII cells is not clear yet. In addition to these caveolae at the cell membranes, numerous small vesicles are also noted in both ATI (but not ATII) cells and pulmonary vascular endothelial cells. Caveolin-1, a 21–24 kDa protein, is the major scaffolding protein that forms the vesicular skeleton of the caveolae. In the alveolar epithelium, caveolin-1 expression appears limited to ATI cells, as ATII cells normally express none or very little *(16)*. Thus for many in vitro studies, caveolin-1 expression was used to discriminate between ATI and ATII cell phenotypes *(17)*.

Several monoclonal antibodies, that discriminate between unidentified ATI and ATII cell surface epitopes, have been produced. Some of these antibodies may recognize the markers described above. The epitope of some antibodies is not yet identified, an example being the rat type I cell-specific antibody, VIIIB2.

Type I cells are also reported to express carbopeptidase membrane-bound enzyme (CP-M), intercellular adhesion molecule-1 (ICAM-1), β2-adrenergic receptor, insulin-like growth factor receptor-2 (IGFR-2), γ-glutamyl transferase (γ-GT), AQP-4, and VAMP-2, a membrane protein associated with caveolae. Whether or not these molecules are expressed in ATI cells exclusively or may also be present in other cell types needs still to be elucidated.

2. Materials

2.1. Isolation of ATII Cells

1. Specimen of distal portions of normal lung tissue (approx 5–30 g) from patients undergoing lung resection.
2. Antibiotics solution 100X: penicillin (10,000 U/mL), streptomycin (10 mg/mL) in 0.9% NaCl (P-0781, Sigma).
3. Balanced salt solution (BSS): 137 mM NaCl (16.0 g), 5.0 mM KCl (0.8 g), 0.7 mM Na$_2$HPO$_4$ • 7 H$_2$O (0.28 g), 10 mM HEPES (4.76 g), 5.5 mM glucose (2.0 g), 20 mL antibiotics solution 100X. Titrate to pH 7.4 at 37°C. BSS can be stored for 3 wk in a refrigerator at 4°C (numbers in brackets are given for a final volume of 2.0 L).
4. Sterile Petri dishes, beakers (100 mL), Pasteur pipets, graded pipets, and centrifuge tubes (15 and 50 mL).
5. Dissection kit, including curved scissors and tweezers.
6. Tissue chopper (McIlwain).
7. Sterile cell strainers: nylon (40 μm and 100 μm mesh) and gauze (*see* **Note 1**).
8. Enzyme solutions: 1.5 mL trypsin type I (10,000 BAEE units/mg protein, T-8003, Sigma) by reconstitution of 1 g in 10 mL BSS. Three hundred μL elastase (44.5 mg protein/mL, 4.7 units/mg protein, Worthington) by reconstitution of 10 mg in 10 mL BSS.

2.2. Purification of ATII Cells

1. Centrifuge with swingout rotor.
2. Shaking water bath.
3. Dulbecco's modified Eagle's medium / Ham's nutrient mixture F-12 (DME/F12, 1:1 mixture) supplemented with 10% FBS and 1% antibiotics solution 100X.
4. Small airways growth medium (SAGM, Cambrex).
5. DNase type I (D-5025, Sigma). Prepare aliquots of 10,000 to 15,000 units in 1 mL BSS.
6. Inhibition solution: a mixture of 30 mL DME/F12, 10 mL FBS, and 1 mL DNase aliquot. Let the solution warm up to room temperature and use.
7. Adhesion medium: 22.5 mL DME/F12, 22.5 mL SAGM, 1 DNase aliquot. Use at 37°C.
8. Coating solution: Into 10 mL SAGM (in a 15-mL centrifuge tube, on ice), add 100 μL collagen type I solution (5 mg/mL in 0.01 N acetic acid) and 100 μL fibronectin solution (at 1 mg/mL). Amount of coating solution needed to coat Transwell® filters (6.5 mm in diameter) is 0.2 mL, Transwell® filters (12 mm in diameter) 0.5 mL, Transwell filters (24 mm in diameter) 1.5 mL, and chamber slides 0.4 mL.

9. PBS 10X: The final PBS 1X contains 130 mM NaCl (7.59 g), 5.4 mM KCl (0.4 g), 11 mM glucose (1.98 g), 10.6 mM HEPES (2.53 g), 2.6 mM Na$_2$HPO$_4$ • 7 H$_2$O (0.7 g) in distilled water, with numbers in brackets being given for a final volume of 100 mL PBS 10X. This 10X solution can be stored for 6 wk at 4°C in a refrigerator.
10. Light Percoll® solution (1.040 g/mL): To make 40 mL, mix 4 mL PBS 10X, 10.88 mL Percoll solution, and 25.12 mL distilled water.
11. Heavy Percoll® solution (1.089 g/mL): To make 40 mL, mix 4 mL PBS 10X, 25.96 mL Percoll solution, 10.04 mL distilled water, and one drop of phenol red.
12. Wash buffer for the magnetic beads: To make 100 mL, mix 10 mL PBS 10X, 0.5 g BSA, 54.5 mg ethylenediaminetetraacetic acid (EDTA), and bring up the volume to 100 mL using distilled water. The wash buffer can be stored for 4 wk at 4°C in a refrigerator.
13. Anti-CD-14 and anti-fibroblast-conjugated Dynabeads® (Dynal) or MicroBeads® (Miltenyi Biotech) can be used for the removal of macrophages and fibroblasts.
14. 0.4% Trypan blue in PBS.
15. Harris' hematoxylin.
16. Lithium carbonate solution: add 2 mL of a saturated lithium carbonate solution (1 g in 100 mL water) to 158 mL distilled water.
17. Ethanol and xylene.

2.3. Primary Culture of Isolated Human Alveolar Epithelial Type II Cells to Exhibit Type I Cell-Like Phenotype

1. Cell culture medium: *see* **Subheading 2.2.**
2. Cell culture incubator with 5% CO$_2$.
3. Cell culture-treated polyester inserts (Transwell Clear®, Corning, pore size 0.4 µm or equivalent). Other cell culture plasticware may be used depending on the experimental requirements (*see* **Note 2**).
4. Epithelial Voltohmmeter (EVOM, World Precision Instruments) equipped with STX-2 chopstick electrodes.

3. Methods

3.1. Isolation of ATII Cells

1. Sterile techniques and materials should be used throughout the whole procedure.
2. Transfer the lung tissue to a Petri dish containing BSS in the laminar flow hood, trim away the major bronchi and blood vessels and cut the parenchyma into small pieces (approx 1 cm^3) (*see* **Note 3**).
3. Mince the lung pieces with the tissue chopper (gauge 0.6 mm) and subsequently transfer them to a centrifuge tube filled with approx 35 mL BSS. Gently mix the contents of the tube and pour the contents into a small sterilized beaker.
4. Pass the minced tissue through a 100-µm mesh cell strainer. Discard the filtrate and add the combined tissues to a new centrifuge tube, filled with 35 mL BSS. Repeat the washing procedure at least three times to remove blood cells and mucus.

5. Fill an Erlenmeyer flask with approx 40 mL BSS, transfer the minced and filtered tissue pieces into the flask, add the enzyme solutions (trypsin and elastase) and incubate for 40 min on the shaking water bath at 37°C.

6. In the meantime, clean up the tissue chopper, and prepare the inhibition solution and adhesion medium.

7. The enzymatic activity is stopped using 40 mL inhibition solution. Thoroughly triturate the digested mixture for approximately 5 min using a 25-mL pipet (*see* **Note 4**).

3.2. Purification of ATII Cells

1. Filter the triturate through the gauze filters and then through a 40-µm cell strainer in tandem to obtain a crude alveolar cell suspension as the final filtrate. Wash the filters with an appropriate amount of BSS, collect and distribute the filtrates in four centrifuge tubes. After centrifugation at 300*g* for 10 min at room temperature, resuspend the cell pellet in the adhesion medium and transfer the cell suspension to the tissue culture–treated plastic Petri dishes (10–15 mL each). Incubate at 37°C in a CO_2 incubator for at least 90 min in order to let macrophages attach to the surface of the Petri dishes.

2. During the cell panning, the permeable Transwell supports need to be incubated with the coating solution. The coating of the filters (or other plastic surfaces) at 37°C in an incubator (5% CO_2 and 95% relative humidity) should take at least 2 h.

3. After the cell panning, the nonadherent cells are gently collected and the Petri dishes are rinsed one more time with BSS. Centrifuge the cell suspension at 300*g* for 10 min at room temperature.

4. During the centrifugation, the Percoll gradient is prepared (*see* **Note 5**).

5. Resuspend the cell pellet in 3 mL DME/F12 and layer the crude cell suspension on top of the Percoll gradient. The preparation is centrifuged at 300*g* for 20 min at 4°C using a swingout rotor to produce an ATII cell enriched layer at the interface between the heavy and light Percoll gradient (*see* **Note 6**).

6. Using a Pasteur pipet, transfer the enriched type II cells to a fresh centrifuge tube filled with approx 40 mL BSS and pellet the type II cells by centrifugation (300*g* for 10 min at room temperature).

7. When Dynabeads® are used, 10 µL of bead-solution (appropriate for 10^6 cells) are washed twice with 2 mL of the wash buffer during the centrifugation. The buffer can be removed by attaching the bead-containing microcentrifuge tube to the magnet for at least 2 min and decant the fluid.

8. When MicroBeads® are used, the separation column (but not the beads) is washed with 500 µL wash buffer.

9. Resuspend the cell pellet in an appropriate volume of magnetic beads (2–5 µL MicroBeads or 10 µL Dynabeads per 10^6 cells) and add appropriate volume of BSS to bring the total volume to 3 mL. Divide the cell suspension into two microcentrifuge tubes and incubate for 20 min at room temperature under slow but constant mixing (e.g., by using an over-end-over rotator or orbital shaker).

10. After the incubation, use the appropriate magnetic system (either magnet or magnetic column) to separate fluids from the magnetic particle–bound cells.

11. Resuspend the resulting cell suspension in a total volume of 10 mL SAGM, set aside 50 μL for cell counting and viability assessment, and centrifuge the rest at 300g for 10 min (*see* **Note 7**).

12. Number and viability of the cells are tested for the aliquots of the mixture comprised of 50 μL trypan blue solution and 400 μL BSS mixed with the 50 μL cell suspension. Exclusion of dye by healthy cells and uptake of dye by damaged cells can be visualized by the blue color of the dead cells (*see* **Notes 8** and **9**).

13. In addition to a mere viability assessment, ATII cells can be readily identified by the modified Papanicolaou method described by Dobbs *(18)*. For this procedure, prepare microscope slides of isolated cells at 2×10^5 cells/cm^2 and air-dry overnight. Incubate the slides in Harris' hematoxylin for 3 to 4 min. Rinse the slides two to three times with distilled water and incubate with lithium carbonate solution for 2 min. Rinse with distilled water again and incubate sequentially in (v/v) 50, 80, 95, and 100% ethanol for 90, 15, 15, and 30 s, respectively. Finally, incubate the slides in a xylene:ethanol (1:1) mixture for 30 s, followed by further incubation in xylene alone for 1 min. Inspect the cells with a light microscope at ×1000 magnification (*see* **Note 10**).

14. During centrifugation of the cell suspension, remove the coating solution from the Transwell filters or respective plastic surfaces by gentle suction and dry off the culture support in the laminar flow hood.

3.3. Culture of ATI Cell-Like Phenotype

1. Resuspend the purified cell pellet in SAGM to a final concentration that will allow seeding at 0.6×10^6 cells/cm^2 onto the appropriate culture support (*see* **Note 11**).

2. Incubate the cells at 37°C, 5% CO_2, and 95% relative humidity with the cell culture medium in the apical and basolateral chambers of the Transwell. The first exchange of medium [or transfer of the monolayers to air-interfaced culture (AIC) condition] can be performed at 48 h postseeding (*see* **Note 12**).

3. For AIC conditions, (*see* **Note 13**), the apical bathing medium has to be removed completely by gentle suction and the basolateral fluid has to be replenished to a volume that no hydrostatic pressure is obtained. The appropriate basolateral fluid volumes for the various Transwell sizes for the AIC are 400 μL (6.5 mm), 650 μL (12 mm), and 1.3 mL (24 mm), respectively.

4. The cells are cultured for 6–10 d with the culture medium changed every 48 h. The cells will spread to form a confluent monolayer and assume type I cell-like morphology. A concomitant increase in transepithelial electrical resistance (TEER) is measured using an EVOM. By day 8 in culture, the cell monolayers usually exhibit TEER values of >1500 Ω • cm^2 when cultured under LCC conditions.

4. Notes

1. Gauze filters can be easily prepared by taping a gauze compressed to the bottom of a truncated 50-mL centrifuge tube and subsequent sterilization.

2. Cell culture on permeable filter supports allows access to both sides of the monolayer. This is crucial for studies of drug or ion transport. When cultured in flasks or cluster wells, only the apical surface of the cells can be accessed.

3. In case of sanguineous tissues, it is important to rinse the tissue multiple times (e.g., more than three) before the chopping step to reduce the number of blood cells.

4. The tissue suspension should appear as a cloudy liquid after a successful trituration, because of multiple singular cells released from the tissue.

5. The heavy Percoll solution is carefully layered under the light Percoll solution to obtain the gradient. The added phenol red aids identification of the light/heavy Percoll interface, from which the ATII cells are harvested. The discontinuous gradient is cooled on ice before layering the cell suspension on the top of the light Percoll.

6. It is important to use a "soft" stopping method of centrifugation mode (e.g., turning off the brake that speeds up the deceleration) for the Percoll gradient step.

7. Although the yield of ATII cells may vary, depending on size and quality of the tissue specimen, between 8×10^5 and 2×10^6 cells per gram of tissue should be obtainable after purification.

8. Uptake of trypan blue dye is time-dependent. If the staining procedure takes too long, viable cells may also begin to take up the dye. We usually assess the cell viability within 10 min of trypan blue addition.

9. It is helpful to monitor the purification method before and after each purification step for comparison with successful versus unsuccessful cell preparations.

10. The lamellar bodies of ATII cells are stained blue with modified Papanicolaou staining procedures. Contaminating cells (fibroblasts, endothelial cells, bronchial cells, macrophages, and so on) are not stained by this method.

11. The seeding density is important. If the seeding density is too high, the cells do not have enough space for spreading (and lower TEER values are obtained), whereas lower seeding density may result in non-confluent monolayers (with no significant TEER value developing for up to 14 d of culture).

12. When feeding cells cultured on Transwell filter inserts, it is important to first remove the basolateral fluid. When replenishing, the apical medium is added before refilling the basolateral compartment. The reason for this is to prevent the development of a hydrostatic pressure in the basolateral to apical direction, which could detach the cells from the filter.

13. The development of tight type I cell-like monolayers depends on the cell-culture conditions. Here, the choice of culture medium may be important as well as changing to AIC conditions. In AIC, the apical side of the monolayers is exposed to the surrounding air and the cells are only fed from the basolateral side (through the filter). For rat AECs, it has been reported that AIC results in a prolongation of the ATII cell phenotype *(19)*. One aspect of AIC conditions is that secreted components are not washed away by the medium, but stay on top of the monolayer.

Acknowledgments

This work was supported in part by the ZEBET, the Hastings Foundation, and research grants HL38658 and HL64365 from the National Institutes of Health.

References

1. Meaney, C., Florea, B. I., Ehrhardt, C., et al. (2002) Bronchial epithelial cell cultures, in *Cell Culture Models of Biological Barriers* (Lehr, C. M., ed.), Taylor & Francis, London, pp. 211–227.
2. Kim, K. J., Borok, Z., and Crandall, E. D. (2001) A useful in vitro model for transport studies of alveolar epithelial barrier. *Pharm. Res.* **18,** 253–255.
3. Elbert, K. J., Schäfer, U. F., Schäfers, H. J., Kim, K. J., Lee, V. H. L., and Lehr, C. M. (1999) Monolayers of human alveolar epithelial cells in primary culture for pulmonary absorption and transport studies. *Pharm. Res.* **16,** 601–608.
4. Corti, M., Brody, A. R., and Harrison, J. H. (1996) Isolation and primary culture of murine alveolar type II cells. *Am. J. Respir. Cell Mol. Biol.* **14,** 309–315.
5. Kikkawa, Y. and Yoneda, K. (1974) The type II epithelial cell of the lung. I. Method of isolation. *Lab. Invest.* **30,** 76–84.
6. Dobbs, L. G. and Mason, R. J. (1978) Stimulation of secretion of disaturated phosphatidylcholine from isolated alveolar type II cells by 12-O-tetradecanoyl-13-phorbol acetate. *Am. Rev. Respir. Dis.* **118,** 705–733.
7. Goodman, B. E. and Crandall, E. D. (1982) Dome formation in primary cultured monolayers of alveolar epithelial cells. *Am. J. Physiol.* **243,** C96–C100.
8. Shen, J., Elbert, K. J., Yamashita, F., Lehr, C. M., Kim, K. J., and Lee, V. H. (1999) Organic cation transport in rabbit alveolar epithelial cell monolayers. *Pharm. Res.* **16,** 1280–1287.
9. Crapo, J. D., Barry, B. E., Gehr, P., Bachofen, M., and Weibel, E. R. (1982) Cell number and cell characteristics of the normal human lung. *Am. Rev. Respir. Dis.* **126,** 332–337.
10. Fehrenbach, H. (2001) Alveolar epithelial type II cell: defender of the alveolus revisited. *Respir. Res.* **2,** 33–46.
11. Williams, M. C. (2003) Alveolar type I cells: molecular phenotype and development. *Annu. Rev. Physiol.* **65,** 669–695.
12. Hermans, C. and Bernard, A. (1999) Lung epithelium-specific proteins—Characteristics and potential application as markers. *Am. J. Respir. Crit. Care Med.* **159,** 646–678.
13. Cheek, J. M., Evans, M. J., and Crandall, E. D. (1989) Type I cell-like morphology in tight alveolar epithelial monolayers. *Exp. Cell Res.* **184,** 3753–3787.
14. Kasper, M. and Singh, G. (1995) Epithelial lung cell marker: current tools for cell typing. *Histol. Histopathol.* **10,** 155–169.
15. Dobbs, L. G., Gonzalez, R. F., Allen, L., and Froh, D. K. (1999) HTI56, an integral membrane protein specific to human alveolar type I cells. *J. Histochem. Cytochem.* **47,** 129–137.

16. Newman, G. R., Campbell, L., von Ruhland, C., Jasani, B., and Gumbleton, M. (1999) Caveolin and its cellular and subcellular immunolocalisation in lung alveolar epithelium: implications for alveolar type I cell function. *Cell Tissue Res.* **295,** 111–120.
17. Fuchs, S., Hollins, A. J., Laue, M., et al. (2003) Differentiation of human alveolar epithelial cells in primary culture—Morphological characterisation and expression of caveolin-1 and surfactant protein-C. *Cell Tissue Res.* **311,** 31–45.
18. Dobbs, L. G. (1990) Isolation and culture of alveolar type II cells. *Am. J. Physiol.* **258,** L134–L147.
19. Dobbs, L. G., Pian, M. S., Maglio, M., Dumars, S., and Allen, L. (1997) Maintenance of the differentiated type II cell phenotype by culture with an apical air surface. *Am. J. Physiol.* **273,** L347–L354.

15

A New Approach to Primary Culture
of Human Gastric Epithelium

Pierre Chailler and Daniel Ménard

1. Introduction

The gastric mucosa assumes several digestive and protective functions including enzyme and mucus secretion, as well as sterilization of the luminal content through the release of hydrochloric acid. Distinct epithelial populations located in specific compartments of the gastric pit-gland units mediate these secretory functions. In addition, previous studies from our laboratory *(1–5)* and others suggest that local growth factors and extracellular matrix (ECM) proteins may cooperatively regulate epithelial morphogenesis, differentiation, and homeostasis at the level of the stomach. Owing to the absence of a normal or cancer cell line that could serve as an in vitro model to delineate the molecular mechanisms involved, gastric epithelial biology mostly relied on primary culture of epithelial cells freshly isolated from the stomach mucosae of animal species. The protocols used were based on two different approaches, i.e., microsurgical/scraping techniques and enzymatic treatments. The first requires laboratory expertise and, unless gradient fractionation or elutriation is used to purify glandular cells, these are rapidly overgrown during the first 2 d of culture by dividing mucous cells originating from the pits (foveolae) of gastric units *(6–8)*. The second approach, which commonly uses a combination of collagenase and hyaluronidase, generates cultures that are initially composed of mucus-secreting cells mainly, with a low proportion of acid-secreting parietal cells *(9,10)*. Such cultures are also devoid of glandular zymogenic cells, named chief or pepsinogen cells. These represent the major cell type found at the base of gastric glands and they are highly specialized for the secretion of pepsinogens. Their absence in epithelial cultures is particularly deleterious for the advancement of gastric

From: *Methods in Molecular Medicine, vol. 107: Human Cell Culture Protocols, Second Edition*
Edited by: J. Picot © Humana Press Inc., Totowa, NJ

physiology in humans. Indeed, human chief cells not only contribute to the digestion of dietary proteins, but also to triglyceride lipolysis through the secretion of a bile salt–independent and acid-tolerent gastric lipase, which is absent in rodents.

Also consistent with the complex nutritional requirements of epithelial cells, supplements must be added to support the attachment and survival of gastric epithelium in primary culture. Either surgical or enzymatic dissociation procedures yield a crude suspension of single dispersed cells with a few multicellular aggregates that necessarily require the presence of a biological substratum. Hence, the conventional use of fetal bovine serum (FBS) plus fibronectin or collagen I as supplements, appears to optimize the multiplication and differentiation of mucous cells at the expense of other epithelial cell types *(11,12)*. It is, thus, obvious that primary cultures prepared from total and mixed gastric populations using the above techniques are not representative of the integral gastric epithelium. So far, limited applications have been found for the study of mucous cell proliferation *(13–15)*, mucin biosynthesis *(16,17)*, and bacterial pathogen binding *(18)*.

In the past, several investigators *(19–21)* have noted in their extensive review of culture methodologies used for intestinal epithelial cells, that microexplants or multicellular clumps of epithelial cells have a much greater capacity to adhere and survive on plastic than single dispersed cells. Recently, a new nonenzymatic method based on a procedure to recover cells grown on Matrigel with *Matrisperse Cell Recovery Solution* was used to dissociate the human fetal intestinal epithelium from its surrounding lamina propria *(22)*. We successfully applied this strategy to the human fetal stomach *(23,24)* and showed that gastric epithelial cells can easily be dissociated using this commercial agent, as reviewed herein. Primary cultures were generated by seeding viable multicellular aggregates prepared by mechanical fragmentation of the material collected after *Matrisperse* treatment. We further demonstrate that this simple and convenient technique is particularly efficient, allowing for the first time the maintenance of gastric epithelial primary cultures on plastic without a biological matrix, as well as the persistent presence of functional glandular chief cells able to synthesize and secrete gastric digestive enzymes, i.e., pepsinogen and gastric lipase. As exemplified briefly, the new primary culture system will allow one to verify the influence of individual growth factors and ECM components, as well as their combinatory effects on gastric epithelial cell behavior.

2. Materials

2.1. Epithelium Isolation

1. Dissection medium: Leibovitz L-15 medium plus gentamycin and nystatin, 40 µg/mL each (all from Gibco BRL / Life Technologies, Burlington, ON, Canada).

2. Matrisperse Cell Recovery Solution® (Becton Dickinson Biosciences, Bedford, MA).
3. Hank's balanced salt solution (HBSS from Gibco BRL) or phosphate-buffered saline (PBS) composition in grams per liter: 8 g NaCl, 0.2 g KCl, 0.73 g anhydrous Na_2HPO_4, 0.2 g KH_2PO_4, pH 7.4.

2.2. Cell Culture Methodology

1. Culture medium: Dulbecco's modification of essential medium (DMEM) and Ham F-12 medium (1:1) supplemented with penicillin (50 U/mL), streptomycin (50 μg/mL) (all from Gibco BRL), and 10% (v:v) FBS (Cellect Gold FBS from ICN Pharmaceuticals Canada, Montreal, QC, Canada).
2. Plastic multiwell plates, 6 or 24 well.
3. Alternative media and additives for consideration (**Subheading 3.2.2.**): DMEM, Ham F-12, RPMI-1640, Opti-MEM (all from Gibco BRL), glutamine, HEPES, insulin, transferrin, and EGF.

2.3. Cell Characterization

2.3.1. Cell Growth

1. Cell counting equipment.
2. ^{3}H-thymidine (specific activity 80 Ci/mmol; Amersham Canada, Oakville, ON, Canada).

2.3.2. Dye Staining

1. Glass cover slips (round 13 mm, Electron Microscopy Sciences, distributed by Cedarlane, Hornby, ON, Canada).
2. Formaldehyde.
3. 1% periodic acid.
4. Schiff's reagent (Fisher Scientific, contains basic fuschin plus sodium metabisulfite).
5. Regaud's fluid: Dissolve 0.48 g of potassium bichromate in 20 mL of distilled water (0.3% solution) and add 5 mL of 37% formaldehyde.
6. Bowie stain solution: Dissolve 1 g of Biebrich scarlet in 250 mL of distilled water and filter using Whatmann paper (qualitative 4). Dissolve 2 g of ethyl violet in 500 mL of distilled water. Filter into Biebrich scarlet solution until neutralization (color shifts from red to violet). Filter. Heat-dry the precipitate and dissolve in ethanol to obtain a 1% solution.
7. Acetone.
8. Xylene.
9. Permount cover slip mounting medium (Fisher Scientific).

2.3.3. Immunocytochemistry

1. Formaldehyde or methanol.
2. Immunocytochemistry quenching solution: 100 m*M* glycine in PBS, pH 7.4.
3. Immunocytochemistry blocking solution: either use 2% BSA in PBS or 5% Blotto (Nestlé powdered milk) in PBS.

Table 1
Description of Primary Antibodies

Category	Antibody	Dilution for use	Source
Growth factor receptors	EGF/TGFα-R mouse monoclonal	1:100	Upstate Biotechnology Inc., Lake Placid, NY
	IGFI-R mouse monoclonal	1:25	Oncogene Research distributed by Cedarlane, Hornby, ON
	HGF-R rabbit polyclonal	1:200	Santa Cruz Biotechnology, Santa Cruz, CA
	KGF-R rabbit polyclonal	1:100	Kindly given by Dr. M. Terada, National Cancer Center Research Institute, Tokyo, Japan
Zymogen products of gastric chief cells	Pg5 polyclonal antisera	1:150	From A. DeCaro, F. Carrière, and R. Verger, INSERM Marseille, France
	HGL polyclonal antisera	1:4000	From A. DeCaro, F. Carrière, and R. Verger, INSERM Marseille, France
Intermediate filaments	epithelial keratin-18 mouse monoclonal	1:2000	Sigma-Aldrich
	mesenchymal vimentin mouse monoclonal	1:200	Sigma-Aldrich
Junctional proteins	E-cadherin (clone 36) mouse monoclonal	1:800	Transduction Laboratories, Lexington, KY
	ZO-1 of tight junctions rabbit polyclonal	1:500	Zymed Laboratories, San Francisco, CA

4. Antihuman antibodies: all are diluted in 0.2% BSA-PBS (*see* **Table 1**).
5. Secondary antibodies: FITC-conjugated sheep anti-rabbit and anti-mouse IgG antibodies (Chemicon, Temecula, CA) diluted 1:50 and 1:30, respectively, in 0.2% BSA-PBS.

2.3.4. Western Blotting

1. Homogenization buffer: Tris-HCl, 4% sodium dodecyl sulfate (SDS), 2% β-mercaptoethanol, 20% glycerol, and 0.08% bromophenol blue.
2. Nitrocellulose membranes.
3. Primary antibodies against HGL, Pg5, and keratin-18 (*see* **Subheading 2.3.3.** and **Table 1**).
4. Western-Light Plus Chemiluminescent Detection System (Tropix, Bedford, MA).

2.3.5. Determination of Enzymatic Activities

For these purposes, cells were seeded in six-well plates and collected by trypsinization after selected intervals along with their corresponding media. Cells from each well were resuspended in 250 µL of either citrate-phosphate buffer (for lipase assay) or glycine-HCl (for pepsin assay), then lysed by freeze-thawing-refreeze cycle in liquid nitrogen.

2.3.5.1. LIPASE ACTIVITY

1. Glycerol [^{14}C]trioleate (Amersham Canada; specific activity 50 mCi/mmol).
2. Fatty-acid-free BSA (Sigma-Aldrich).
3. Citrate-phosphate buffer pH 6.0: Dissolve 0.1 M citric acid and 0.2 M Na_2HPO_4 in water. Mix 37.4 mL of 0.1 M citric acid solution with 62.6 mL of 0.2 M Na_2HPO_4 to obtain a final pH 6.0 solution.
4. Triton X-100.
5. Methanol:chloroform:heptane (1.41:1.25:1).
6. Carbonate-borate buffer pH 10.5. This buffer extracts free fatty acids. Prepare 100 mL each of 0.1 M KOH and 0.1 M Na_2CO_3, and 200 mL of 0.1 M boric acid. Mix KOH and Na_2CO_3 solutions, then add boric acid until pH 10.5 is reached.

2.3.5.2. PEPSIN ACTIVITY

1. Glycine-HCl buffer pH 3.0: Dissolve 0.1 M glycine in water and adjust pH with HCl.
2. Acid-denatured and dialyzed hemoglobin: Prepare a 6% hemoglobin (Sigma-Aldrich) solution and dialyze against 0.1 N HCl for 24 h at 4°C. Centrifuge at 10,000g for 15 min. Collect the supernatant and store 3-mL aliquots at –80°C. Upon thawing, dilute at 2% concentration in glycine-HCl buffer.
3. Trichloroacetic acid.

2.4. Addition of Growth Factors and Biological Substratum

1. Human recombinant growth factors: epidermal growth factor (EGF; Becton Dickinson) and transforming growth factor-β1 (TGFβ1; R&D Systems, Minneapolis, MN).
2. ECM proteins: collagen-I (purified from rat tail; Becton Dickinson), Matrigel reconstituted basement membrane (Becton Dickinson), laminin-1 (LN-1, Gibco BRL), and laminin-2 or merosin (LN-2, Gibco BRL).

3. Methods

The methods described below outline: (1) the procedure of epithelium isolation; (2) the culture methodology; (3) the characterization of cultured cells; and (4) the effects of culture supplements, i.e., growth factors and biological substratum.

3.1. Dissection and Epithelium Dissociation

3.1.1. Specimens

Tissues from 16 fetuses varying in age from 17 to 20 wk of gestation [post-fertilization ages estimated according to Streeter *(25)*] were obtained from normal elective pregnancy terminations. No tissue was collected from cases associated with known fetal abnormality or fetal death. Studies were approved by the Institutional Human Subject Review Board. The stomach was brought to the culture room, immersed in dissection medium, and prepared within a few minutes at room temperature.

3.1.2. Dissection and Dissociation

The gastric epithelium was dissociated using a new nonenzymatic technique based on a procedure to recover cells grown on Matrigel. As opposed to enzymatic treatment, this technique allows dissociation of epithelial tissue as intact sheets or large multicellular fragments (*see* **Fig. 1**).

1. Excise cardia and pyloric antrum regions from the stomach.
2. Cut body and fundus tissues into explants (3× 3 mm^2).
3. Rinse twice with dissection medium.
4. Immerse explants in ice-cold and nondiluted Matrisperse Cell Recovery Solution for 16–20 h, depending on the age of each specimen (*see* **Note 1**).
5. Dissociate the epithelium by gentle agitation on ice for approx 1 h (*see* **Note 2**).
6. Collect Matrisperse with detached material in a 50-mL tube. Add PBS or HBSS, agitate manually, and collect fluid. Repeat several times (6–8) until no detachment occurs.

3.2. Cell Culture Methodology

3.2.1. Seeding

The resulting material was then fragmented mechanically into multicellular clumps and seeded in culture vessels.

1. Centrifuge the epithelial preparation at low speed, 100*g*, 7 min.
2. Resuspend in culture medium.
3. Tissue is fragmented mechanically into multicellular clumps (approx 10–15 cells) by repeated pipetting cycles: 5–10 cycles with a 5-mL serological pipet and 5–10 cycles with a 1-mL micropipet. Avoid the formation of air bubbles in the suspension. It is important to minimize cell death and single cell dispersion (*see* **Note 3**).
4. This material is seeded in plastic 6-well multiwell plates (5×10^5 cells in 3 mL) or 24-well plates (1.5×10^5 cells in 1 mL) and left undisturbed for at least 24 h to allow attachment.

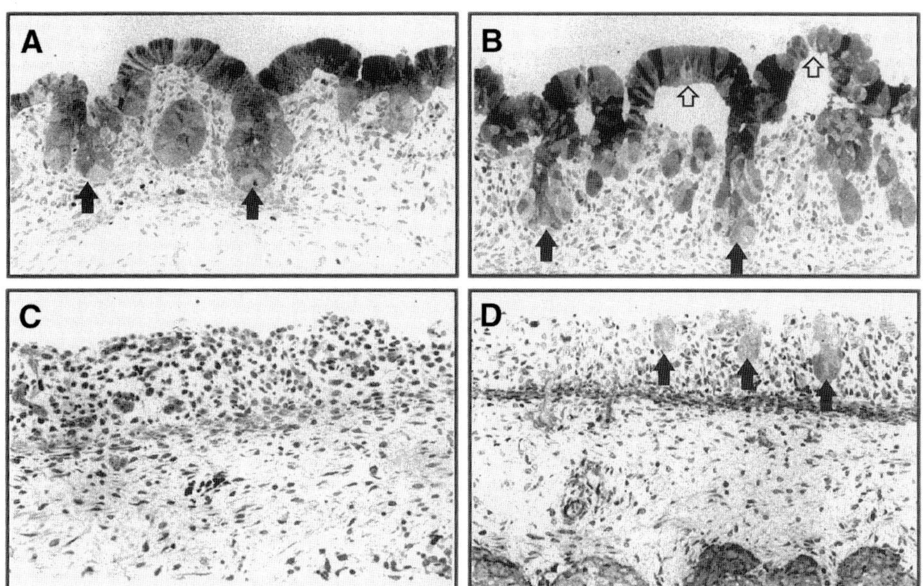

Fig. 1. Dissociation of human fetal gastric epithelium during Matrisperse treatment. **(A)** Optical microscopy of intact explant from 17-wk stomach showing the epithelium and the surrounding lamina propria. Filled arrows indicate the base of forming glands (20×). **(B)** Explant from 17-wk stomach treated for 10 h (midtreatment). Blank arrows illustrate the initial detachment of surface epithelium in large fragment or intact sheet (20×). **(C)** Explant from 17-wk stomach treated for 18 h where epithelial tissue is completely detached (40×). **(D)** Explant from 20-wk stomach treated for 20 h. In some samples prepared from older specimens, a few glandular epithelial cells remained not dissociated, as indicated by filled arrows (20×). Reproduced by kind permission from **ref. *23***.

3.2.2. Culture

Culture medium with nonattached material was discarded after 24 h. Usually, it was necessary to rinse the culture surface with a flow of fresh medium once or twice in order to remove cell debris and attached mucus. Medium was renewed every 48 h thereafter. Cultures were routinely maintained in DMEM/ F-12 mixture supplemented with 10% FBS only (no hormones or matrix coating) and characterized after selected intervals using phase-contrast microscopy. After spreading of epithelial colonies, a confluent monolayer of polyhedral and irregularly shaped cells was obtained (*see* **Fig. 2A,B**).

The overall quality of primary cultures strictly depended on the initial density of epithelial aggregates. Furthermore, several medium formulations were tested for their efficiency to support cell proliferation and monolayer formation:

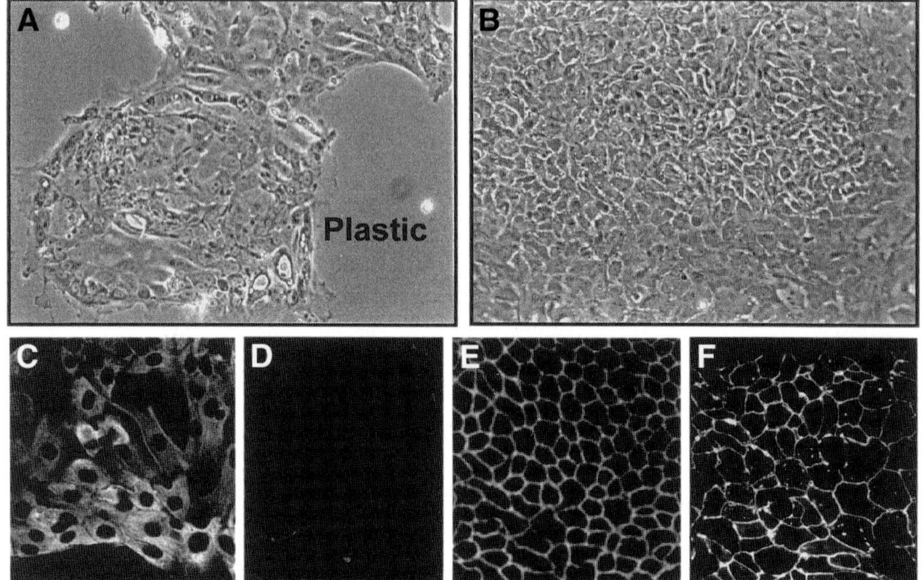

Fig. 2. Epithelial status of human gastric primary cultures. **Upper panel:** Phase-contrast morphology of fetal gastric epithelial cells after 1.5 d (**A**) and after 4 d (**B**) (magnification 10×). **Lower panel:** Indirect immunofluorescence of epithelial keratin-18 (**C**), mesenchymal vimentin (**D**), tight junction ZO-1 (**E**), and E-cadherin (**F**). Adapted from **refs. *23,24*.**

1. DMEM.
2. Ham F-12.
3. DMEM/F-12 mixture.
4. RPMI 1640.
5. Opti-MEM.

As expected from cell-culture references *(26)*, DMEM/F-12 was optimal for reaching confluence rapidly, i.e., after 3–4 d. DMEM alone and Opti-MEM rather supported the formation of more compact epithelial colonies, particularly in the presence of 4 m*M* glutamine and 20 m*M* HEPES. Consequently, the adequate medium formulation should be selected according to the necessity of rapid confluence or a more polarized epithelial morphology in specific experimental studies. It is also worthy of mention that the concentration of FBS can be reduced to 4–5% when the following growth factors are added: 5 µg/mL insulin, 5 µg/mL transferrin, and 10 ng/mL EGF. Under all previous conditions, a low number of cells spontaneously detached from the culture substratum during the first week of culture without altering the integrity of the

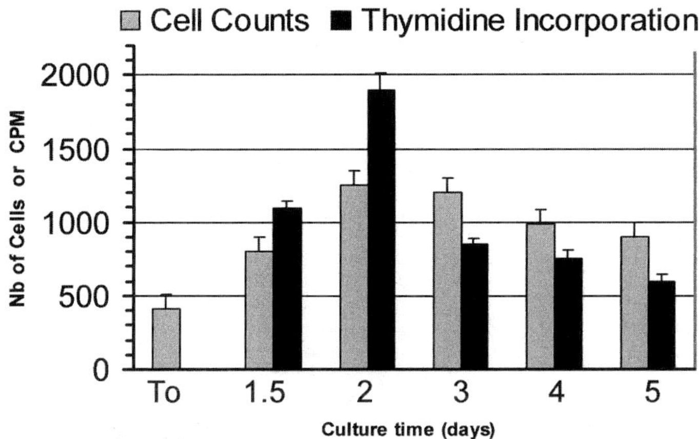

Fig. 3. Growth kinetics. Cell counts were determined at different culture intervals and final results expressed as the number of cells $\times 10^3$ per well. The levels of ^{3}H-thymidine incorporation are in CPM per well. Note that the number of cells and radioprecursor uptake rapidly increase by twofold during the first 48 h. Adapted from **ref. 23**.

monolayer. Thereafter, cell detachment became progressively more intense. Such results may suggest that the cell adhesion kinetics could be ameliorated with the use of ECM protein coatings, as discussed in **Subheading 3.4.**

3.3. Cell Characterization

3.3.1. Cell Growth

Cell number and viability were evaluated after each day by hematocytometer counting and trypan blue exclusion technique performed on trypsin-dissociated cells *(26)* (*see* **Fig. 3**). The DNA synthesis rate was determined by incubating cells seeded in 24-well plates with 2 µCi/mL of ^{3}H-thymidine during the last 12 h of each interval. Specimens were rinsed, radioactivity incorporated into total DNA was precipitated by trichloroacetic acid treatment and then solubilized with 0.1 M NaOH + 2% Na_2CO_3 for 30 min. One-milliliter aliquots were counted in a β-liquid scintillation system (*see* **Fig. 3**).

3.3.2. Dye Staining

In these experiments, suspensions were seeded on sterile glass cover slips deposited in culture vessels. At the end of culture intervals, specimens were rinsed with HBSS pH 7.4 and then processed. PAS staining revealed that a significant proportion of cultured cells, but not all, were mucous cells (*see* **Fig. 4A**). Bowie staining provided the first evidence of the presence of viable chief cells in our primary culture system (*see* **Fig. 4B**).

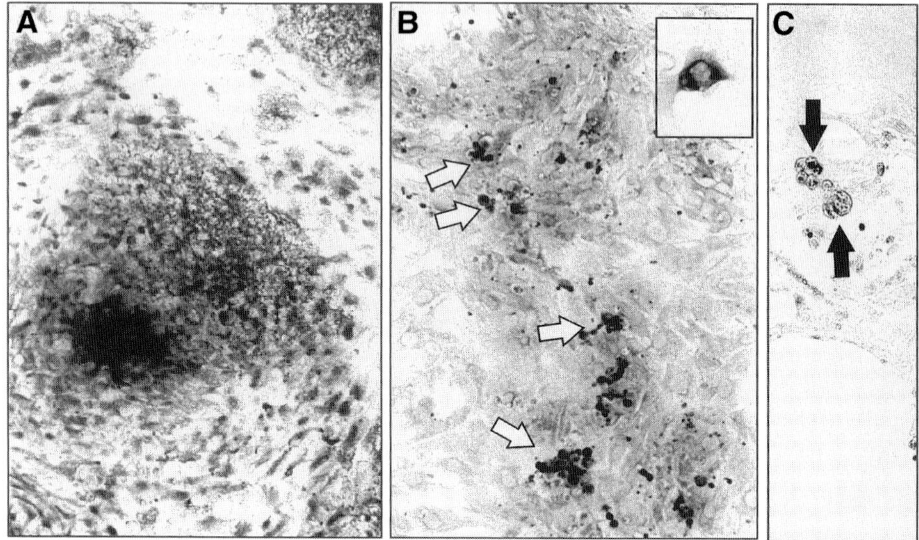

Fig. 4. Histochemical analysis of differentiation markers. **(A)** PAS staining showing intense reactivity in a majority (50–60%) of cells. Dark staining at the center of the specimen reveals the presence of mucus (16×). **(B)** Bowie staining of pepsinogen-containing cells with their cytoplasm appearing dark violet (white arrows) (16×). Inset illustrates positive cells at higher magnification (40×). **(C)** Bowie staining of parietal cells that appear pink (black arrows) (16×). Adapted from **ref. 23**.

3.3.2.1. Periodic Acid Schiff (PAS) for Identification of Mucus-Producing Cells

1. Fix specimens in 3.7% formaldehyde for 15 min at room temperature.
2. Incubate in 1% periodic acid for 10 min.
3. Rinse 5 min in running tap water.
4. Incubate 15 min in Schiff reagent.
5. Wash in water, allow to dry completely, and mount with Permount.

3.3.2.2. Bowie Staining *(27)* of Pepsinogen-Containing Cells

1. Fix specimens 12 min in Régaud's fluid.
2. Rinse 5 min in water.
3. Incubate for 5–10 s in Bowie solution.
4. Rinse in water and PBS.
5. Dehydrate specimens in acetone (two rapid dippings), soak 30 s in acetone:xylene (1:1), differentiate in xylene for 1 min, and mount with Permount.

3.3.3. Immunocytochemistry

Several protein markers previously associated with the functional development of human fetal gastric epithelial cells were analyzed using indirect

Table 2
Expression of Gastric Epithelial Markers by Functional Category

Junctions / Cytoskeleton	Growth Factor Receptors	Integrin Subunits	Secretory Products
E-cadherin	**EGF-R**	$\beta1$	PAS-pos.
ZO-1	**IGFI-R**	$\alpha6$	mucin-6
Keratin-18	**HGF-R**	$\alpha3$	HGL
(vimentin)	**KGF-R**	$\alpha2$	Pg5
		$\beta4$	

Proteins detected in all cultured cells are written in bold. Secretory products were present in the intracellular granules of cell subpopulations, whereas $\alpha2$ and $\beta4$ integrin subunits were absent in a low number of cells. Vimentin immunoreactivity was almost negative (*see* **ref.** *23* for detailed results).

immunofluorescence. Assays were performed on cells cultured on sterile glass cover slips. The epithelial nature of the cells was confirmed by the presence of keratin-18, E-cadherin, ZO-1, and the absence of vimentin staining (*see* **Fig. 2C–F**). **Table 2** summarizes our data (*see* **refs.** *23*, *24*, and *37* for detailed analyses).

1. Fix specimens in fresh 3% formaldehyde for 12 min or in methanol for 10 min.
2. If necessary, free aldehyde residues are quenched by a 15-min incubation in quenching solution.
3. Nonspecific binding is prevented by adding blocking solution for 15 min.
4. Incubate specimens inside a humidified chamber with choice of mouse and rabbit primary antibodies for 60 min.
5. After rinsing, add appropriate dilutions of anti-mouse and anti-rabbit secondary antibodies for 45 min.
6. Rinse and mount on glass slides.
7. Specimens were examined on a Reichert Polyvar microscope equipped for epifluorescence and photographed with Kodak TMAX film (400 ASA).

3.3.4. Western Blotting

1. Dissociate cultures seeded in six-well plates.
2. Lyze cells in homogenization buffer.
3. Separate proteins (180–200 µg aliquots) by sodium dodecyl sulfate-polyacrylamide gel electrophoresis (SDS-PAGE), transfer onto nitrocellulose membranes, and then incubate with primary antibodies against HGL, Pg5, and keratin-18.
4. Membranes are finally processed with the Western-Light Plus Chemiluminescent Detection System.
5. Autoradiograms exposed in a linear range are quantified by densitometry and the signals are normalized to those of the keratin-18 control.

3.3.5. Determination of Enzymatic Activities

Previous studies from our laboratory (cited in **24**) have already established that human fetal gastric explants can be maintained fully functional in organ culture for several days. We also unravelled a nonparallelism of synthesis and secretion for both gastric enzymes in the last system: pepsin activity increases over a 5-d period whereas lipase activity decreases. In order to determine whether isolated gastric epithelial cells (and chief cells, in particular) retain their intrinsic capacity to produce and secrete digestive enzymes in primary culture, enzymatic activities were measured in cells and culture fluid as described previously for human fetal stomach *(28)*.

3.3.5.1. LIPASE ACTIVITY

Lipolytic activity attributed to HGL in the presence of triglycerides was determined using glycerol [^{14}C]trioleate as substrate and fatty-acid-free BSA as carrier of released fatty acids *(29)*.

1. The assay system contains: 1.2 μmol labeled triglyceride, 10 μmol citrate-phosphate buffer pH 6.0, 0.1 μmol BSA, 2 μ*M* Triton X-100, and 100 μL of cell homogenate (prepared in citrate phosphate buffer as described in **Subheading 2.3.5.**) in a final volume of 200 μL.
2. Incubate 60 min at 37°C.
3. Stop reaction and separate free [^{14}C]oleic acid by liquid–liquid partition: add 3.25 mL of methanol:chloroform:heptane (1.41:1.25:1 by volume) and vortex vigorously. Immediately add 1.05 mL of carbonate-borate buffer pH 10.5 and vortex during 45 s.
4. Centrifuge at 1800*g* for 20 min.
5. Count a 1-mL aliquot of the superior fraction of supernatant in a β-scintillation system.
6. The specific activity of HGL is expressed as nanomoles (nmol) of free fatty acids (FFA) released per minute per milligram of protein.

3.3.5.2. PEPSIN ACTIVITY

Pepsin activity resulting from activation of pepsinogen at acid pH was measured by the method of Anson and Mirsky *(30)*.

1. The assay tube contains: 900 μL of glycine-HCl and 100 μL of cell homogenate (prepared in glycine-HCl as described in **Subheading 2.3.5.**) to which is added 1 mL of 2% hemoglobin as substrate.
2. Reaction is carried out at 37°C for 10 min.
3. Stop with 6.2% trichloroacetic acid, vortex, and keep on ice for 1 h.
4. Centrifuge samples at 20,000*g* for 20 min.

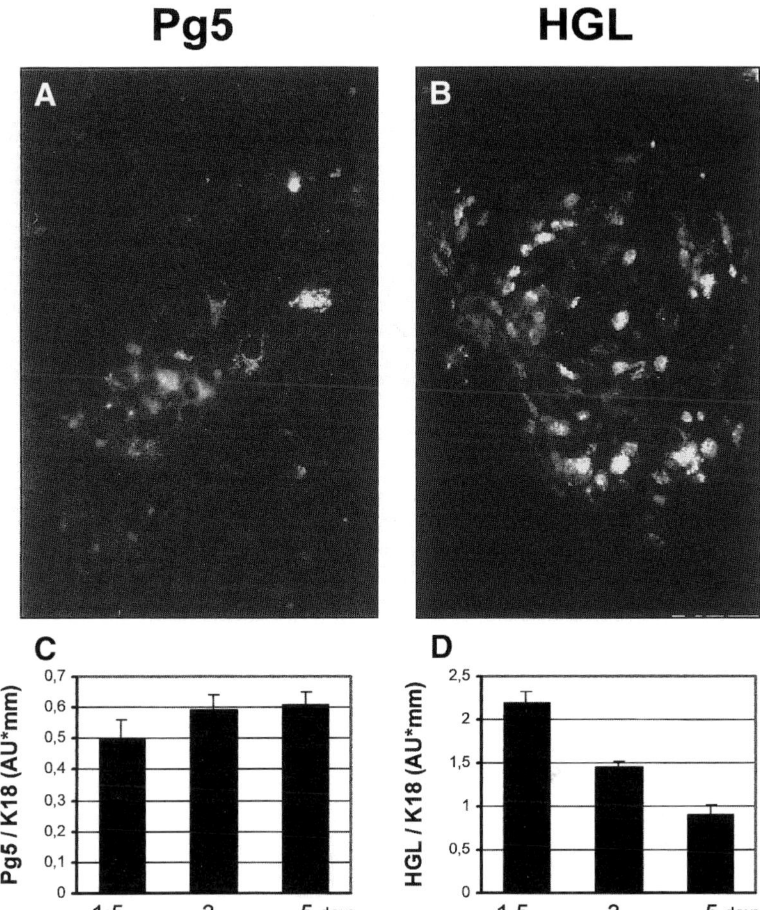

Fig. 5. Immunodetection of human pepsinogen (Pg5) and human gastric lipase (HGL). **(A)** Anti-Pg5 immunoreactive cells observed at 1.5 d. A subpopulation of cultured cells exhibited an intense granular staining (40×). **(B)** Similarly, a subpopulation were positive for anti-HGL staining at 1.5 d (40×). **(C)** Densitometric analysis of Western blot experiments ($n = 3$) revealing the progressive increase in pepsinogen reactivity after normalization with the keratin-18 control. **(D)** Densitometric analysis ($n = 3$) of HGL signals shows that the amount of HGL protein progressively decreased in cultured cells. Adapted from **ref. 23**.

5. Quantitate the free amino acid products generated by pepsin activity in the supernatant by spectrometry (280 nm) using a L-tyrosine standard.
6. Pepsin specific activity is expressed in units (μmoles/min) per milligram of protein.

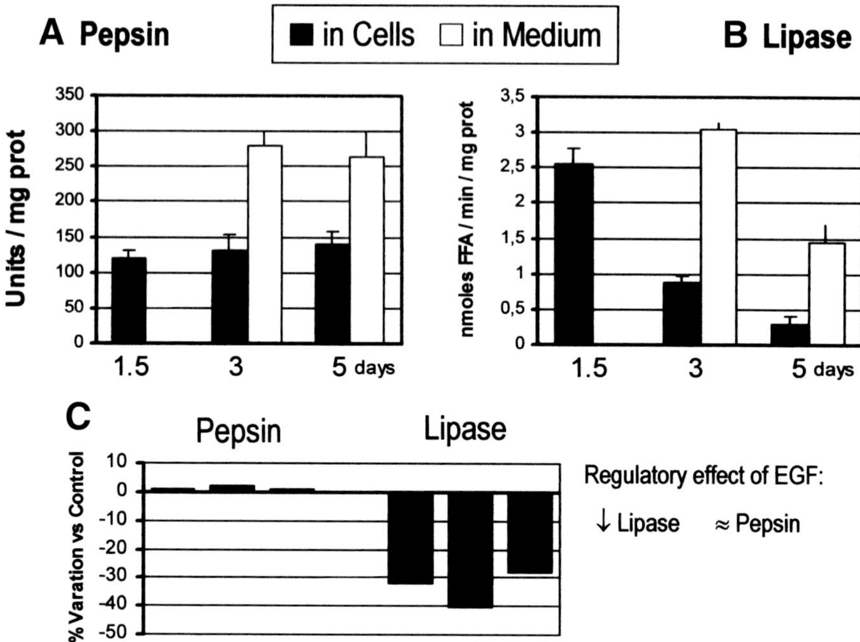

Fig. 6. Measurements of pepsin and lipase activities: variations during culture and upon EGF supplementation. **Upper panel:** Synthesis and secretion profiles of Pg5 (**A**) and HGL (**B**). Specific activities were measured in cells and media between culture intervals 1.5, 3, and 5 d. **Lower panel:** (**C**) Effect of 100 ng/mL EGF on intracellular lipase and pepsin activities after 1.5 d in three independent experiments. Adapted from **refs. *23*** and ***24***.

Protein content of the homogenates was measured by the method of Lowry et al. *(31)*.

In primary culture, pepsin activity remained fairly constant (*see* **Fig. 6A**), whereas intracellular activity of HGL dramatically decreased and its extracellular activity was high on d 3 and reduced on d 5 (*see* **Fig. 6B**). This illustrates that the digestive enzymes coexpressed by glandular chief cells are differentially regulated, corroborating previous data obtained in organ culture *(2,28,33)*.

3.4. Addition of Growth Factors and Biological Substratum

Homeostasis of digestive epithelia, maintenance of cell polarity, as well as differentiation of specialized cell lineages, are governed by cell-to-cell interactions and extracellular signals present in the cell microenvironment. The latter are represented by growth factors and by ECM components such as laminins (LNs), collagen-IV, and heparan sulfate found at the level of the epithelial basement membrane *(34–36)*. Obviously, the current culture system is the first that allows one to discriminate their individual effects on the gastric epithelium because it does not contain hormonal supplements, mesenchymal constituents, or a biological substratum.

3.4.1. EGF Supplementation

Growth factors known to be trophic for the gastric mucosa such as EGF, TGFα, IGF-I, and IGF-II play a major role in gastric physiology because of their effects on mucosal repair and/or inhibition of gastric acid secretion. Their receptors are ubiquitously expressed along the foveolus-gland axis, they stimulate epithelial cell proliferation in organ culture, and they downregulate HGL expression (mRNA, protein, enzyme activity) without affecting Pg5 (**refs. 2, 5, 33**, and review *24*). In total accordance with the last results, EGF triggered the same effects in the current primary culture system (*see* **Fig. 6C**): at the 100 ng/mL concentration, it specifically decreased HGL activity after 1.5 d.

3.4.2. Biological Matrix Coatings

Basement membrane LNs (LN-1, LN-2, LN-5, and LN-10) provide cues for cell polarity and promote the expression of tissue-specific genes in differentiating epithelial cells. In addition, their localization along the foveolus-gland axis in the developing human gastric mucosa *(1,3)* either suggests a role in differentiation of epithelial cell lineages or gland morphogenesis. TGFβ1 is also one of a few growth factors associated with basement membranes and it is now recognized as a regulator of cell–matrix interactions. As described in **ref. 37** and illustrated below (*see* **Fig. 7**), these mediators had profound effects on cell polarity and HGL expression in gastric primary cultures.

1. For this purpose, culture flasks were precoated with ECM proteins: collagen-I (20 µg/mL), LN-1, LN-2, or their mixture (10 µg/mL total), and Matrigel (a complex reconstituted basement membrane, 3.4 mg/mL).
2. Individual ECM components (*see* **Fig. 7A**) could not prevent the spontaneous decrease of cellular HGL activity and mRNA that occurs during a 5-d period; stromal-type collagen-I even had a negative influence compared to TGFβ1 and LN-1/LN-2 mixture (individual LN-1 and LN-2 exerted similar effects; not shown).
3. By contrast, a thin coating of Matrigel initially increased and then maintained HGL levels (activity, mRNA) during 5 d (*see* **Fig. 7B**). Moreover, a combination of LNs plus TGFβ1, which are purified constituents of Matrigel, was able to upregulate HGL synthesis through a powerful synergism (*see* **Fig. 7B**). Matrigel and LNs+TGFβ1 exerted beneficial effects on many aspects of human gastric epithelial morphology such as epithelial cell polarity, accumulation of intracellular granules in apical cytoplasm, and maintenance of E-cadherin at cell–cell contacts (*see* **Fig. 7C** and **ref. 37** for detailed results).

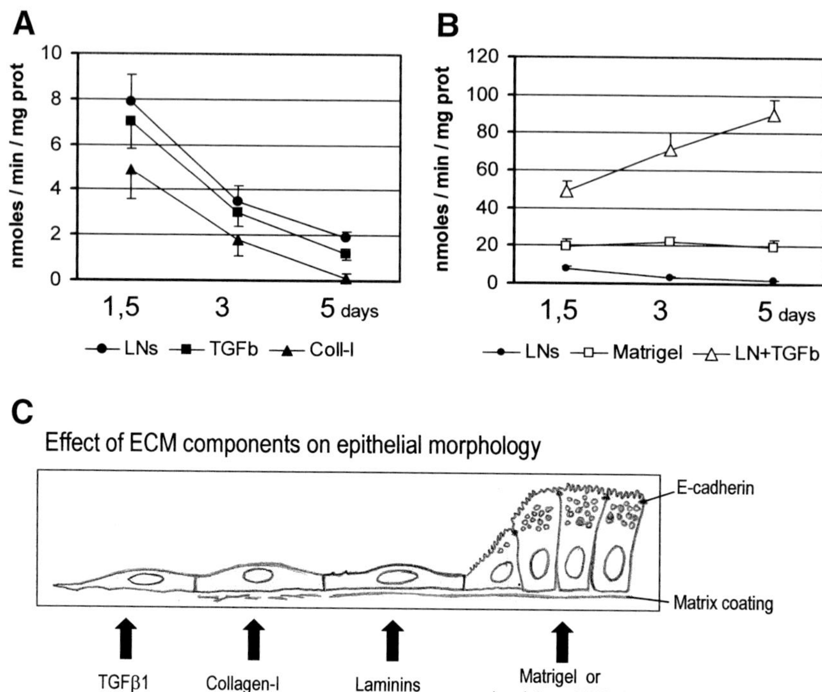

Fig. 7. Regulation of lipase activity by ECM components. In (**A**), specific activities were measured in cells between culture intervals 1.5, 3, and 5 d in the presence of either 5 ng/mL TGFβ1 (plus 0.1% BSA) or thin coatings of LNs (LN-1/LN-2 mixture) and collagen-I (Coll-I). In (**B**), comparison on a different scale between the effects of LNs, Matrigel, and a LNs+TGFβ1 combination. In (**C**), illustration of ECM protein bioactivity on gastric epithelial cultures.

Taken altogether, the new and future data generated with this primary culture system of human gastric epithelium using a convenient nonenzymatic dissociation technique will certainly contribute to a better understanding of the intrinsic mechanisms by which mucosal growth factors and ECM cooperatively and directly regulate gastric epithelial functions. Alternatively, the morphological and functional characteristics of primary cultures suggest applications for the study of pathogen binding to gastric epithelial barrier and for the study of regeneration and restitution associated with ulcer wounding.

4. Notes

1. Period of incubation in Matrisperse solution increases with age: 16 h is usually sufficient for a 16-wk specimen, 18 h for 17-wk, 20 h for 18-wk, and 24 h for 20-wk stomach. Larger explants will require a longer incubation. It is worth men-

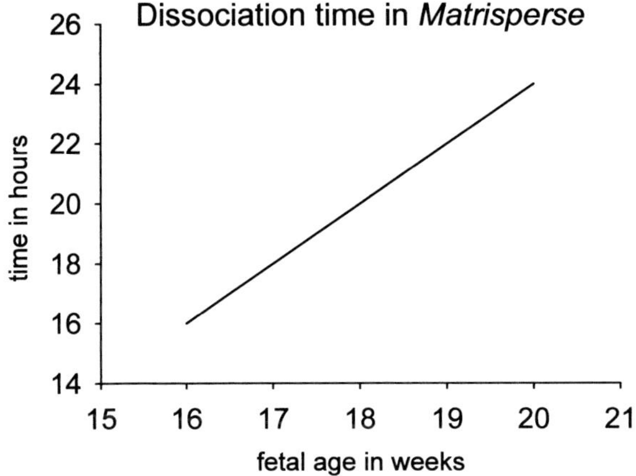

Fig. 8. Dissociation time in Matrisperse.

tion that we noted no significant alteration of culture viability and functionality when incubation was prolonged for a few hours.

2. Gentle agitation on a rocker platform (approx 30 cycles per min) is recommended and detachment of epithelial sheets is generally visualized after 45–50 min. They appear thin and translucent compared to intact explants. More vigorous manual agitation will accelerate the process at this stage. However, do not agitate the suspension by hand too strenuously in order to prevent the induction of apoptosis (anoikis) in dissociated cells.

3. Hand-to-hand adaptation of the technique aims at minimizing single-cell dispersion. Epithelial cells isolated from gastrointestinal organs attach and survive poorly on plastic compared to multicellular aggregates *(19–21)*. The abundance of damaged cells during initiation of the culture is possibly deleterious and their number can be reduced by simply performing a 5-min sedimentation after **step 3.1.2.6**. Discard the cells/debris that remain in suspension and continue to **Subheading 3.2.1.**

Acknowledgment

This work was supported by an IRSC Grant MOP-36495 to Daniel Ménard.

References

1. Tremblay, E. and Ménard, D. (1996) Differential expression of extracellular matrix components during the morphogenesis of human gastric mucosa. *Anat. Rec.* **245,** 668–676.
2. Tremblay, E., Monfils, S., and Ménard, D. (1997) Epidermal growth factor influences cell proliferation, glycoproteins and lipase activity in human fetal stomach. *Gastroenterology* **112,** 1188–1196.

3. Chénard, M., Basque, J.-R., Chailler, P., Beaulieu, J.-F., and Ménard, D. (2000) Expression of integrin subunits correlates with differentiation of epithelial cell lineages in developing human gastric mucosa. *Anat. Embryol.* **202,** 223–233.

4. Chailler, P., Basque, J.-R., Corriveau, L., and Ménard, D. (2000) Functional characterization of the keratinocyte growth factor system in human gastrointestinal tract. *Pediatr. Res.* **48,** 504–510.

5. Tremblay, E., Chailler, P., and Ménard, D. (2001) Coordinated control of fetal gastric epithelial functions by insulin-like growth factors and their binding proteins. *Endocrinology* **142,** 1795–1803.

6. Sanders, M.J., Amirian, D.A., Ayalon, A., and Soll, A.H. (1983) Regulation of pepsinogen release from canine chief cells in primary monolayer culture. *Am. J. Physiol.* **245,** G641–G646.

7. Rattner, D. W., Ito, S., Rutten, M. J., and Silen, W. (1985) A rapid method for culturing guinea pig gastric mucous cell monolayers. *In Vitro (Cell. Dev. Biol.)* **21,** 453–462.

8. Chew, C. (1994) Parietal cell culture: new models and directions. *Annu. Rev. Physiol.* **56,** 445–461.

9. Terano, A., Ivey, K. J., Stachura, J., et al. (1982) Cell culture of rat gastric fundic mucosa. *Gastroenterology* **83,** 1280–1291.

10. Terano, A., Mach, T., Stachura, J., Sekhon, S., Tarnawski, A., and Ivey, K. J. (1983) A monolayer culture of human gastric epithelial cells. *Dig. Dis. Sci.* **28,** 595–603.

11. Fukamachi, H., Ichinose, M., Ishihama, S., et al. (1994) Fetal rat glandular stomach epithelial cells differentiate into surface moucous cells which express cathepsin E in the absence of mesenchymal cells in primary culture. *Differentiation* **56,** 83–89.

12. Fujiwara, Y., Arakawa, T., Fukuda, T., Higuchi, K., Kobayashi, K., and Tarnawski, A. (1995) Role of extracellular matrix in attachment, migration, and repair of wounded rabbit cultured gastric cells. *J. Clin. Gastroenterol.* **21(Suppl.),** S125–S130.

13. Nakagawa, S., Yoshida, S., Hirao, Y., Kasuga, S., and Fuwa, T. (1985) Biological effects of biosynthetic human EGF on the growth of mammalian cells in vitro. *Differentiation* **29,** 284–288.

14. Rutten, M. J., Dempsey, P. J., Solomon, T. E., and Coffey, R. J., Jr. (1993) TGF-alpha is a potent mitogen for primary cultures of guinea pig gastric mucous epithelial cells. *Am. J. Physiol.* **265,** G361–G369.

15. Takahashi, M., Ota, S., Shimada, T., et al. (1995) Hepatocyte growth factor is the most potent endogenous stimuklant of rabbit gastric epithelial cell proliferation and migration in primary culture. *J. Clin. Invest.* **95,** 1994–2003.

16. Yoshida, S., Kasuga, S., Hirao, Y., Fuwa, T., and Nakagawa, S. (1987) Effect of biosynthetic human epidermal growth factor on the synthesis and secretion of mucin glycoprotein from primary culture of rabbit fundal mucosal cells. *In Vitro (Cell. Dev. Biol.)* **23,** 460–464.

17. Boland, C. A., Kraus, E. R., Scheiman, J. M., Black, C., Deshmukh, G. D., and Dobbins, W. O. (1990) Characterization of mucous cell synthetic functions in

a new primary canine gastric mucous cell culture system. *Am. J. Physiol.* **258,** G774–G787.

18. Wagner, S., Beil, W., Mai, U.E., Bokemeyer, C., Meyer, H. J., and Manns, M. P. (1994) Interaction between Helicobacter pylori and human gastric epithelial cells in culture: effect of antiulcer drugs. *Pharmacology* **49,** 226–237.

19. Quaroni, A. and May, R. J. (1980) Establishment and characterization of intestinal epithelial cell cultures. *Methods Cell Biol.* **218,** 403–427.

20. Kedinger, M., Haffen, K., and Simon-Assman, P. (1987) Intestinal tissue and cell cultures. *Differentiation* **36,** 71–85.

21. Evans, G. S., Flint, N., and Potten, C. S. (1994) Primary cultures for studies of cell regulation and physiology in intestinal epithelium. *Annu. Rev. Physiol.* **56,** 399–417.

22. Perreault, N. and Beaulieu, J.-F. (1998) Primary cultures of fully differentiated and pure human intestinal epithelial cells. *Exp. Cell Res.* **245,** 34–42.

23. Basque, J.-R., Chailler, P., Perreault, N., Beaulieu, J.-F., and Ménard, D. (1999) A new primary culture system representative of the human gastric epithelium. *Exp. Cell Res.* **253,** 493–502.

24. Basque, J.-R. and Ménard, D. (2000) Establishment of culture systems of human gastric epithelium for the study of pepsinogen and gastric lipase synthesis and secretion. *Microsc. Res. Tech.* **48,** 293–302.

25. Streeter, G. L. (1920) Weight, sitting head, head size, foot length and menstrual age of the human embryo. *Carnegie Inst. Contrib. Embryol.* **11,** 143–179.

26. Freshney, R. I. (1987) *Culture of Animal Cells: a Manual of Basic Technique.* 2nd ed., Wiley-Liss, New York, pp. 246–247.

27. Bowie, D. J. (1936) A method for staining the pepsinogen granules in gastric glands. *Anat. Rec.* **64,** 357–365.

28. Ménard, D., Monfils, S., and Tremblay, E. (1995) Ontogeny of human gastric lipase and pepsin activities. *Gastroenterology* **108,** 1650–1656.

29. Lévy, E., Goldstein, R., Freier, S., and Shafrir, E. (1981) Characterization of gastric lipolytic activity. *Biochim. Biophys. Acta* **664,** 316–326.

30. Anson, M. L. and Mirsky, A. E. (1932) The estimation of pepsin with hemoglobin. *J. Gen. Physiol.* **10,** 342–344.

31. Lowry, O. H., Rosebrough, N. F., Farr, A. L., and Randall, R. J. (1951) Protein measurement with Folin phenol reagent. *J. Biol. Chem.* **193,** 265–275.

32. Chomczynski, P. and Sacchi, N. (1987) Single-step method of RNA isolation by acid guanidium thiocyanate-phenol-chloroform extraction. *Anal. Biochem.* **162,** 156–159.

33. Tremblay, E., Basque, J.-R., Rivard, N., and Ménard, D. (1999) Epidermal growth factor and transforming growth factor-α downregulate human gastric lipase gene expression. *Gastroenterology* **116,** 831–841.

34. Beaulieu, J.-F. (1997) Extracellular matrix components and integrins in relationship to human intestinal cell differentiation. *Prog. Histochem. Cytochem.* **31,** 1–78.

35. Montgomery, R. K., Mulberg, A. E., and Grand, R. I. (1999) Development of the human gastrointestinal tract: twenty years of progress. *Gastroenterology* **116,** 702–731.

36. Podolsky, D. K. and Babyatsky, M. W. (1995) Growth and development of the gastrointestinal tract, in *Textbook of Gastroenterology*, 2nd ed. (Yamada, T., ed.), Lippincott, Philadelphia, PA, pp. 546–577.
37. Basque, J.-R., Chailler, P., and Ménard, D. (2002) Laminins and TGF-β maintain cell polarity and functionality of human gastric glandular epithelium. *Am. J. Physiol. Cell Physiol.* **282,** C873–C884.

16

Isolation and Culture of Human Colon Epithelial Cells Using a Modified Explant Technique Employing a Noninjurious Approach

Hamid A. Mohammadpour

1. Introduction

Colorectal cancer, originating in tissues of epithelial origin, is one of the most prolific among human malignant diseases. The prevalence of this disease creates a need for colon epithelial cell lines to facilitate the study of the etiology of this disease and the biochemical and molecular characteristics of epithelial cells. Several approaches for the isolation of colon epithelial cell cultures have previously been reported *(1–8)*. Methods vary, ranging from enzymatic, mechanical disruption, and explant procedures. The enzymatic and mechanical approaches employed to isolate individual cell types are not always practical because they generally result in a homogeneous cell population of fibroblasts *(2,3)*. This is owing to the fact that fibroblasts adhere to a matrix more efficiently, preventing other cell types from propagating and because few epithelial cells survive harsh treatments *(2)*. The method described herein is based on a noninjurious approach using minimal primary tissue manipulation to isolate colon epithelial cells in culture *(1)*. The colon epithelial cell lines are developed through long-term observation of untreated explants. Epithelial cell lines are obtained through the use of cloning rings. The isolation of colon epithelial cancer cells is obtained by differential attachment of the epithelial cells and/or the use of cloning rings.

The protocols below describe the procedures required to: (1) prepare conditioned media; (2) transport and wash human colon tissues; (3) explant human colon tissues; and (4) isolate and propagate human colon epithelial cells in culture.

From: *Methods in Molecular Medicine, vol. 107: Human Cell Culture Protocols, Second Edition*
Edited by: J. Picot © Humana Press Inc., Totowa, NJ

2. Materials

2.1. Tissue-Culture Facility

1. Laminar flow hood, BioSafety level two
2. A humidified, incubator set at 37°C.
3. An inverted phase-contrast microscope.
4. Hemocytometer.
5. Scalpel.
6. Forceps.
7. Tissue-culture flasks and dishes.
8. Assorted cloning rings.
9. 70-µm cell strainers.

2.2. Solutions

1. Hank's balanced salt solution (HBSS) containing penicillin (50 U/mL), strepto-mycin (50 µg/mL), and amphotericin B (0.5 µg/mL).
2. Fibroblast growth media (FGM) consisting of RPMI-1640 supplemented with 2 m*M* glutamine, penicillin (50 U/mL), streptomycin (50 µg/mL), amphotericin B (0.5 µg/mL), 10% fetal bovine serum (FBS), insulin (5 µg/mL), and transferrin (5 µg/mL).
3. Colon tissue transport media consisting of RPMI-1640 supplemented with 2 m*M* glutamine, penicillin (50 U/mL), streptomycin (50 µg/mL), and amphotericin B (0.5 µg/mL).
4. Epithelial growth media (EGM) consisting of RPMI-1640 supplemented with 3–5% fibroblast conditioned media, 2 m*M* glutamine, penicillin (50 U/mL), strep-tomycin (50 µg/mL), amphotericin B (0.5 µg/mL), 10% FBS, insulin (5 µg/mL), and transferrin (5 µg/mL). This may be filtered using a 0.22-µm sterile filter to insure sterility has been maintained.
5. Cell culture freezing media.
6. Trypan blue (0.4%).
7. 0.05% trypsin-ethylenediaminetetraacetic acid (EDTA).
8. Conditioned media.
9. Sterile grease.
10. Phosphate-buffered saline (PBS).

2.3. Cells and Tissue

1. Fibroblasts at low passage/population doubling level (PDL) (available from American Type Tissue Culture).
2. Human colon tissue.

3. Methods

3.1. Preparation of Conditioned Media using Fibroblasts (see Note 1)

The culture of fibroblasts should be conducted under sterile conditions.

1. Rapidly thaw a vial of frozen cells (PDL <30, approx 1×10^6 cells) using a 37°C water bath.
2. Transfer the cells into a T-75 cm^2 tissue-culture flask containing FGM.
3. Incubate the cells in the presence of 5% CO_2 and humidity at 37°C overnight.
4. Replace the media the next day and continue to change the media every 2–3 d.
5. Allow the cells to grow to 80% confluency.
6. Expand the cells:
   a. Wash the cells with PBS.
   b. Add 2 mL of 0.05% trypsin/EDTA and incubate at room temperature until the cells dislodge from the plate.
   c. Add 10–12 mL of FGM and gently pipet up and down to disperse the cells.
   d. Seed the cells in new flasks at a ratio of 1:3 in FGM.
7. Allow the cells to grow to 80% confluency.
8. Remove the growth media and wash the cells once with serum-free FGM.
9. Maintain the cells in serum-free FGM for 48–72 h (*see* **Note 2**).
10. Collect the media and replace it with fresh serum-free FGM every 48–72 h.
11. Filter the collected media using a cell strainer and then centrifuge at 800*g*, 4°C for 5 min to pellet any cell debris.
12. Remove the supernatant and aliquot into 15-mL sterile tubes and store at –20°C. This is the conditioned media to be added as a supplement to the epithelial growth media.
13. Repeat this procedure several times or until the cells become senescent.
14. This procedure can be repeated as needed (*see* **Note 3**).

3.2. Transporting and Washing Human Colon Tissues

Colon tissue is obtained from patients undergoing colon resection due to colon cancer or other gastrointestinal diseases or abnormalities. The following steps should be carried out using sterile techniques to minimize contamination.

1. Immediately place the colon tissue in a sterile plastic beaker containing 4°C colon tissue transport media.
2. Transport the tissue on ice to the laboratory for processing.
3. Transfer the tissue into a sterile Petri dish. Remove and discard any damaged areas using a scalpel, gently holding the tissue with a forceps.
4. Remove the connective tissue and membranes from the colon exposing the mucosa layer below, which contains epithelial layers.
5. Dice into 2–3-mm sections and transfer the sections into 50-mL conical tubes.
6. Add 37°C HBSS to cover the sections. Swirl the tube gently to remove any debris or fecal materials that might have been trapped in the tissue.
7. Allow the sections to settle and remove HBSS using a pipet.
8. Repeat this procedure at least 10 times to ensure the sections are clean, followed by several washes with EGM (*see* **Note 4**).

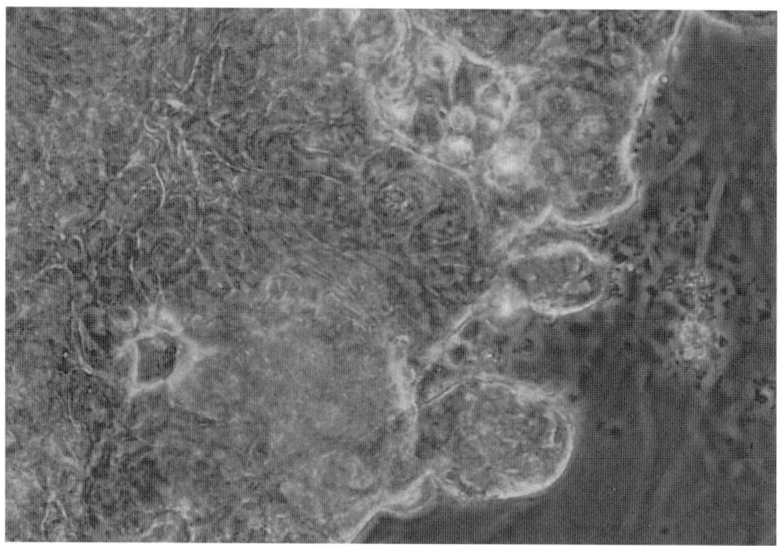

Fig. 1. Extensions or appendages growing outward from explanted colon tissue sections.

3.3. Explanting

The explanting procedures should be conducted under sterile conditions.

1. Transfer 10–15 washed tissue sections (2–3 mm) into a 100-mm Petri dish containing EGM (*see* **Note 5**). Place the section with the mucosal layer in contact with the surface of the dish.
2. Carefully transfer the dish to the incubator with humidity and 5% CO_2 at 37°C.
3. Do not disturb the dishes for several days, allowing the sections to adhere to the plate.
4. After a few days, gently remove the dishes from the incubator and replace 50% of the media with fresh media without disturbing the adhered sections (*see* **Note 6**).
5. Remove sections that are not adhered to the plate.
6. Return dishes to the incubator.
7. Change 50% of the media 2–3 times per week without disturbing the sections.
8. Over time, you will observe (using an inverted microscope) the presence of extensions or appendages growing outward from the sections (*see* **Fig. 1**). Initially, these extensions will be made up of fibroblasts and it may require weeks (≥6) of observation before obvious epithelial cell growth is apparent. This outgrowth of fibroblasts and other cells will sometimes occur in clusters and sometimes as individual cells. This observation may not be true for all the explanted sections (*see* **Note 7**).
9. Continue changing the media as before until epithelial cell layers are evident under the microscope. The morphology of these epithelial cells can be described as

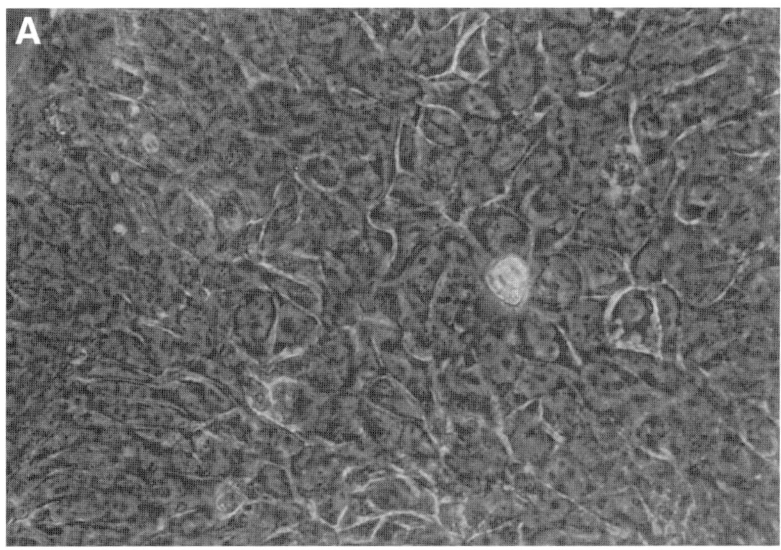

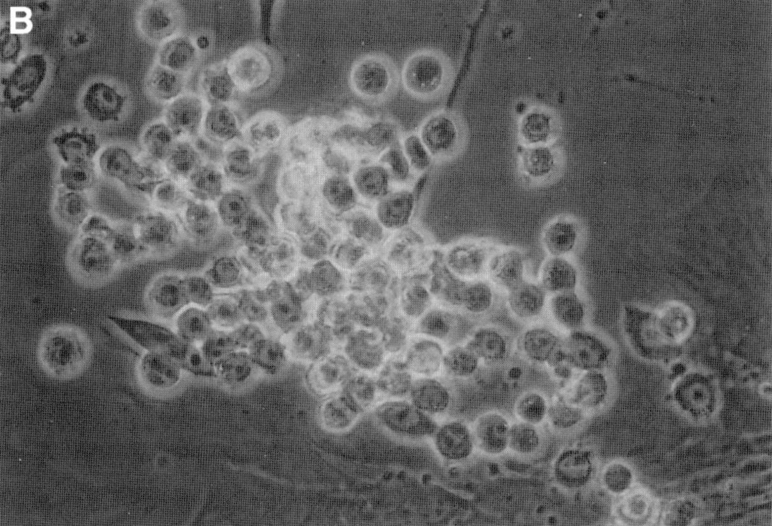

Fig. 2. **(A)** Epithelial cells showing columnar or brick-like morphology. **(B)** Raised clusters of epithelial cells surrounded by mucus.

columnar or brick-like (*see* **Fig. 2A**) or as raised clusters covered with clear mucus (*see* **Fig. 2B**) rather than the elongated, spindle shape common of fibroblasts.

10. Isolate the epithelial cells when you see colonies or fields that can be removed from other cells successfully either by cloning rings (*see* **Fig. 3A**) or by serial replating (*see* **Fig. 3B**).

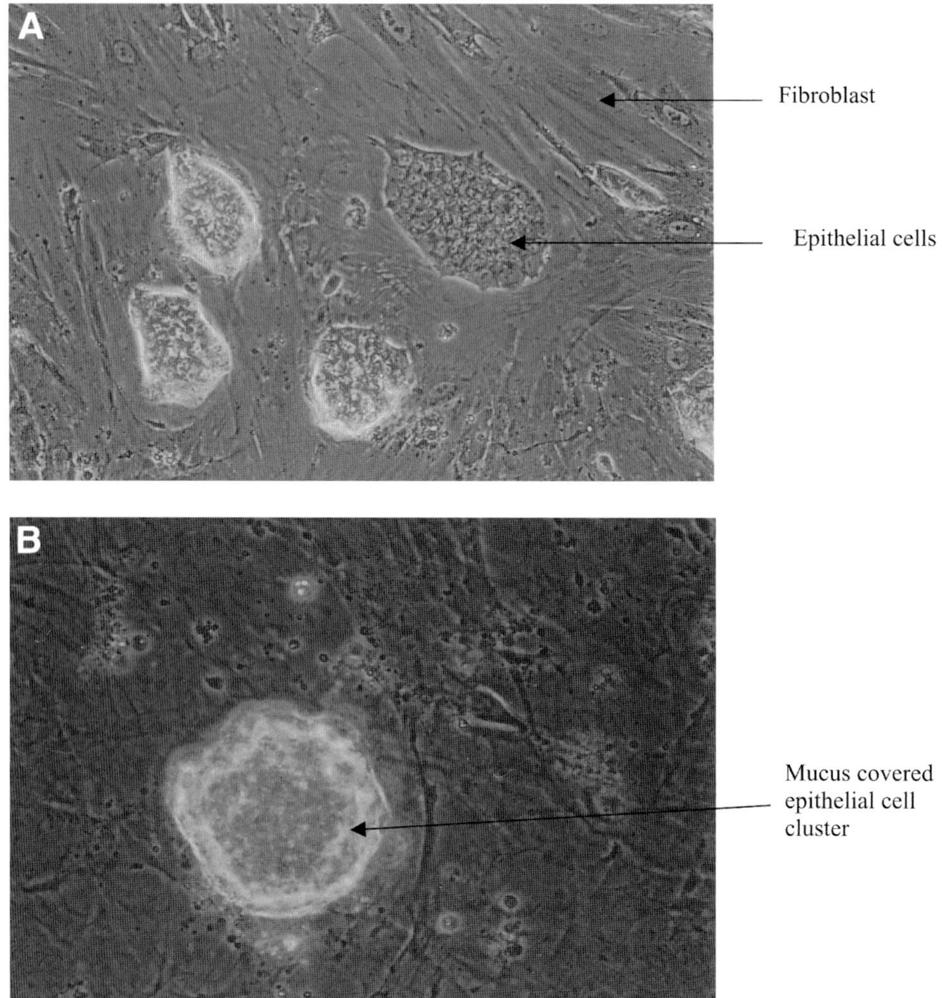

Fig. 3. **(A)** Epithelial cells growing as colonies adhered to the surface of the plate. The epithelial cells are surrounded by fibroblasts and can be isolated through the use of cloning rings. **(B)** Epithelial cells in the form of a cluster surrounded by mucus and loosely associated with fibroblasts. The epithelial cells can be separated from the fibroblasts by Serial Replating.

3.4. Isolating Colon Epithelial Cells from Other Cell Populations

The isolation of colon epithelial cells should be conducted under sterile conditions.

3.4.1. Isolating Colon Epithelial Cells Using Cloning Rings

Epithelial cells can be isolated from a population derived from explants using a cloning ring when the cells are forming distinct colonies and are adhered to the surface (*see* **Fig. 3A**).

1. Remove the media from the Petri dish and carefully wash the cells with 37°C PBS.
2. Identify and mark those areas containing normal epithelial cells with a marker on the outside of the Petri dish.
3. Select a cloning ring smaller than the identified area to reduce the transfer of other cell types during the isolation procedure (*see* **Note 8**).
4. Using a sterile Q-tip, apply a small amount of sterile grease to the outer edge of the cloning ring.
5. Gently place the ring on top of the patch selected for isolation and apply even pressure to anchor the ring. Avoid spreading the grease or damaging the cells.
6. Add a few drops of 0.05% trypsin-EDTA in the cloning ring and gently rotate the plate to cover the cells.
7. Allow to sit at room temperature until the cells appear to be dislodged from the plate.
8. Add a few drops of EGM to inactivate the trypsin.
9. Using a pipet, remove the dislodged cells and place into a conical tube containing several milliliters of EGM.
10. Pipet the cells up and down and filter through a cell strainer to make a single-cell suspension.
11. Centrifuge (4°C) at 800*g* for 5 min to remove media containing trypsin.
12. Remove the supernatant and add fresh EGM. Resuspend the cell pellet gently and determine cell number using trypan blue.
13. Plate the cells at a density of 0.5–1.0 million cells per T-75 cm^2 flask and continue cell maintenance in EGM.

3.4.2. Separating Epithelial Colon Cells from Other Cell Populations Without the Use of a Cloning Ring and/or Enzymatic Digestion

The presence of mucus surrounding clusters of cancer cells often causes the epithelial cells to form a loose association with the surface of the plate and/or fibroblasts (*see* **Fig. 3B**). This property permits the separation of cancer cells from other cell populations without the use of cloning rings or enzymatic disruption.

1. Gently tap the side of the dish with the palm of your hand to dislodge the clusters.
2. Transfer the media containing the clusters to a centrifuge tube.
3. Centrifuge at 4°C and 800*g* for 5 min.
4. Discard the supernatant.
5. Add 2–3 mL (depending on the size of the pellet) of 0.05% trypsin-EDTA and incubate at room temperature with occasional agitation to break up cell clumps and mucus.

6. Add approx 10 mL EGM and filter through a cell strainer, centrifuge as before, and remove the media.
7. Resuspend the cell pellet in EGM and plate into a new flask.
8. Expand the colon epithelial cells when the cell density reaches 75–80% confluency:
   a. Wash the cells with PBS.
   b. Add 2 mL of 0.05% trypsin/EDTA and incubate at room temperature until the cells dislodge from the plate.
   c. Add 10–12 mL of EGM and gently pipet up and down to disperse the cells.
   d. Seed the cells in new flasks at a ratio of 1:3 in EGM.

3.4.3. Isolating Epithelial Colon Carcinoma Cells by Serial Replating

If the isolation of pure cultures is not obtained by the above procedures, a serial replating procedure, which takes advantage of the differential attachment of transformed epithelial cells, can be applied.

3.4.3.1. SERIAL REPLATING

1. Wash the cells with 37°C PBS.
2. Add 1–2 mL of warm (37°C) 0.05% trypsin-EDTA , and incubate at room temperature until the cells are detached from the plate.
3. Add 5–10 mL EGM, pipet the cells up and down and filter the cells through a cell strainer to remove any clumps.
4. Centrifuge the suspension at 4°C and 800g for 5 min and remove the supernatant.
5. Add fresh EGM and resuspend the pellet gently.
6. Count the cells and replate at a density of 0.5–1.0 million cells in a T-75 cm^2 flask and return to the incubator.
7. Incubate the cells for approx 1 h. Most fibroblasts adhere to the surface during the incubation because of the fact that they attach at a faster rate than other cells such as epithelial cancer cells. This differential attachment allows for the isolation of the nonadherent cells (*see* **Note 9**).
8. Remove the media containing suspended cells, which includes cancer cells, and transfer into a new T-75 cm^2 flask.
9. Repeat **steps 7** and **8** until all of the fibroblasts have been removed. This is an effective method for isolating pure epithelial cancer cells from a mixed population of cells.
10. When fibroblasts are no longer present, return the flask to the incubator and allow the cells to grow to 80% confluency.
11. Continue to culture the pure epithelial cells (*see* **Fig. 4**).
    a. Wash the cells with PBS.
    b. Add 2 mL of 0.05% trypsin/EDTA and incubate at room temperature until the cells dislodge from the plate.
    c. Add 10–12 mL of EGM and gently pipet up and down to disperse the cells.
    d. Seed the cells in new flasks at a ratio of 1:3 in EGM.

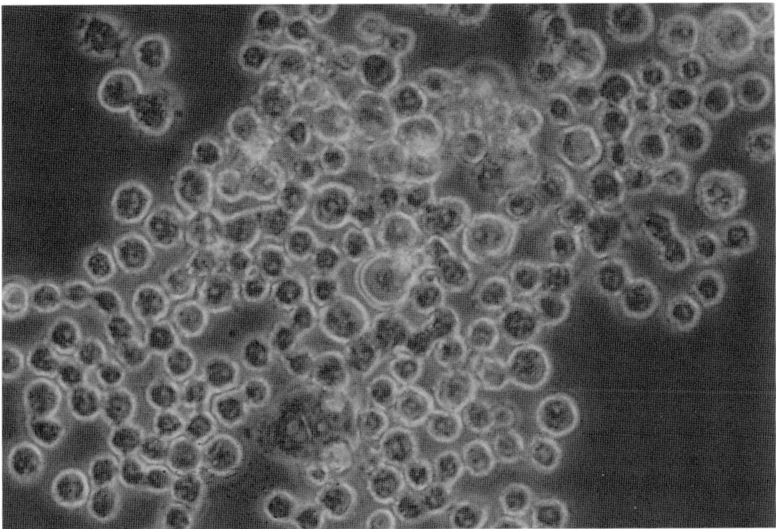

Fig. 4. Pure epithelial cells originating from colon tissue explants and isolated by Serial Replating.

3.5. Freezing and Storage of the Cells

At various PDLs, aliquots of the cultured cells can be frozen and stored in the vapor phase of liquid nitrogen for future use.

1. Grow cells to 75–80% confluency in a tissue-culture flask.
2. Wash the cells three times with PBS and add 2 mL of 0.05% trypsin-EDTA. Incubate at room temperature to dislodge the cells.
3. Inactivate the trypsin with RPMI-1640 media containing 10% FBS.
4. Centrifuge at 800g for 5 min.
5. Remove the supernatant and resuspend the cells in cold freezing media (*see* **Note 10**).
6. Aliquot 1 mL (approx 1×10^6 cells) of cell suspension into cryogenic tubes and freeze in the vapor phase of a liquid nitrogen storage tank (*see* **Note 11**).

4. Notes

1. To ensure successful cell isolation and growth, it is essential to provide the cells with necessary growth factors. There are several ways of accomplishing this, including the addition of individual growth factors to the media. This requires extensive knowledge of various growth factors, however, and is the most effective when the exact cellular requirements are known. Although growth media formulations have improved dramatically, many cells require additional factors for optimum growth. Two well-accepted methods of providing the necessary requirements

are the use of conditioned media and the use of feeder cells. Of the two, the author considers conditioned media to be more effective for the successful propagation of cultured cells.

2. In the absence of serum, the cells are stimulated to produce and excrete growth factors. This growth factor–enriched media is used as a supplement to provide necessary factors for the propagation of pure cultures of low cell number.

3. Fibroblasts isolated from the obtained colon tissues may also be used to produce conditioned media. Transfer several pieces of colon tissue to a 25-mL Erlenmeyer flask containing 5–10 mL of 0.05% trypsin-EDTA. Stir gently for 10–30 min at 37°C. Periodically check the solution for cell dissociation. It should not take more than 30 min to dissociate enough fibroblast cells from the tissue. Inactivate the trypsin by adding 10 mL of FGM. Filter through a cell strainer. Centrifuge at 800g for 5 min and discard the supernatant. Determine cell viability using trypan blue and plate the cells in a T75 cm^2 flask at 50% density in FGM. Incubate overnight in the CO$_2$ incubator. Change the media to remove nonviable cells and cellular debris. In a few days, fibroblasts start to propagate. Continue changing the media every 2–3 d until the cells are 80% confluent. These cells can be used to harvest conditioned media as described in **Subheading 3.1.**

4. This step is crucial to the success of the isolation procedures. Contamination will result in a failed effort and contaminated cultures should be discarded. Attempts to wash the contaminated cultures will not be successful.

5. In order to increase the probability that the explant is successful it is essential to plate several individual dishes containing 10–15 tissue sections each because most of the explants either do not adhere to the matrix or result in fibroblasts.

6. The media contains intestinal growth factors produced by the cells and is necessary for cellular growth.

7. Fibroblasts grow faster than other cells, if there is a cluster of epithelial cells growing under the sections, they will not be obvious early on and time is required for them to grow out from under the sections. Removing the sections after observing fibroblast growth is premature and may remove the growing epithelial cells underneath.

8. The use of a cloning ring smaller than the patch of selected cells ensures that the cells you are isolating are pure.

9. Often fibroblasts adhere within 30 min of incubation, therefore, observe the cells at approx 15-min intervals. The cancer cells are surrounded by mucus that retards the attachment of the cells to the surface. This principle allows for the removal of the suspended and loosely attached epithelial cells after the fibroblasts have adhered to the surface of the plate.

10. Dimethyl sulfoxide (DMSO) is toxic to cells and personal experience suggests that greater cell viability can be attained when cold freezing media is added to chilled cells. This procedure should be done rapidly to prevent cell death.

11. Cancer cells can be stored in a –80°C freezer, although cell viability will decrease over time, normal cells should be stored in the vapor phase of liquid nitrogen.

Acknowledgments

The author wishes to thank Byron McGregor, M.D. and Laura Briggs, Ph.D. for their assistance in the preparation of this manuscript. Their input and guidance was invaluable.

References

1. Mohammadpour, H., Hall, M. R., Pardini, R. S., Khaiboullin, S. F., Manalo, P., and McGregor, D. B. (1999) An atraumatic method to establish human colon carcinoma in long term culture. *J. Surg. Res.* **82,** 146–150.
2. Baton, A., Sakamoto, K., and Shamsuddin, A. M. (1992) Long-term culture of normal human colonic epithelial cells in vitro. *FASEB J.* **6,** 2726–2734.
3. Moyer, M. P., Dickson, P. S., Culpepper, A. L., et al. (1990) In vitro propagation and characterization of normal, preneoplastic, and neoplastic colonic epithelial cells, in *Colon Cancer Cells* (Moyer, M. P. and Poste, G., eds.), Academic, San Diego, CA, pp. 85–136.
4. Owens, R. B., Smith, H. S., Nelson-Rees, W. A., and Springer, F. L. (1976) Epithelial cell cultures from normal and cancerous human tissues. *J. Natl. Cancer Inst.* **56,** 843–849.
5. Moyer, M. P. and Aust, J. B. (1982) Human colon cells: Culture and in vitro transformation. *Science* **224,** 1445–1447.
6. McGregor, D. B., Morris, L. L., Manalo, P., et al. (1989) Pentagastrin stimulation of human colon carcinoma. *Arch. Surg.* **124,** 470–472.
7. Danes, B. S. (1985) Long-term-cultured colon epithelial cell lines from individuals with and without colon cancer genotypes. *J. Natl. Cancer Inst.* **75,** 261–267.
8. Staufer, J. S., Manzano, L. A., Balch G. C., et al. (1995) Development and characterization of normal colonic epithelial cell lines derived from normal mucosa of patients with colon cancer. *Am. J. Surg.* **169,** 161–169.

17

Isolation and Culture of Human Hepatocytes

Martin Bayliss and Graham Somers

1. Introduction

The liver performs a wide range of physiologically important metabolic functions including the synthesis and secretion of albumin, fibrinogen, and other plasma proteins, the synthesis of cholesterol and bile acids, and the metabolism of drugs, steroids, and amino acids. In addition, the liver has a central role in energy metabolism as the major store of glycogen and the site of gluconeogenesis and synthesis of fatty acids and triglycerides. The liver is, therefore, a vital organ; however, it is sometimes difficult to study specific liver functions in vivo owing to interfering influences from other organs, e.g., the kidney, gut, and lungs, which metabolize drugs and the muscle involvement in glucose homeostasis. A number of in vitro preparations such as tissue homogenates or subcellular fractions are available to study tissue-specific functions, although experiments in intact cells, with cell membrane and cellular organelles are postulated to reflect the in vivo situation more closely. Hepatocytes represent 60–65% of the number of cells in the liver but occupy some 80% of the liver volume because of their large volume *(1)*. Nonparenchymal cells such as Kuppfer cells, lipocytes, and endothelial cells (6%) make up the remaining tissue. Intercellular spaces account for approx 14% of the liver volume *(1)*. Isolated human hepatocytes have been used extensively for a variety of techniques investigating cellular metabolism *(2,3)* and the induction *(4–9)*, metabolism *(10–14)*, and toxicology of xenobiotics *(15,16)*.

The preparation of isolated hepatocytes from a heterogeneous tissue such as the liver involves several steps. Initially, the cells must be dissociated from the fibroconnective skeleton. Tissue dispersion requires the disruption of reticulin fibrils along with adhesion proteins such as fibrinonectin and laminin, which

From: *Methods in Molecular Medicine, vol. 107: Human Cell Culture Protocols, Second Edition*
Edited by: J. Picot © Humana Press Inc., Totowa, NJ

constitute the framework of the liver lobule *(1)*. Hepatocyte isolation also requires the disruption of cell–cell adhesions, i.e., the junctional complexes, and finally, the recovery of the hepatocytes and their preparation for further use *(1)*.

Enzymatic hepatocyte isolation was first introduced in 1967 *(17)*. Prior to this time, mechanical and chemical dissociation of the liver parenchyma had been used. This latter isolation yielded very poor quality hepatocytes. The chemical and mechanical dissociation methods are discussed at length in a review by Berry et al. *(18)*. The development of a technique utilizing enzymes to digest the liver providing a high yield of functionally active rat hepatocytes led to a rapid increase in the utility of the isolated hepatocyte as a model system for the study of liver functions *(19)*. The perfusion technique originally developed for the preparation of hepatocytes from small animal species has, over the last 20 yr, been adapted and applied to human material.

Human hepatocytes have been prepared from a variety of liver samples, including a whole liver *(20,21)*, a portion of whole liver *(20)*, end of lobe wedge biopsy samples *(1,22–24)*, and small biopsy fragments *(25,26,27)*. The majority of human liver made available for research is in the form of large biopsies, which are usually obtained from patients undergoing partial hepatectomy. This material is usually 5–10 g in size and normally has only one cut surface. Therefore, the wedge biopsy perfusion method for preparing hepatocytes appears to be the most suitable method for utilizing the available material. The preparation of human hepatocytes using this method has produced yields of between $0.1–30 \times 10^6$ hepatocytes/g wet liver with cell viability in the range of 70–95% as determined by trypan blue exclusion. However, cell yields and viability vary depending on the size of the biopsy, the ability of the operator to cannulate the exposed vessels, and the resultant perfusion of the tissue.

The lack of human liver tissue has prompted developments in maximizing the use of human hepatocytes by developing isolation, culture, co-culture, and cyropreservation techniques. The limited availability of human material requires a system that can provide a high yield of functional isolated hepatocytes rapidly utilizing all the available material.

A major consideration when using human material to prepare hepatocytes is the variation in the functional activity between cell preparations. This variability may be caused in part to the length of time taken in obtaining the sample, sample storage conditions, or the sample history, as well as the genetic and environmental factors affecting the donor. Information regarding the patient history, social history, dietary habits, and genetic polymorphism are important for interpreting results. For example, a liver sample removed from a patient who had taken a barbiturate overdose can have drug-metabolizing enzymes that are significantly induced *(28)*. These enzymes remain elevated in isolated hepatocytes, which are cultured for several days. Furthermore, studies by Houssin et al. *(29)* have also suggested that variable intracellular ATP and glycogen content in hepatocyte populations is attributed to the nutritional status of the donor.

Isolated human hepatocytes used in suspension with no attachment to any form of extracellular matrix are viable for only a few hours *(30)*. Hepatocyte survival and functionality can be extended considerably when cells are maintained in monolayer culture conditions in the presence of an extracellular matrix and/or a supplemented media. Human hepatocytes in monolayer culture have been shown to maintain plasma protein production, glycolysis, and urea synthesis for extended periods of time from periods of several days up to 35 d *(31–34)*.

Drug metabolizing enzymes, on the other hand, particularly the cytochromes P450, are maintained during the initial stages of culture, but decline with time *(35,36)*. However, these enzymes can be elevated in culture to some extent by the use of inducing agents such as phenobarbital, 3-methycholanthrene, and rifampicin *(5,7–9,37,38)*.

A number of factors including the extracellular matrix used, seeding densities, culture media, media supplements, and cell–cell interactions need to be considered when culturing human hepatocytes. Culture configurations that allow greater cell–cell contact have been suggested to have more influence over the quality of the resultant hepatocytes than the constitution of the medium. The encapsulation of hepatocytes in gel matrices (sandwich configuration) or production of cell aggregates (spheroids) have been suggested. The isolation and culture of hepatocytes and the evaluation of many of the above factors has been discussed and reviewed by a number of workers *(7,8,39–42)*.

It should be emphasized that a number of different conditions for isolating and culturing human hepatocytes have been published and are dependent on the type of studies to be carried out with the resultant cell preparation. This chapter represents one of the accepted practices at the time of writing.

2. Materials

1. In the U.K., human liver samples can be used for hepatocyte isolation and culture following approval from the local Ethical Review Board. These samples are often obtained as surgical waste following partial hepatactomy (*see* **Notes 1** and **3**). Where fresh tissue cannot be obtained, cryopreserved hepatocytes or hepatic cell lines may be considered (*see* **Notes 13** and **14**).
2. A good supply of cell culture grade water should be used.
3. The following reagents are required for cell culture. Bovine serum albumin (BSA) (Fraction V, fatty-acid free), collagen type I (soluble rat tail or calf skin), collagenase (EC 3.4.24.3) from *Clostridium histolyticum* (specific activity, 0.17–0.44 Wunsch U/mg lyophilisate), deoxyribonucleate 5′ oligonucleotidohydrolase (DNase I; EC 3.2.21.1) from bovine pancreas, Earle's balanced salt solution

(EBSS; 100 mL 10X concentrate, calcium and magnesium free), ethylenediamine-tetraacetic acid (EDTA), ethylene glycol *bis*-(β-aminoethyl ether) N,N,N′,N′-tetra acetic acid (EGTA), 4-(2-hydroxyethyl) piperazine-2-ethanesulfonic acid (HEPES), L-glutamine (200 mM), insulin (from bovine pancreas), sodium bicarbonate solution (7.5%), trypsin inhibitor (from soyabean, Type II-S), Williams' Medium E (WME) without phenol red or L-glutamine, trypan blue [4,4′-*bis*(8-amino-3, 6-disulpho-1-hydroxy-2-naphthalazo)3,3′-dimethylbiphenyl tetrasodium salt] formulated as a 0.5% w/v solution in 0.85% saline, penicillin/streptomycin (lyophilized; 10,000 U/10,000 µg/mL). All other chemicals and solvents should be of Analar grade or equivalent.

4. Perfusion buffer: EBSS (magnesium and calcium free); 117 mM NaCl (6.8 g/L), 5.4 mM KCl (0.4 g/L), 0.9 mM NaH$_2$PO$_4$•H$_2$O (0.125 g/L), 26.16 mM NaHCO$_3$ (2.2 g/L), 5.56 mM glucose (1 g/L), and 0.05 mM phenol red (0.02 g/L). Add EBSS (10X concentrate) to cell-culture grade water (870 mL). Sodium bicarbonate (7.5% solution; 30 mL) should be added to give a final concentration of 26 mM and the pH adjusted to 7.4. Commercial EBSS (10X concentrate) can be purchased and stored for up to 6 mo. Solutions should be prepared on the day of isolation.

5. Chelating solution: A stock solution of EGTA (25 mM) should be prepared in 0.1 M NaOH and the pH adjusted to 7.4. This stock can be stored at 4°C for up to 3 mo. An aliquot of the stock EGTA (25 mM) solution (10 mL) is then added to 490 mL EBSS solution (pH 7.4) to give a final concentration of 0.5 mM EGTA, prepare on the day of isolation.

6. Enzyme solution: Collagenase (0.24 U/mL) is dissolved in EBSS (pH 7.4) (100 mL). Standardization on a fixed number of units allows for the varying specific activities of different batches of enzyme. Trypsin inhibitor (10 mg) and calcium chloride (2 mM) are also added. Prepare on the day of isolation.

7. Hepatocyte dispersal buffer: HEPES (10 mM) containing NaCl (142 mM) and KCl (7 mM) should be adjusted to pH 7.4. BSA should be added slowly to give a 1% w/v solution. Stir slowly and continuously. Prepare on the day of isolation.

8. Culture medium: L-Glutamine (20 mM) should be added to WME to give a final concentration of 4 mM. It is recommended that the culture medium be purchased from a commercial supplier. For discussion of alternative media, *see* **Note 10**.

9. Attachment media: WME containing L-glutamine (4 mM), penicillin/streptomycin solution (100 U/100 µg/mL), insulin (0.25 U/mL), and 10% fetal calf serum (FCS). The penicillin/streptomycin, insulin are prepared as X100 stock solutions in WME without L-glutamine and stored at –20°C in aliquots (1 mL). The stock solutions described above are added to WME in a final volume of 100 mL. A variety of attachment factors have been used to culture human hepatocytes and are discussed in **Note 9**.

10. Incubation media is the same as the attachment media excluding FCS.

11. Culture plates can be purchased commercially precoated with collagen (0.4–1.4 µg type I collagen/cm^2). Alternatively, plastic culture plates or flasks can be coated with soluble collagen. A stock solution of collagen is prepared by dissolving either

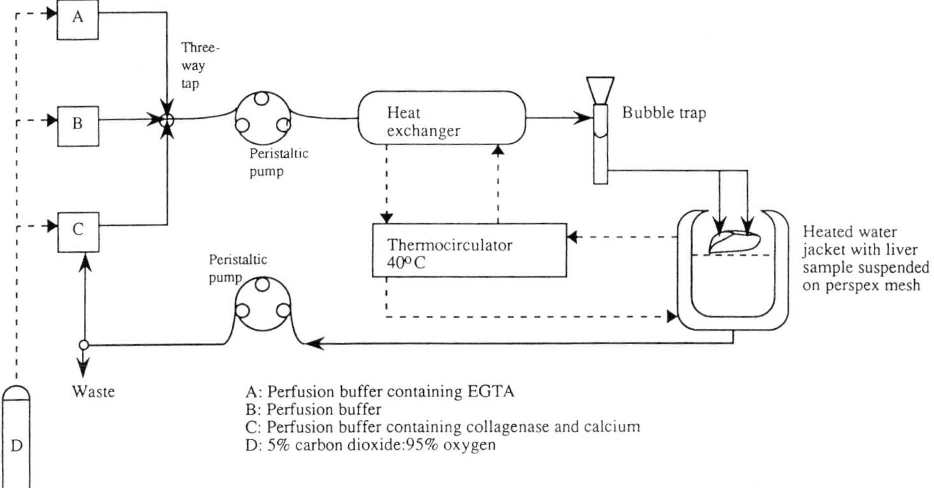

Fig. 1. Apparatus used for the perfusion of human liver biopsy samples. Reproduced with kind permission of D. Cross.

soluble rat tail or calf skin collagen type I in acetic acid (0.1 *M*). This solution is diluted with sterile water and 100 µg applied to a 35-mm culture well. This gives a coating of 5–10 µg collagen/cm^2. Plates are air-dried in sterile conditions, preferably in a laminar flow cabinet. The dried plates are washed well with either culture quality water or media to remove the acetic acid.

12. Perfusion apparatus: The equipment used for the perfusion of liver samples to prepare isolated hepatocytes is shown in **Fig. 1**. All perfusion equipment should either be autoclaved or washed with 70% ethanol and rinsed with sterile distilled water prior to use. The perfusion system consists of three reservoirs each of which contains one of the three solutions used to perfuse the liver biopsy sample. Reservoir 1 contained chelating solution [perfusate (EBSS, pH 7.4) with EGTA], reservoir 2 contained perfusate only, and reservoir 3 contained enzyme solution (collagenase and calcium ions). The reservoirs are linked via a three-way tap system. The flow of perfusate is maintained by a Watson-Marlow 502S peristaltic pump (Watson-Marlow Ltd., Falmouth, Cornwall, U.K.). The temperature of the perfusate is maintained at 37°C by passage through a heat exchanger. The water jacket around the heat exchanger is maintained at 40°C by a recirculating thermoregulator (Harvard Apparatus Ltd, Edenbridge, Kent, U.K.) to accommodate a 3°C loss in perfusate temperature between the heat exchanger and the polyethylene catheters used to cannulate the liver sample. A bubble trap is placed between the heat exchanger and the cannulae. Silicon tubing (1.6 mm id and 1.6 mm wd) is used to connect the system.

All perfusion buffers are continually gassed with a mixture of oxygen and carbon dioxide (95%:5% v/v) and the pH of each solution is monitored throughout the perfusion using a pH stick electrode (Jenway Ltd., Dunmow, Essex, U.K.).

Similar types of apparatus allowing constant perfusion can be designed according to space limitations as long as the main components (water-jacketed temperature-controlled reservoirs, constant flow, nonpulsatile pump, and bubble trap) are present.

13. Other equipment requirements include 18-gauge catheters (Becton Dickinson, NJ), boulting cloth (64 μm; Henry Simon Ltd., Stockport, Cheshire, U.K.), refrigerated bench top centrifuge (Hereaus Omnifuge 2.0RS), Improved Neubauer counting chamber (Weber Scientific International, Ltd., Lancing, Sussex, U.K.), a standard laboratory and an ID03 inverted laboratory binocular microscope (×400 and ×320 magnification, respectively; Carl Zeiss Ltd., Welwyn, Hertfordshire, U.K.), a thermostatically controlled CO_2 incubator, an autoclave, a class II laminar flow cabinet and a sterile form of dispensing media, i.e., an automatic Pipettus system (Flow Laboratories, Rickmansworth, Hertfordshire, U.K.).

3. Methods

Safety note. When working with human material, a solution of medical disinfectant, i.e., Virkon solution (2%), should always be available to disinfect spills. All waste media should be disinfected prior to disposal and any tissue remaining after the perfusion must be autoclaved before disposal. Local safety rules for the handling of human material need to be carefully observed.

Human hepatocytes are isolated from liver using a three-stage perfusion modified from the method of Oldham et al. *(43)*, which was derived from the method of Strom et al. *(44)*. The major modifications were the substitution of EBSS for HEPES as the perfusion buffer and an additional washing step to remove residual EGTA prior to collagenase perfusion, adopted later by other workers *(45,46)*.

1. Sterile techniques should be employed during the isolation of human hepatocytes for culture. All solutions should be either filtered (0.22 μm) or autoclaved and instruments autoclaved when possible.
   Human liver wedge biopsy (5–10 *g*) samples should be obtained with only one cut surface where possible. Polyethylene catheters (18 gauge) should be placed into the lumen of the exposed vessels. It is essential to use several catheters (preferably four) and obtain a good perfusion of the tissue sample.
2. The tissue sample should be perfused initially with the chelating solution containing EGTA (reservoir 1) at a flow rate of 8 mL/min/cannula for up to 8 min (*see* **Note 4**). This should result in rapid blanching of the liver sample. Areas not perfused will not yield isolated hepatocytes. The perfusate should be allowed to drain to waste.

3. The second step should be to perfuse with EBSS (reservoir 2) for a further 10 min to remove the chelating agent from the samples; EGTA inhibits the action of collagenase in the subsequent perfusion. The perfusate should be allowed to drain to waste.

4. The third stage should be to perfuse with enzyme solution (reservoir 3). The sample should be perfused with collagenase in the presence of calcium ions and trypsin inhibitor. Commercial collagenase preparations may be contaminated with trypsin, which if active during perfusion is thought to be detrimental to cell-surface receptors. The perfusion of the liver sample with collagenase solution should continue for up to 40 min or until the surface structure of the liver sample obviously changes owing to digestion of the tissue. This solution should be continuously recirculated. The perfusion buffers should be continuously gassed with oxygen, carbon dioxide (95%, 5% v/v), and the pH of each buffer monitored (*see* **Note 2**). Adjustments should be made to keep the pH constant as required. Perfusion of the liver acidifies the buffers which will be observed by a change in the color of the phenol red indicator.

5. Following digestion, the liver sample should be removed carefully into ice-cold dispersal buffer containing DNase I (4 mg/100 mL) *(47)*. DNase helps prevent cell clumping by acting on nucleic acids released by damaged cells during the isolation procedure. Care should be taken not to rupture the capsule of the perfused sample during transfer to the dispersal buffer. Once in the buffer, the capsule surrounding the liver should be ruptured with forceps and the hepatocytes gently combed from the tissue mass. The cell suspension should be filtered through prewetted boulting cloth, and the filtrate gassed with oxygen and carbon dioxide (95% : 5% v/v). The cell suspension should be centrifuged at 50 g and 4°C for 2 min. Round-bottom tubes (20 mL; Nunc, Gibco Ltd.) rather than conical shaped tubes should be used to minimize cell damage. The supernatant containing cell debris and nonviable cells should be removed by aspiration and the cell pellet resuspended in dispersal buffer containing DNase I by gentle swirling of the tube. The centrifugation procedure should be repeated. The third and final wash of the hepatocyte preparation should be repeated in the absence of DNase I. After centrifugation, resuspend the viable hepatocyte preparation in pregassed WME (30–100 mL) without phenol red and albumin, but containing L-glutamine and maintain at 4°C.

6. Plasma membrane integrity and hepatocyte number should be determined using trypan blue exclusion *(1,18,48,49)* in the absence of supplemented protein (*see* **Note 5**). Ensure that no albumin is present at this stage. An aliquot of trypan blue solution (50 µL) should be mixed with the human hepatocyte suspension (250 µL) and allowed to stand for 2 min at room temperature. The cell suspension should then be analyzed using an improved Neubauer counting chamber. The sample should be mixed and applied to both grids using a Pasteur pipet. With the cover slip in the correct position, the chamber has a depth of 0.1 mm. The cells should be counted in the central grid in the squares surrounded by triple lines. The area of this grid is 1 mm^2 and the volume is therefore 0.1 mL. The number of

cells in 1 mL is therefore the number of cells counted in the grid $\times 10^4$. Cells will always be found overlapping the outer lines of the grids and the usual practice is to include the top and left-hand lines in the count and ignore those cells over lapping the bottom and right-hand lines. Any dilution used (e.g., the addition of trypan blue) needs to be taken into account.

Cellular yields should be in the order of 1×10^7 cells/gm tissue.

Viable cells are recognized by their ability to exclude trypan blue from the nucleus *(1,48,49)*. Therefore, cell viability is determined by counting the number of stained and unstained cells:

$$\% \text{ Viability} = \frac{\text{Number of cells unstained}}{\text{Total cell number}} \times 100\%$$

All counts should be completed in duplicate. The counting chambers should be viewed using a standard laboratory 16 binocular microscope ($\times 400$ magnification). Typically 70–90% of isolated human hepatocytes exclude trypan blue.

Suspensions of isolated human hepatocytes can be used for short-term experiments, i.e., up to 4 h. If longer term experiments are needed, the cells require culturing.

7. If viability has been established, only hepatocyte preparations of greater than 90% should be used for culture. The final cell suspension should be centrifuged and the medium aspirated. The isolated hepatocytes should then be resuspended in the attachment medium. The final cell suspension should be in the region of 5×10^5 cells/mL.

8. For cell culture, all cell manipulations should be performed under sterile conditions in a class II laminar flow cabinet. All liquid handling should be performed under sterile conditions using a Pipettus system or similar equipment.

9. Isolated hepatocytes sediment rapidly on standing, therefore, to achieve a homogeneous cell suspension, the cell mixture should be swirled prior to pipeting. The hepatocyte suspension (2 mL) should be dispersed into the wells of the culture plates; either commercially precoated or "in house" prepared plates should be used. A seeding density of approx $10–15 \times 10^4$ cells/cm^2 (*see* **Note 7**) can be used as recommended by a number of workers *(8,28,37,41,44,45)*. Depending on the size of culture plate used, the initial hepatocyte suspension can be diluted or concentrated to achieve the seeding density.

10. After plating with hepatocytes, the culture plates should be placed in a CO_2 incubator at 37°C to allow hepatocytes to attach to the extracellular matrix. The incubator should contain a humidified atmosphere of 5% carbon dioxide in air. Attachment of human hepatocytes can take up to 16 h (*see* **Note 8**).

11. Culture plates containing hepatocytes should be viewed under the inverted microscope. Flattening of the hepatocytes should indicate cell attachment. Once hepa-

tocyte attachment is complete, the media above the attached monolayer should be removed by aspiration using a sterile glass pipet or the Pipettus system. The culture plates should be tilted slightly and care must be taken not to damage the hepatocyte monolayer. The monolayer should then be covered with serum-free incubation media and the plates returned to the CO_2 incubator. If the hepatocyte culture is to be kept longer than 24 h, the media should be replaced every 24 h. The morphology of the hepatocyte monolayers and the attachment index should be inspected and determined using an inverted microscope.

12. For xenobiotic metabolism or induction studies using cultured hepatocytes, compounds should be added to the media after cell attachment (*see* **Notes 11** and **12**). Metabolism experiments are usually of up to 24 h duration. However, induction experiments can last significantly longer. On completion of the incubation the media should be removed and rapidly frozen. The monolayer should be washed with phosphate-buffered saline (PBS). The hepatocyte monolayer can then be disrupted by either a detergent solution (Triton X-100, 1% solution in PBS) or mechanical action. The supernatant containing the disrupted cell contents should also be rapidly frozen.

4. Notes

The preparation and culture of animal hepatocytes, particularly rat, has been extensively evaluated and refined *(42)*. This evaluation and refinement has led to the development of standard protocols for the isolation and culture of rat hepatocytes *(50,51)*. In contrast, the limited availability of human material for the preparation and culture of hepatocytes means that human systems are still under evaluation and, as such, the protocols used for the isolation and culture of human hepatocytes are diverse *(8,9)*.

The protocol presented here is just one of many that have been reported for human hepatocytes. Different protocols may give different results as observed with rat hepatocytes, although these differences have not yet been fully evaluated for human hepatocytes. Alternative protocols and approaches are reviewed in the following sections.

4.1. Isolation of Human Hepatocytes

A number of factors are important in ensuring a functional hepatocyte preparation with a high yield.

1. Perfusion of the tissue by various media is critical in the isolation of high-quality hepatocytes. It is essential that the perfused tissue has only one cut surface and that the capsule surrounding the tissue is otherwise intact. Careful placement of the cannulae is crucial in achieving the optimum perfusion of the tissue sample.
2. It is important to monitor pH and temperature. The advantage of the bicarbonate buffering system reported here is that oxygenation of the perfusate with a mixture

of oxygen and carbon dioxide (95% : 5%) is combined with a very high buffering capacity. A disadvantage of the system is the formation of microbubbles. It is important not to allow bubbles to enter the tissue as this will prohibit the perfusion of the tissue. A small drop in temperature of 1–2°C in perfusion temperature can prolong the perfusion time and increase the cellular damage caused by the perfusion. The temperature at the liver should be monitored and maintained at 37°C.

3. The size of the liver wedge biopsy sample to be perfused should be considered. Gomez et al. *(31)* reported that smaller human liver samples up to 5 g gave better yields than samples over 5 g. Furthermore, these workers have indicated that the quality of the collagenase used for the perfusion is important. Collagenase with a high specific activity provides a better yield of viable hepatocytes. This increase in yield was thought to be caused by a reduced perfusion time, which, in turn, leads to a reduction in cell destruction. Other workers *(26)* have shown that hepatocyte yield and viability can be increased if either hyaluronidase (0.05%) or dispase (500 U/mL) is added to the hepatocyte dissociation medium. However, for the preparation of rat hepatocytes, hyaluronidase is thought to be detrimental *(50)*.

4. The use of a calcium chelator to remove calcium ions that are essential for junctional complex desmosome integrity is recommended. Perfusion with EGTA brings about cleavage of the junctional complexes. It has been suggested that perfusion without a chelator reduces hepatocyte yield *(18)*.

5. Trypan blue exclusion or leakage of certain cytoplasmic enzymes are relatively quick and simple assays of cell viability. Care must be taken with the trypan blue assay because it has been reported that a decrease in media pH can lead to increased dye uptake in rat hepatocytes *(48)*, which is reversible. This phenomena has not been reported for human hepatocytes. Furthermore, it should be noted that staining intensity can vary depending on the batch of trypan blue or media components *(48,49)*. Protein binds trypan blue strongly and, thus, there should be no protein in the medium when analyzing cell viability by this method, otherwise a falsely high value for viability will be obtained.

6. A review detailing a number of methods for the isolation of hepatocytes has recently been completed *(1)*.

4.2. Culture of Human Hepatocytes

7. Confluent monolayers have been established with seeding densities of $4-18 \times 10^4$ cells/cm^2 *(8,26–28,32,37,41,44–46,52)*. In general, seeding densities should be sufficient to allow cell growth and to enable the maintenance of cell–cell contacts. Generally, cell densities of $10-15 \times 10^4$ cells/cm^2 should be sufficient.

8. The attachment efficiency (number of cells attached/number of cells seeded) should be determined and be in the region of 70–80% *(31)*. Attachment periods of up to 20 h for human hepatocytes have been reported *(46)*. Human hepatocytes prepared from the livers of older donors require increased attachment time *(52)*.

9. The influence of various matrix components on hepatocytes in culture has been widely studied. In the liver, the hepatocyte is surrounded by a matrix that contains collagen, laminin, fibronectin, and heparan sulfate proteoglycan *(1,18)*. An ideal

cell matrix should contain some of these components, be able to support hepatocyte morphology over long periods, allow formation of the monolayer and intercellular contacts, protect cells from sheer forces of fluids, and allow free exchange of salts within the media. However, in general, these organic matrices do not delay the occurrence of phenotypic changes. Moreover, the substrata can enhance alteration of liver gene expression.

Hepatocytes in culture have been shown to have the ability to secrete several types of collagen and noncollagenous glycoproteins *(53,54)*. The use of plastic culture plates coated with an extracellular matrix such as collagen, fibronectin, laminin, collagen gels, collagen gels embedded on a nylon mesh, or a laminin-rich matrix termed Matrigel® has been shown to improve the efficiency of hepatocyte attachment, increasing the duration of functional activity and the morphology of the cells in culture. The use of attachment factors in animal hepatocyte culture is extensively reviewed by Berry et al. *(18)*, Guillouzo et al. *(30)*, Skett *(59)*, and in human hepatocytes by Hamilton *(41)*, Lecluse *(7,8)*, and Silva *(9)*. Human hepatocytes in culture have been reported to attach well to extra cellular matrices such as uncoated plastic *(55–57)*, soluble rat tail or calf skin collagen type I *(26,45,46)*, fibronectin *(31,55)*, collagen gel immobilized hepatocytes *(8,9,41,58)*, and Matrigel *(8,9,41)*.

10. The effects of various hepatocyte culture media and supplements on the maintenance of hepatocyte function have been the subject of a number of reviews *(18,30,40,59)*. Many different media have been used and little work has been published comparing media effects on animal hepatocytes. In one recent study, for example, Skett and Roberts *(60)* showed that WME was the best of the commercially available media for rat hepatocytes, but that Chee's medium offered advantages. However, other authors *(7–9,41)* find little difference in the experimental effects of culture medium for human hepatocytes in short-term culture (1–2 wk). As with matrices, the systems employed for human hepatocyte culture are very diverse and tend to be extensions of conditions used for the culture of animal hepatocytes. Media that have been used for human hepatocyte culture include Weymouth's 752, minimal essential media mixed with media 199 (75% 25% v/v), Ham's F12, Leibovitz L-15, WME, and ISOMs.

Once the basic media has been selected, a number of media supplements can be added. These supplements range from natural compounds including hormones, minerals, and vitamins to nonphysiological factors such as dimethyl sulfoxide (DMSO).

a. Serum—composition unknown but contains proteins, growth factors.

b. Hormones—insulin, glucagon, glucocorticoids, thyroxine, sex steroids.

c. Growth supplements—selenium, zinc, EGF, HGF.

d. Nonphysiological factors—DMSO, ammonium chloride, metyrapone

The role of each of these supplements is, at best, unclear, even for the well-evaluated rat hepatocyte culture *(59)*. However, a number of workers have established long-term human hepatocyte cultures by the use of a number of different medium additives *(32–34)*.

11. Studies of the functionality of human hepatocytes in culture are also limited when compared to data on animal hepatocytes *(28,31,53)*. However, workers have shown that gluconeogenesis and glycolysis could be stimulated in human hepatocytes by physiological concentrations of glucagon and insulin. The gluconeogenic rate in human hepatocytes was similar to that estimated for the fasted human liver, whereas basal glycolysis is higher in cultured human hepatocytes than in vivo.

    Urea synthesis from ammonia has been used as an indicator of mitochondrial function. Under basal conditions human hepatocytes produce 2–4 nmoles urea/mg cell protein/minute. Hepatocytes can be stimulated to produce significantly more urea, up to 11 nmoles/min/mg. The maximal rate of ureagenesis decreased by 50% after one day in culture but was then stable for several days thereafter *(31)*.

    Protein production by human hepatocytes has been used as a measure of cell function *(1,37,61,62)*. Human, hepatocytes in culture have been shown to secrete albumin, α-antitrypsin, α-antichymotrypsin, α-acid glycoprotein, serum amyloid A, fibronectin, hepatoglobulin, and α_2-macroglobulin. The ability of cultured human hepatocytes to secrete albumin has been used by many workers as a measure of culture viability *(5,32–34)*. However, a number of other methods have also been reported including measurement of ATP content *(23)* and LDH leakage *(1)*.

12. A primary objective of human hepatocyte cultures is to provide models to study the potential disposition of xenobiotics prior to administration to humans. The cytochromes P450 (CYP450) present in cultured human hepatocytes are thought to be more stable than those in rat hepatocytes *(13,28,63)*. Studies have shown that CYP450 may be maintained for a number of days in primary culture *(8,9,13)*. In addition, human hepatocytes cocultured with nonparenchymal cells maintain a relatively stable cytochrome P450 content for longer periods of time *(28,37,61, 64–67)*. Cytokines produced during inflammation or infection have been shown to affect the expression of the major human CYP450 enzymes in hepatocyte cultures. The interleukins -1B, -4, -6, tumor necrosis factor-α, and interferon-α and -γ in addition to nitric oxide have been shown to downregulate specific isozymes of CYP450 in human hepatocyte cultures *(3,68,69)*. The observed in vitro effects of cytokines on drug-metabolizing enzymes may be important clinically with respect to chronic hepatitis patients who receive high doses of interferon as well as antiviral compounds.

    Investigations and evaluation of the drug-metabolizing activities present in human hepatocyte cultures has demonstrated that these cells in culture appear to be a good model for predicting drug metabolism in vivo *(10–14,45,46,70)*. Human hepatocyte cultures provide an opportunity to study both the routes and rates of metabolism of new drugs and are a key in vitro model to improve the drug development process *(7,8,10–12,14–16,70)*. Several workers have shown that the metabolic routes and rates of a number of drugs in human hepatocyte cultures are similar to those observed in vivo *(10,12–14,45,46,74,75)*. Furthermore, human hepatocyte cultures have also been used to help identify species differences in the metabolism of a number of drugs *(45,46,71–75)*.

Hepatocyte cultures have been used extensively to study CYP450 expression and regulation. The CYP450 can be induced by certain compounds in human hepatocyte cultures. Incubation of human hepatocyte cultures in the presence of either phenobarbital (1.5–32 m*M*), rifampacin (50 μ*M*), 3-methylcholanthrene (25–50 μ*M*), benzanthracene (12.5 μ*M*), and ethanol (200 m*M*) induce both the mRNA and protein of specific cytochrome CYP450 isozymes *(4–9,32,37)*. Furthermore, the human hepatocyte culture system has been used successfully to examine the potential of therapeutic agents such omeprazole, lansoprazole, and pantoprazole to induce cytochromes P450 in vitro *(6,76,77)*. Whether the induction observed in closed in vitro systems can be extrapolated to humans is a key question currently under investigation. However, the contribution of pharmacokinetics to the in vivo disposition of compounds shown to affect the regulation of enzymes in vitro should be carefully considered.

The human hepatocyte culture has also been used to model a number of drug–drug interactions resulting from changes in the metabolism compounds caused by either induction or inhibition of the enzymes responsible for their metabolism *(16,78–80)*.

The main focus of many researchers has been the cytochrome CYP450 drug-metabolizing enzymes. However, the conjugating enzymes have also been shown to be altered in human hepatocyte culture. UDP glucuronyltransferase increased during culture whereas sulfotransferase activity decreased. Glutathione content remained relatively stable *(32,81)*. Again, the phase II enzyme content of the cells is affected by medium composition and other culture conditions.

In addition to metabolism studies human hepatocytes in culture have been evaluated as screens for hepatotoxicity *(16,61,62)* or to elucidate mechanisms of toxicity *(52,82)*.

4.3. The Cryopreservation of Human Hepatocytes

13. Owing to the infrequent supply of human material, methods of cryopreserving hepatocytes prior to culture have been developed. Numerous researchers have published results showing that cryopreserved hepatocytes retain their metabolic capabilities *(83–86)*. Human hepatocytes have been cryopreserved in mixtures of buffers containing a cryoprotectant. DMSO generally appears to be the cryoprotectant of choice. Hepatocytes are then frozen in a controlled manner and then ideally stored at <–150°C. At present, cyropreserved human hepatocytes tend to have poor attachment rates (approx 50%) in culture *(86)*.

4.4. Immortalized Cell Lines

14. In the absence of freshly isolated or cryopreserved native hepatocytes, immortalized hepatic cell lines such as the human-derived hepatoma cell line HEP G2 have also been utilized. These cell types have been shown to express both phase I and II drug-metabolizing activity albeit lower than that expressed in normal liver cells

(87). Cell lines have also been used with some success for the study of drug metabolism, induction, or cytotoxicity *(88–90)*.

References

1. Puviani, A., Ottolenghi, C., Tassinari, B., Pazzi, P., and Morsiani, E. (1998) An update on high yield hepatocyte isolation methods and on the potential clinical use of isolated liver cells. *Comp. Biochem. Physiol.* **121,** 99–109.
2. Bodin, K., Andersson, U., Rystedt, E., et al. (2002) Metabolism of 4β-hydroxy-cholesterol in humans. *J Biolog. Chem.* **277(35),** 31,534–31,540.
3. Alexander, B. (1998) The role of nitric oxide in hepatic metabolism. *Nutrition* **14(4),** 376–390.
4. Kedderis, G. (1997) Extrapolation of in vitro enzyme induction data to humans in vivo. *Chemico–Biolog. Interact.* **107,** 109–121.
5. Meunier, V., Bourrie, M., Julian, B., et al. (2000) Expression and induction of CYP1A1/1A2 and CYP3A4 in primary cultures of human hepatocytes: a 10 year follow up. *Xenobiotica* **30(6),** 589–607.
6. Masubuchi, N., Li, A., and Okazaki, O., (1998) An evaluation of the cytochrome P450 Induction potential of pantoprazole in primary human hepatocytes. *Chemico-Biolog. Interact.* **114,** 1–13.
7. LeClyse, E., Madan, A., Hamilton, G., Carrol, K, Dehann, R, and Parkinson, A. (1999) Expression and regulation of cytochrome P450 enzymes in primary cultures of human hepatocytes. *J. Biochem. Molec. Toxicol.* **14(4),** 2000.
8. LeClyse, E. (2000) Human hepatocyte culure systems for the in vitro evaluation pf cytochrome P450 expression and regulation. *Europ. J. Pharmaceut. Sci.* **13,** 343–368.
9. Silva, J., Morin, P., Day, S., et al. (1998) Refinemant of an in vitro cell model for cytochrome P450 Induction. *Drug Metab. Disposition* **26(5),** 460–496.
10. Cross, D. and Bayliss, M. (2000) A commentry on the use of hepatocytes in drug metabolism studies during drug discovery and development. *Drug Metab. Rev.* **32(2),** 219–240.
11. Iwatsubo, T., Hirota, N., Ooie, T., et al. (1997) Prediction of in vivo drug metabolism in the human liver from in vitro metabolism data. *Pharmacol. Therapeut.* **73(2),** 147–171.
12. Lave, T., Coassolo, P., and Reigner B. (1999) Prediction of hepatic metabolic clearance based on interspecies allometric scaling techniques and in vitro-in vivo correlations. *Clinical Pharmacokinet.* **36(3),** 211–231.
13. Kern, A., Bader, A., Pichmayr, R., and Sewing, A. (1997) Drug metabolism in hepatocyte sandwich cultures of rats and humans. *Biochem. Pharmacol.* **54,** 761–772.
14. Zuegge, J., Schneider, G., Coassolo, P., and Lave, T. (2001) Prediction of hepatic metabolic clearance: Comparison and assesment of prediction models. *Clin. Pharmacokinet.* **40(7),** 553–563.
15. Guillouzo, A., Morel, F., Langouet, S., Maheo, K., and Rissel, M. (1997) Use of hepatocyte cultures for the study of hepatotoxic compounds. *J. Hepatol.* **26(suppl. 2),** 73–80.

16. Buchan, P., Wade, A., Ward, C., Oliver, S., Stewart, A., and Freestone, S. (2002) Frovatriptan: A review of drug drug interactions. *Headache* **42,** S63–S73.
17. Howard, R. B., Christensen, A. K., Gibbs, F. A., and Pesch, L. A. (1967) The enzymatic preparation of isolated intact parenchymal cells from rat liver. *J. Cell Biol.* **35,** 675–684.
18. Berry, M. N., Edwards, A. M., and Barritt, G. J. (1991) Isolated hepatocytes preparation; properties and applications, in *Laboratory Techniques in Biochemistry and Molecular Biology*, Vol. 21 (Burdon, R. H. and van Knippenberg, P. H., eds.), Elsevier, Amsterdam, pp. 1–460.
19. Berry, M. N. and Friend, D. S. (1969) High yield preparation of isolated rat liver parenchymal cells. *J. Cell Biol.* **43,** 506–520.
20. Donini, A., Baccarani, U., Piccolo, G., et al. (2001) Hepatocyte isolation using human livers discarded from transplantation:Analysis of cell yield and function. *Transplant. Proc.* **22,** 654–655.
21. Hewitt, W., Corno, V., Eguchi, S., Kamlot, A., Middleton, Y., Beeker, T., Demetriou, A., Roizga, J. (1997) Isolation of human hepatocytes from livers rejected for whole organ transplantation. *Transplant. Proc.* **29,** 1945–1947.
22. Caruana, M., Battle, T., Fuller, B., and Davidson, B. (1999) Isolation of human hepatocytes after hepatic warm and cold ischaemia: A practical approach using university of winsconsin solution. *Cryobiology* **38,** 165–168.
23. Olinga, P., Merema, M., Hof, I., et al. (1998) Effect of human liver source on the functionality of isolated hepatocytes and liver slices. *Drug Metab. Disposition* **26(1),** 5–11.
24. Olinga, P., Merema, M., Hof, I., et al. (1997) Effect of cold and warm ischaemia in isolated hepatocytes and slices from human and monkey liver. *Xenobiotica* **28(4),** 349–360.
25. David, P., Viollon, C., Alexandre, E., et al. (1998) Metabolic capacities in cultured human hepatocytes obtained by a new isolating proceedure from non wedge biopsies. *Human and Experiment. Toxicol.* **17,** 544–553.
26. Maekubo, H., Ozaki, S., Mitmaker, B., and Kalant, N. (1982) Preparation of human hepatocytes for primary culture. *In Vitro* **18,** 483–491.
27. Miyazaki, K., Takaki, R., Nakayama, F., Yamauchi, S., Koga, A., and Todo, S. (1981) Isolation and primary culture of adult human hepatocytes. *Cell Tissue Res.* **218,** 13–21.
28. Guillouzo, A., Beaune, P., Gascoin, M. N., et al. (1985) Maintenance of cytochrome P450 in cultured adult human hepatocytes. *Biochem. Pharmacol.* **34,** 2991–2995.
29. Houssin, D., Capron, M., Celier, C., Cresteil, T., Demaugre, F., and Beaune, P. (1983) Evaluation of isolated human hepatocytes. *Life Sci.* **33,** 1805–1809.
30. Guillouzo, A., Morel, F., Ratanasavanh, D., Chesne, C., and Guguen-Guillouzo, C. (1990) Long term culture of functional hepatocytes. *Toxicology in Vitro* **4,** 415–427.
31. Gomez-Lechon, M. J., Lopez, P., Donato, T., et al. (1990) Culture of human hepatocytes from small surgical liver biopsies. Biochemical characterisation and comparison with in vivo. *In Vitro Cell. Dev. Biol.* **26,** 67–74.

32. Ferrini, J.-B., Pichard, L., Domerge, J., and Maurel, P. (1997) Long term primary cultures of adult human hepatocytes. *Chemico–Biolog. Interact.* **10**, 31–45.
33. Katsura, N., Ikai, I., Mikata, T., et al. (2002) Long term culture of primary human hepatocytes with preservation of proliferative capacity and differentiated functions. *J. Surg. Res.* **106**, 115–123.
34. Runge, D., Runge, D., Jager, D., et al. (2000) Serum free, long-term cultures of human hepatocytes: Maintainence of cell morphology, transcription factors and liver specific functions. *Biochem. Biophys. Res. Commun.* **269**, 46–53.
35. Goodwin, G., Liddle, C., Tapner, M., and Farrell, G. (1997) Time dependent expression of cytochrome P450 genes in primary cultures of well-differentiated human hepatocytes. *J. Lab. Clin. Med.* **129(6)**, 638–648.
36. Bousquet-Melou, A., Laffont, C., Laroute, V., and Toutain, P. (2002) Modelling the loss of metabolic capacities of cultured hepatocytes: application to measurement of Michaelis-Menten kinetic parameters in in vitro systems. *Xenobiotica* **32(10)**, 895–906.
37. Morel, F., Beaune, P. H., Ratanasavanh, D., et al. (1990) Expression of cytochrome P450 enzymes in cultured human hepatocytes. *Europ. J. Biochem.* **191**, 437–444.
38. Morel, F., Beaune, P., Ratansavanh, D., Flinois, J. P., Guengerich, F. P., and Guillouzo, A. (1990) Effects of various inducers in the expression of cytochromes P450 2C8, 9, 10 and 3A in cultured human hepatocytes. *Toxicol. In Vitro* **4**, 458–460.
39. Li, A. and Kedderis, G. (1997) Primary hepatocyte culture as an experimental model for the evaluation of interactions between xenobiotics and drug metabolising enzymes. *Chemico-Biolog. Interact.* **107**, 1–3.
40. Runge, D., Michalopoulos, G., Strom, S., and Runge, D. (2000) Recent advances in human hepatocyte culture systems. *Biochem. Biophys. Res. Commun.* **274**, 1–3.
41. Hamilton, G., Jolley, S., Gilbert, D., Coon, D., Barros, S., and LeClyse, E. (2001) Regulation of cell morphology and cytochrome P450 expression in human hepatocytes by extracellular matrix and cell-cell interactions. *Cell Tissue Res.* **306**, 85–99.
42. Richert, L., Binda, D., Hamilton, G., et al. (2002) Evaluation of the effect of culture configuration on morphology, survival time, anti oxidant status and metabolic capacities of cutured rat hepatocytes. *Toxicol. In Vitro* **16**, 89–99.
43. Oldham, H. G., Norman, S. J., and Chenery, R. J. (1985) Primary cultures of adult rat hepatocytes—a model for the toxicity of histamine H2-receptor antagonists. *Toxicology* **36**, 215–229.
44. Strom, S. C., Jirtle, R. L., Jones, R. S., et al. (1982) Isolation, culture and transplantation of human hepatocytes. *J. Nat. Cancer Inst.* **68**, 771–778.
45. Seddon, T., Michelle, I., and Chenery, R. J. (1989) Comparative drug metabolism of diazepam in hepatocytes isolated from man, rat, monkey and dog. *Biochem. Pharm.* **38**, 1657–1665.
46. Oldham, H. G., Standring, P., Norman, S. J., et al. (1990) Metabolism of temelastine (SKF93944) in hepatocytes from rat, dog, cynomologous monkey and man. *Drug Metab. Disp.* **18**, 146–152.

47. Bellemann, P., Gebhardt, R., and Mercke, D. (1977) An improved method for the isolation of hepatocytes from liver slices. *Anal. Biochem.* **81,** 408–415.

48. Baur, H., Kasperek, S., and Pfaff, E. (1975) Criteria of viability of isolated liver cells. *Hoppe-Seyler's Z. Physiol. Chem.* **356,** 827–838.

49. Jauregui, H. O., Hayner, N. T., Driscoll, J. L., Williams-Holland, R., Lipsky, M. H., and Galletti, P. M. (1981) Trypan blue dye uptake and lactate dehydrogenase in adult rat hepatocytes—freshly isolated cells; cell suspensions, and primary monolayer cultures. *In Vitro* **17,** 1100–1110.

50. Blaauboer, B., Boobis, A. R., Castell, J. V., et al. (1994) The practical applicability of hepatocyte cultures in routine testing. *ATLA* **22,** 231–242.

51. Skett, P. and Bayliss, M. (1996) Time for a consistent approach to preparing and culturing hepatocytes. *Xenobiotica* **26(1),** 1–7.

52. Butterworth, B. E., Smith-Oliver, T., Earle, L., et al. (1989) Use of primary cultures of human hepatocytes in toxicology studies. *Cancer Res.* **49,** 1075–1084.

53. Clément, B., Guguen-Guillouzo, C., Campion, J. P., Glaise, D., Bourel, M., and Guillouzo, A. (1984) Long term co-cultures of adult human hepatocytes with rat liver epithelial cells: Modulation of active albumin secretion and accumulation of extracellular material. *Hepatology* **4,** 373–380.

54. Schuetz, E. G., Schuetz, J. D., Strom, S. C., et al. (1993) Regulation of human liver cytochromes P450 in family 3A in primary and continuous culture of human hepatocytes. *Hepatology* **18,** 1254–1262.

55. Guguen-Guillouzo, C., Campion, J. P., Brissot, P., et al. (1982) High yield preparation of isolated human adult hepatocytes by enzymic perfusion of the liver. *Cell Biol. Int. Rep.* **6,** 625–628.

56. Ballet, F., Bouma, M. E., Wang, S. R., Amit, N., Marais, J., and Infante, R. (1984) Isolation, culture and characterisation of adult human hepatocytes from surgical liver biopsies. *Hepatology* **4,** 849–854.

57. Vons, C., Pegorier, I. P., Ivanov, M. A., et al. (1990) Comparison of cultured human hepatocytes isolated from surgical biopsies or cold-stored organ donor livers. *Toxicol. In Vitro* **4,** 432–434.

58. Koebe, H. G., Pahernik, S., Eyer, P., and Schildberg, F. W. (1994) Collagen gel immobilisation: A useful cell culture technique for long term metabolic studies on human hepatocytes. *Xenobiotica* **24,** 95–107.

59. Skett, P. (1994) Problems in using isolated and cultured hepatocytes for xenobiotic metabolism/metabolism based toxicity testing—solutions? *Toxicol. In Vitro* **8,** 491–504.

60. Skett, P. and Roberts, P. (1994) Effect of culture medium on the maintenance of steroid metabolism in cultured adult rat hepatocytes. *In Vitro Toxicol.* **7,** 261–267.

61. Guillouzo, A., Begue, J. M., Campion, J. P., Gascoin, M. N., and Guguen-Guillouzo, C. (1985) Human hepatocyte cultures: A model of pharmaco-toxicological studies. *Xenobiotica* **15,** 635–641.

62. Ratanasavanh, D., Baffet, G., Latinier, M. F., Rissel, M., and Guillouzo, A. (1988) Use of hepatocyte co-cultures in the assessment of drug toxicity from chronic exposure. *Xenobiotica* **18,** 765–771.

63. Guguen-Guillouzo, C., Baffet, G., Clement, B., Begue, J. M., Glaise, D., and Guillouzo, A. (1983) Human adult hepatocytes: Isolation and maintenance at high levels of specific function in a co-culture system, in *Isolation, Characterisation and Use of Hepatocytes* (Harris, R. A., ed.).
64. Bhatia, S., Bali, U., Yarmush, M., and Toner, M. (1998) Microfabrication of hepatocyte/fibroblast Co-cultures: Role of homotypic cell interactions. *Biotechnol. Progress* **14,** 378–387.
65. Hino, H., Tateno, C., Sato, H., et al. (1999) A long term culture of human hepatocytes which show a high growth potential and express their differentiated phenotypes. *Biochem. Biophys. Res. Commun.* **256,** 184–191.
66. Bhatia, S., Balis, U., Yarmush, M., and Toner, M. (1999) Effect of cell-cell interactions in preservation of cellular phenotype: cocultivation of hepatocytes and non parenchymal cells. *FASEB* **13,** 1883–1900.
67. Bhandari, R., Riccalton, L., Lewis, A., et al. (2001) Liver tissue engineering: a role for co-culture systems in modifying hepatocyte function and viability. *Tissue Eng.* **3,** 345–357.
68. Abdel-Razzak, Z., Loyer, P., Fautrel, A., et al. (1993) Cytokines down-regulate expression of major cytochrome P450 enzymes in adult human hepatocytes in primary culture. *Mol. Pharm.* **44,** 770–775.
69. Donato, M. T., Herrero, E., Gomez-Lechon, M. J., and Castell, J. V. (1993) Inhibition of monooxygenase activities in human hepatocytes by interferons. *Toxicol. In Vitro* **7,** 481–485.
70. Fabre, G., Combalbert, J., Berger, Y., and Cano, J.P. (1990) Human hepatocytes as a key in vitro model to improve pre clinical drug development. *Eur. J. Drug Met. Pharmacokinet.* **15,** 165–171.
71. Hewitt, N., Buhring, K., Dasenbrock, J., Haunschild, J., Ladsletter, B., and Utesch, D. (2001) Studies comparing the in vivo:in vitro metabolism of three pharmaceutical compounds in rat, dog, monkey and human using cryopreserved hepatocytes, microsomes and collagen gel immobilised hepatocyte cultures. *Drug Metab. Disposition* **29(7),** 1042–1050.
72. Bayliss, M., Bell, J., Jenner, W., Park, G., and Wilson, K. (1999) Utility of hepatocytes to model species differences in the metabolism of loxtidine and to predict pharmacokinetic parameters in rat, dog and man. *Xenobiotica* **2(3),** 253–268.
73. Li, C., Chauret, N., Trimble, L., et al. (2001) Investigation of the in vitro metabolic profile of a phosphodeiesterase-IV inhibitor leading to structural optimization. *Drug Metab. Disposition* **29(3),** 232–241.
74. Le Bigot, J. F., Begue, J. M., Kiechel, J. R., and Guillouzo, A. (1987) Species differences in metabolism of ketotifen in rat, rabbit and man: Demonstration of similar pathways in vivo and in cultured hepatocytes. *Life Sci.* **40,** 883–890.
75. Dauphin, J. F., Graviere, C., Bouzard, D., Rohou, S., Chesne, C., and Guillouzo, A. (1993) Comparative metabolism of tosufloxacin and BMY43748 in hepatocytes from rat, dog, monkey and man. *Toxicol. In Vitro* **7,** 499–503.
76. Diaz, D., Fabre, I., Daujat, M., et al. (1990) Omeprazole is an aryl hydrocarbon like inducer of human hepatic cytochrome P450. *Gastroenterology* **99,** 737–747.

77. Curi-Pedrosa, R., Daujat, M., Pichard, L., et al. (1994) Omeprazole and lansoprazole are mixed inducers of CYP1A and CYP3A in human hepatocytes in primary culture. *J. Pharmacol. Expt. Ther.* **269**, 384–392.

78. Li, A., Maurel, P., Gomez-Lechon, M., Cheng, L., and Jurima-Romet, M. (1997) Preclinical evaluation of drug-drug interaction potential: Present status of the application of primary human hepatocytes in the evaluation of cytochrome P450 induction. *Chemico-Biolog. Interact.* **107**, 5–16.

79. Li, A. and Jurima-Romet, M. (1997) Applications of primary human hepatocytes in the evaluation of pharmacokinetic drug-drug interactions:evaluation of model drugs terfenadine and rifampin. *Cell Biol. Toxicol.* **13(4–5)**, 365–374.

80. Ito, K., Iwatsubo, T., Kanamitsu, S., Nakajima, Y., and Sugiyama, Y. (1998) Quantitative prediction of in vivo drug clearance and drug interactions from in vitro data on metabolism, together with binding and transport. *Ann. Rev. Pharmacol. Toxicol.* **38**, 461–469.

81. Iqbal, S., Elcombe, C. R., and Elias, E. (1991) Maintenance of mixed function oxidase and conjugation enzyme activities in hepatocyte cultures prepared from normal and diseased human liver. *J. Hepatol.* **12**, 336–343.

82. Ulrich, R., Bacon, J., Brass, E., Cramer, C., Petrella, D., and Sun, E. (2001) Meatbolic, idiosyncratic toxicity of drugs:overview of the hepatic toxicity induced by the anxiolytic, panadiplon. *Chemico-Biolog. Interact.* **134**, 251–270.

83. Li, A., Gorycki, P., Hengsler, J., et al. (1999) Present status of the application of cryopreserved hepatocytes in the evaluation of xenobiotics:consensus of an international expert panel. *Chemico-Biolog. Interact.* **121**, 117–123.

84. Hegstler, J., Utesch, D., Steinberg, P., et al. (2000) Cryopreserved primary hepatocytes as a constantly available in vitro model for the evaluation of human and animal drug metabolism and enzyme induction. *Drug Metab. Rev.* **32(1)**, 81–118.

85. Hegstler, J., Ringel, M., Biefang, K., et al. (2000) Cultures with cryopreserved hepatocytes: applicability for studies with enzyme induction. *Chemico-Biolog. Interact.* **125**, 57–73.

86. Alexandre, F., Viollon-Abadie, C., David, P., et al. (2002) Cryopreservation of adult human hepatocytes obtained from resected liver biopsies. *Cryobiology* **44**, 103–111.

87. Rodriguez-Antona, C., Donato, M., Boobis, A., et al. (2002) Cytochrome P450 expression in human hepatocytes and hepatoma cell lines:molecular mechanisms that determine lower expression in cultured cells. *Xenobiotica* **32(6)**, 505–520.

88. Knasmuller, S., Parzefall, W., Sanyal, R., et al. (1998) Use of metabolically competant human hepatoma cells for the detection of mutagens and antimutagens. *Mutat. Res.* **402**, 185–202.

89. Reuf, J., Chiapella, J., Darroudi, F., et al. (1996) Development and validation of alternative metabolic systems for mutagenicity testing in short term assays. *Mutat. Res.* **353**, 151–176.

90. Khalil, M., Shariat-Panahi, A., Tootle, R., et al. (2001) Human hepatoma cell lines proliferating as cohesive spheroid colonies in alginate merkedly upregulate both synthetic and detoxificatory liver function. *J. Hepatol.* **297**, 68–77.

18

Glomerular Epithelial and Mesangial Cell Culture and Characterization

Heather M. Wilson and Keith N. Stewart

1. Introduction

The advent of in vitro culture techniques has enabled the culture of homogeneous populations of glomerular mesangial and epithelial cells to aid our understanding of the development of glomerular disease at the cellular level. Advances in our knowledge of the pathogenic mechanisms have made it clear that the response of intrinsic glomerular cells to external stimuli plays an important role in glomerular injury *(1,2)*. Glomerular cells from several mammalian species have been isolated and propagated, and, in some instances, cell lines have been generated by viral or nonviral oncogenic transformation *(3–5)*.

The renal glomerulus is a complex anatomical structure that contains many different cell types, including visceral and parietal epithelial cells, mesangial cells, and endothelial cells. These cells have long been recognized as distinct entities, because they occupy defined anatomical locations in vivo and have distinguishable morphological and cytochemical features. This compartmentalization is, however, lost in culture, as are several of the anatomical characteristics such as endothelial fenestrations and epithelial pedicels *(6,7)*; in addition, cultured cells may undergo dedifferentiation. Despite these limitations, the study of glomerular cells in culture has proven useful, and valuable information has been obtained about their physiology and pathophysiology. Human glomeruli are usually obtained from the normal pole of kidneys surgically removed from patients with renal carcinoma or from donor kidneys that cannot be used for transplantation for technical reasons. Glomeruli are isolated using sieves such that they can be rendered virtually free of tubule contamination *(7,8)*. Thereafter, glomeruli can be seeded into culture flasks, and after a week,

From: *Methods in Molecular Medicine, vol. 107: Human Cell Culture Protocols, Second Edition*
Edited by: J. Picot © Humana Press Inc., Totowa, NJ

cells can be seen growing out of the glomerular core *(5)*. Alternatively, glomeruli can be dissociated by incubation with an enzyme, such as collagenase before culture *(9)*.

The first cells to emerge from explanted glomeruli are epithelial cells, which have a distinctive "cobblestone" appearance *(9)* and, for the first 7–10 d of outgrowth are the most common cell type in the mixed population of glomerular cells. If Bowman's capsule is not stripped from the glomeruli, parietal epithelial cells are the dominant cell type *(10)*; endothelial cells and mesangial cells are present at this stage. Mesangial cells become more evident later in culture and have a stellate appearance. They grow vigorously, in multilayers, whereas epithelial cells grow in a monolayer and are subject to contact inhibition. Mesangial cells, therefore, outgrow the epithelial cells, and after 30 d of growth, the cultures are nearly pure glomerular mesangial cells *(11)*. The difference in growth potential of primary cells in culture can be explained by the "Mosaic Theory" *(12)*, which states that separate populations of cells can be obtained from a mixed population of cells based on their different growth rates or culture requirements.

For most experimental purposes, a homogeneous population of cells is required. It is, therefore, important to assess the purity of the isolated cells, because even a small population of contaminating cells can affect the experimental results. Homogeneity of cells can be improved by using cloning rings and repeated cloning but the yield of cells is less and they do dedifferentiate with passage number.

It is now clear that morphology alone is not sufficiently discriminating to assess cell purity or ensure that a homogeneous population of cells has been isolated *(13)*. Antibodies to specific cell-surface and cytoskeletal markers need to be used to confirm both cell identity and purity. **Table 1** shows the main markers used to differentiate between the glomerular cell types present in culture *(5,11,14–17)*.

The outlined methods are routinely used in our laboratory to obtain and characterize pure populations of glomerular epithelial and mesangial cells (*see* **Note 1**).

2. Materials

1. Wash medium: RPMI-1640 Dutch modification supplemented with 2 m*M* L-glutamine, 100 U/mL penicillin, 100 µg/mL streptomycin, and 2.5 µg/mL amphotericin B (all reagents from Gibco Invitrogen Inc., Paisley, Scotland). Store at 4°C for up to 1 mo.
2. Culture medium: RPMI-1640 Dutch modification or MEM-D-valine (Gibco Invitrogen Inc.) supplemented as in **step 1** with the addition of 10% fetal calf serum (FCS) (Gibco Invitrogen Inc.) and insulin–transferrin–sodium selenite (ITS)

Table 1
Main Cell Markers That Can be Used to Differentiate Between Glomerular Cell Types In Vitro

Antibody to	Visceral Epithelial Cells	Parietal Epithelial Cells	Mesangial Cells	Endothelial Cells
GLEPP1[a]	+	−	−	−
Vimentin[b]	+	−	+	−
CALLA[c]	+	+	−	−
Cytokeratin 18[b]	−	+	−	−
Cytokeratin 19[b]	−	+	−	−
α-SMA[d]	−	−	+	−
Myosin (smooth)	−	−	+	−
VLA-1 integrin	−	−	+	−
von Willebrand (factor VIII antigen)	−	−	−	+
HLA-DR	−	−	−	+

+ = positive ; − = negative.

[a]GLEPP1, Glomerular Epithelial protein 1.

[b]GVEC and GPEC can dedifferentiate in culture and are not always distinguished specifically by vimentin and cytokeratin; cytokeratin can be expressed by both GVEC and GPEC in culture.

[c]cALLA, common acute lymphoblastic leukemia antigen.

[d]α-SMA, Smooth muscle actin.

medium supplement (Sigma-Aldrich, Poole, Dorset, U.K., I-1884). Reconstitute 1 vial of ITS in 50 mL sterile distilled water, and add 1 mL of this stock to 100 mL of medium to give a final concentration of 5 µg/mL insulin, 5 µg/mL transferrin, and 5 ng/mL sodium selenite. Store the stock solution of ITS at 4°C and protect from light.

3. Fibronectin (Sigma-Aldrich F-2006): reconstitute to 1 mg/mL in sterile distilled water, and dilute to 10 µg/mL in RPMI-1640 wash medium. Store at –20°C in aliquots. Coat flasks at a concentration of 5 µg/cm^2, and leave solution to bind for at least 30 min. Decant excess fibronectin solution.

4. Type I collagen (tissue-culture grade, Sigma-Aldrich C-7661): reconstitute to 1 mg/mL in 0.1 *M* acetic acid, leave for 1–3 h until it is dissolved. Transfer this to a glass bottle with chloroform at the bottom. Do not shake or stir the collagen after this point. Store in the dark at 4°C. Coat the flasks at a concentration of 10 µg/cm^2, and leave to bind for 4 h at 37°C or overnight at 4°C. Expose the coated flask to UV radiation if you suspect the solution is not sterile; do not filter sterilize.

5. Phosphate-buffered saline (PBS): 1.5 m*M* (0.2 g/L) KH$_2$PO$_4$, 8.1 m*M* (1.15 g/L) Na$_2$HPO$_4$, 2.7 m*M* (0.2 g/L) KCl, 140 m*M* (8.0 g/L) NaCl. Filter sterilize through a 0.22-µm filter before use. Store at room temperature.

6. Trypsin/EDTA 10X (Sigma T-4174): Dilute trypsin/EDTA 1:10 with sterile PBS, and store in aliquots at –20°C.
7. Tris-buffered saline (TBS): 0.05 M Tris-HCl, pH 7.3, containing 0.15 M NaCl. Store at room temperature.
8. Veronal acetate buffer: Dissolve 1.47 g sodium barbitone and 0.97 g sodium acetate (trihydrate) in 200 mL water, pH to 9.2, using 0.1 M HCl, and make up to 250 mL with water. Store at 4°C for no longer than 1 mo.
9. Scot's tap water substitute: Dissolve 2 g of sodium bicarbonate and 20 g of magnesium sulfate in 1 L of distilled water, and store at room temperature.
10. Stainless-steel sieves of mesh size 250, 200, 150, 106, and 63 μm (Endecotts Ltd., London, U.K.). A diameter size of 7.5 cm or greater is best for processing large amounts of tissue.
11. Lux chamber slides (eight-well Permanox) from Gibco-Invitrogen Inc.
12. Rabbit anti-mouse Ig binding antibody (Dako Z0259, Ely, Cambridgeshire, U.K.). Dilute 1/20 in 1% BSA/TBS immediately before use.
13. Alkaline phosphatase anti-alkaline phosphatase (APAAP) complexes (Dako D0651). Dilute 1/40 immediately before use.
14. Substrate solution: Dissolve 25 mg naphthol-AS-MX-phosphate disodium salt (Sigma-Aldrich N-5000), 12 mg of levamisole (Sigma-Aldrich L-9756), and 25 mg of fast red TR salt (Sigma-Aldrich F-2768) in 50 mL of veronal acetate buffer, and filter through Whatman no. 1 filter paper. Prepare freshly as required.
15. Mayer's hematoxylin (Sigma-Aldrich).
16. Apathy's aqueous mounting medium (BDH Laboratory Supplies, Poole, Dorset, U.K.).

3. Methods

3.1. Isolation of Whole Glomeruli

Human glomeruli are most often isolated from the nonaffected pole of nephrectomy specimens from patients with renal cell carcinoma (*see* **Notes 2** and **3**). Aseptic conditions in a laminar flow hood should be adopted throughout.

1. Place the tissue in a sterile Petri dish, and cover it with RPMI-1640 wash medium; no FCS should be used during the isolation procedure, because it may initiate clotting owing to blood products present in the tissue. Using a scalpel, remove surrounding capsule and any fat.
2. Cut the cortex away from the medulla, and chop the cortex into 1–2 mm² pieces. Press this through a sieve of mesh size 250 μm, into a Petri dish, using the barrel from a 5-mL syringe (*see* **Note 4**). This results in the separation of glomeruli from renal tubules, interstitium, and vasculature. Wash the retained tissue with a generous amount (50–100 mL) of RPMI-1640 wash medium. Collect the glomerular filtrate from the Petri dish into sterile containers on ice.
3. Separate the glomeruli from the tubular fragments by passing it through a 150-μm sieve. This also strips the Bowman's capsule from most of the glomeruli (*see*

Note 5). A further 50 mL of RPMI wash medium are used to rinse the tissue retained on the sieve. Collect the filtrate into sterile containers.

4. Pass the filtrate through a 106-μm sieve, which retains the glomeruli. Because there is a substantial volume of filtrate, it is advantageous to hold the sieve above a beaker to allow the filtrate to pass through quickly. Pour the filtrate through the sieve into a funnel inserted in the beaker to prevent any "splashback" that may occur. Rinse the retained glomeruli with approx 50 mL RPMI-1640 wash medium to eliminate any tubular fragments that may still be present.

5. Collect the glomeruli retained on the 106-μm sieve by inverting the sieve over a Petri dish and washing with RPMI culture medium. Transfer the glomeruli at a concentration of approx 15–20 glomeruli/mL to a fibronectin-coated culture flask (*see* **Fig. 1A** and **Note 6**).

6. Culture the glomeruli at 37°C in a 5% CO_2 incubator.

3.2. Isolation of Glomerular Epithelial Cells

There are two types of epithelial cells within the glomerulus, namely, the visceral epithelial cells (GVEC) and the parietal epithelial cells (GPEC). GVEC, also known as podocytes, are located on the outer side of the glomerular basement membrane and are crucial in the urine filtration process where GPEC form the parietal sheet of Bowman's capsule. It is difficult to differentiate between GVEC and GPEC in culture as phenotypic changes take place. In glomeruli in vivo, GPEC rapidly proliferate and form crescents in proliferative nephritis *(18)*, whereas GVEC rarely undergo cell division. Cytokeratin is present only in the GPEC, whereas vimentin is restricted to GVEC *(19)*. These characteristics are lost in culture and it has been suggested that the GPEC phenotype is adopted *(20)*. When epithelial cells are derived from decapsulated glomeruli they are likely to represent GVEC whereas GPEC grow at the periphery of glomeruli that have Bowman's capsule present. However, it is unlikely to achieve pure cultures of a specific epithelial cell type without the use of cloning rings.

1. Isolate and culture the glomeruli as described in **Subheading 3.1.** Once the glomeruli have adhered to the flask, change the medium (RPMI-1640 culture medium) every 4–5 d (*see* **Note 5**).

2. Epithelial cells can be seen growing out of the glomeruli around 7–10 d. This time may vary and generally it takes longer to see cellular outgrowth in glomeruli prepared from kidneys from older patients (*see* **Note 7**).

3. Once sufficient epithelial cells are identified (these have a polygonal, cobble-stone appearance) (*see* **Fig. 1B**), pour off the unbound glomeruli and rinse the bound glomeruli/cells in PBS. Trypsinize the cells/glomeruli off the flask (*see* **Subheading 3.4.**) and pass them through a 63-μm sieve. Collect the epithelial cells in the filtrate; glomeruli are retained on the sieve. Rinse the sieve several times with a volume of around 10 mL RPMI culture medium to stop the action of trypsin and ensure maximum epithelial cell recovery.

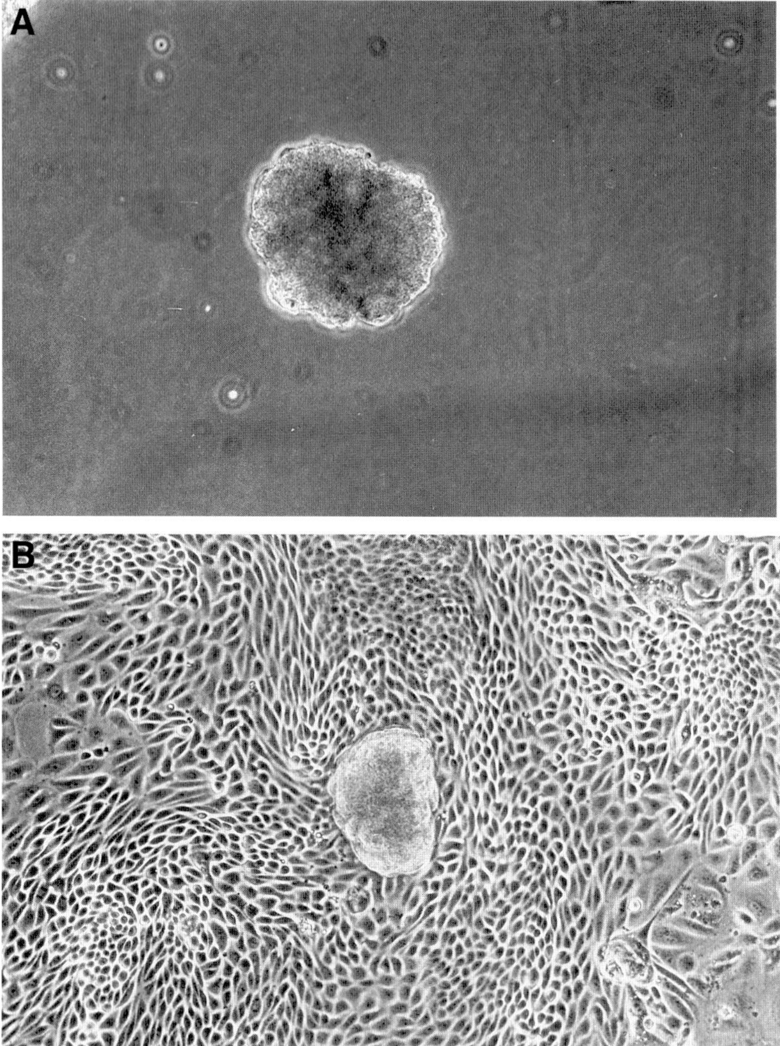

Fig. 1. Glomerular cell culture. (**A**) Decapsulated glomerulus at d 0. (**B**) Cellular growth at d 7; polygonal cells characteristic of epithelial cells.

4. Pellet the cells that have passed through the sieve by centrifugation at 200*g* for 5 min, and then resuspend in RPMI culture medium. Plate the cells into tissue culture flasks which have been coated with type I collagen at a concentration of 10 µg/cm^2 (*see* **Subheading 2.1.** and **Note 8**).
5. After approx 1 wk when the epithelial cells have reached confluence (*see* **Note 9**), passage using trypsin/EDTA (*see* **Subheading 3.4.**). Plate the trypsinized cells

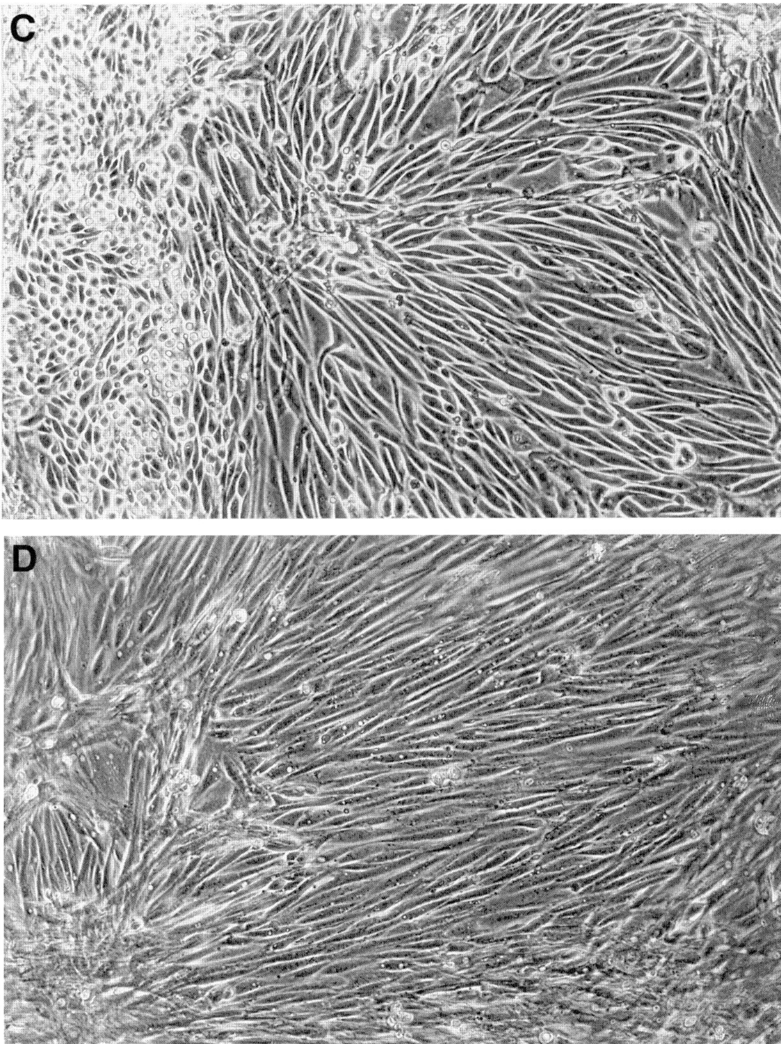

Fig. 1. *(continued)* (**C**) Stellate-shaped cells growing in multilayers characteristic of contractile mesangial cells at d 14; a few epithelial cells are still present. (**D**) Mesangial cell outgrowth at d 28; no epithelial cells are evident.

onto plastic tissue-culture grade dishes or flasks without type I collagen at this point. At confluence, the cells will be homogeneous and display a cobblestone appearance (*see* **Fig. 2**).

6. Characterize the cells before use (*see* **Subheading 3.5.**)

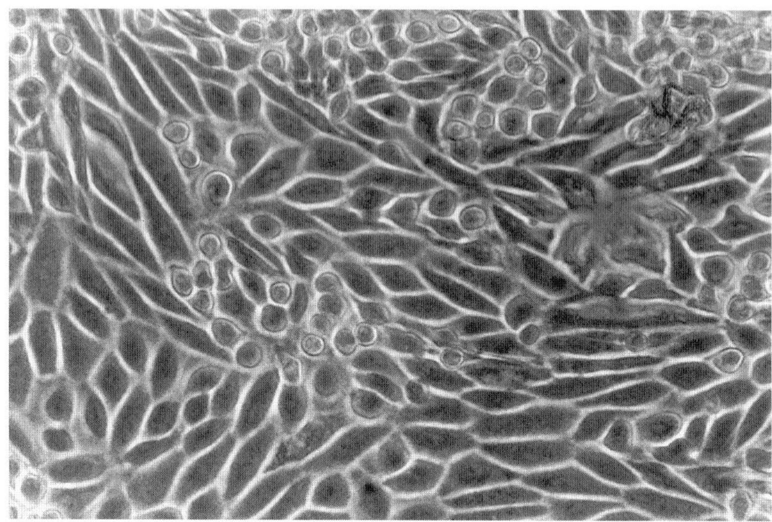

Fig. 2. Epithelial cell preparation grown on type I collagen. At confluence, epithelial cells display a cobblestone-like appearance.

3.3. Glomerular Mesangial Cell Culture

1. Glomeruli are isolated and cultured as described in **Subheading 3.2.** Once the glomeruli have adhered to the flask, the medium should be changed every 4–5 d.
2. The first cells to grow out of the glomerular core are epithelial cells (*see* **Fig. 1B**). After a further 2 wk in culture, a mixed population of glomerular cell types is observed (*see* **Fig. 1C**). Continue cultures for 2–3 wk to allow the mesangial cells to overgrow the glomerular epithelial cells (*see* **Fig. 1D**). From wk 3 onward, grow the cells in MEM-D-valine culture medium to inhibit any fibroblast growth (*see* **Note 10**) *(22)*.
3. Once the cells have grown to confluence, split them 1:1 using trypsin/EDTA (*see* **Subheading 3.4.**), and continue to culture in uncoated tissue-culture grade flasks.
4. Cells should be used between passages 3 and 7 (*see* **Notes 7** and **11**) after being fully characterized.

3.4. Trypsinization of Adherent Cells

To remove the epithelial or mesangial cells from tissue-culture flasks:

1. Pour the medium off the cells, and wash out the flask with sterile PBS.
2. Pour off the PBS, add 5 mL of 1% trypsin/EDTA solution, and then incubate the flask at 37°C for 4–7 min.
3. Tap the flask lightly on the bench to loosen the cells before adding fresh medium. Split the cell suspension into two flasks to continue the culture, or centrifuge and count the cells before using for experiments (*see* **Note 12**).

3.5. Validation of Glomerular Cell Cultures

3.5.1. Cell Morphology

Cell morphology should be assessed throughout the culture period using an inverted microscope with phase-contrast illumination. Cell morphology at different stages of glomerular culture is shown in **Fig. 1**.

1. Epithelial cells: Epithelial cells are homogeneous in appearance *(9)*, polygonal, and form cobblestone-like monolayers (*see* **Fig. 2**). They are closely packed and adhere tightly to each other at the edge of the growing monolayer. Epithelial cells are subject to contact inhibition (*see* **Note 9**) and are sensitive to puromycin amino nucleotide.
2. Mesangial cells: Mesangial cells are elongated and stellate-shaped, and are not subject to contact inhibition (*see* **Fig. 1D**). They grow in multilayers to form characteristic hills and valleys *(5,23)*.

3.5.2. Alkaline Phosphatase Antialkaline Phosphatase (APAAP) Immunohistochemistry Technique

The morphological assessment of cell cultures should be confirmed by immunohistochemistry using currently available monoclonal antibodies (MAbs) (*see* **Table 1**). The APAAP immunohistochemical technique is a very sensitive method *(24)* and is particularly useful when using MAbs. In this technique, soluble complexes of alkaline phosphatase and mouse monoclonal anti-alkaline phosphatase are used to amplify the primary antibody–antigen interaction via a bridging antibody (rabbit anti-mouse) that links the primary antibody (mouse) to the APAAP complex. An intense immunohistochemical staining with low nonspecific backgrounds can be obtained. The enzyme conjugate is detected using a naphthol-phosphatase derivative as substrate, and the naphthol compound produced is visualized by forming a diazonium salt with Fast red to give a colored product. Levamisole is included in the substrate to inactivate the endogenous phosphatase activity of cells.

1. The cells to be tested should be passaged, and 0.4 mL of cells at 1×10^6/mL seeded onto eight-well Permanox Lux chamber slides (Gibco-Invitrogen) and incubated overnight in a CO_2 incubator.
2. Remove the slides from the incubator, and without removing the upper chamber structure, aspirate tissue-culture supernatant. Gently wash the cells twice with 0.8 mL of TBS/well. Discard the final rinse of TBS, and allow the slides to air-dry for 20 min or overnight at room temperature.
3. Fix the slides by adding 0.3 mL of acetone at 4°C and incubating slides on ice for 10 min. (Note: Do not use more than 0.3 mL of acetone, because this dissolves the upper structure of the eight-well chamber.)
4. Remove the acetone and allow the slides to air-dry at room temperature for 10–15 min.

Table 2
**Source and Working Dilutions of Antibodies Used
to Characterize Glomerular Cell Cultures**

MAbs	Clone / source[a] / product code	Dilution
Glomerular epithelial protein-1	5C11/ Biogenex / 336M	1/100
Vimentin	LN-6/ Sigma / V2258	1/200
Common ALL Antigen (CD10)	SS2/36/ Dako / M727	1/50
Cytokeratin 18	CY-90/ Sigma / C8541	1/500
Cytokeratin 19	A53-BA2/ Sigma / C6930	1/50
α-Smooth muscle actin	1A4/ Sigma / A2427	1/200
Myosin (smooth)	hSM-V/ Sigma / M7786	1/500
VLA-1 integrin	TS2/7/ Serotec / MCA1133	1/50
von Willebrand (factor VIII antigen)	F8/86/ Dako / M0616	1/50
HLA-DR	LN-3/ Novacastra / NCL-LN3	1/50

[a]Locations of companies: Biogenex (San Ramon CA); Dako (Ely Cambridgeshire, U.K.); Sigma-Aldrich (Poole, Dorset, U.K.); Serotec (Kiddington, U.K.); and Novacastra Laboratories (Newcastle upon Tyne, U.K.).

5. Prepare the antibodies at this stage, diluting them to the optimal concentration in 1% BSA/TBS (w/v) (*see* **Table 2**).
6. Rehydrate the slides by adding 0.8 mL of TBS to each well (two washes of 5 min). Ensure that the slide does not dry out at any stage after this point.
7. Remove the TBS from the wells, and add 200 μL of the primary MAb diluted to its optimal concentration in 1% BSA/TBS. Incubate the slides for 1 h at room temperature.
8. Discard the antibody, and wash the slides three times for 2 min in TBS.
9. To each well, add 200 μL rabbit anti-mouse immunoglobulin bridging antibody (Dako Z0259) diluted 1/20 in 1% BSA/TBS. Incubate for 30 min at room temperature.
10. Wash as in **step 8**.
11. Add 200 μL of mouse APAAP complexes (Dako D0651) at a 1/40 dilution, and incubate for a further 30 min at room temperature.
12. Wash the cells in TBS, three times for 2 min each, followed by a rinse in distilled water.
13. Prepare the substrate solution as described in **Subheading 2., item 14**. To each well of the Permanox Lux slide, add 0.5 mL of the substrate solution.
14. Incubate the slides at room temperature for 10–15 min. During this time, a red color develops if the cells are positive for that marker.
15. Rinse in distilled water.
16. Rinse in tap water and remove the upper chamber structure from the slides according to the manufacturer's instructions.

17. Stain the cells lightly in Mayer's hematoxylin for approx 10 s, and blue the nuclei in Scot's tap water substitute.
18. Wash the slides well in running tap water.
19. Mount the slides in Apathy's aqueous mounting medium.
20. Assess slides for positivity (red) by microscopy, and score on a scale of negative (–) to +++. It is also important to assess the percentage of cells positive for each marker to determine the purity of the epithelial or mesangial cell cultures.

4. Notes

1. It should be noted that the methods described are for isolation of human cells, and sieve sizes and culture conditions should be modified should rat cells (or other mammalian cells) be required *(4,21,25–27)*.
2. All work undertaken with human tissue must adhere to local Ethical Committee guidelines for confidentiality and consent.
3. Appropriate precautions should be used when handling human tissue, e.g., sterile, disposable gloves should be worn at all times; it is recommended that those engaged in cell isolation procedures should be immunized against hepatitis B.
4. It has been suggested that for rodent kidneys *(27)* the number of viable GVECs is increased if mechanical pressing of kidney tissue through the top sieve is avoided and replaced by simply cutting the kidney into small pieces followed by washing through the sieves with medium.
5. If parietal epithelial cells are required, a 200-μm sieve should be used in place of the 150-μm sieve at **Subheading 3.1., step 3** as Bowman's capsule is retained but tubular contamination is minimized.
6. Two techniques can be used to initiate glomerular cell culture. In addition to the methods described in this chapter, glomeruli can be dissociated by incubation with collagenase type I (Sigma-Aldrich C-0130) at a concentration of 1 mg/mL for 20 min at 37°C. After agitation with a Pasteur pipet, glomeruli remnants can be separated from single cells by passing them through a 63-μm sieve. Glomerular fragments and single cells are plated separately for the culture of mesangial and epithelial cells, respectively. Although this improves the plating efficiency of glomeruli, great care must be taken not to "overdigest" the glomeruli, because this may damage epithelial cells.
7. The age of the patient and functional capacity of the tissue will determine how quickly (a) the cells establish themselves in culture (cells from a young, healthy kidney will grow more rapidly) and (b) the number of passages the cells can undergo before reaching senescence.
8. The composition of the extracellular matrix may exert major effects on the phenotypic properties of cells. Attention must be given to the modulatory influences of the matrix on each cell type. Although fibronectin and collagen greatly improve the initial adherence of mesangial and epithelial cells, they can, if necessary, be cultured in the absence of such matrices. Commercially coated type I collagen tissue culture plates (Stratech Scientific Ltd., Luton, U.K.) are now available and can be used for culturing epithelial cells at the early stages.

9. Epithelial cells are subject to contact inhibition and should be passaged as soon as they reach confluence to reduce cell death. Ideally, cells should be used at, or before, passage three, because their proliferative activity decreases suddenly around this time. Epithelial cells can sometimes adopt a spindle-like structure after passage; this is usually owing to dedifferentiation, and their use for experimentation should be considered carefully.

10. One easy and reliable way to check the cultures for fibroblast contamination is by growing cells in medium containing D-valine substituted for L-valine, a condition in which fibroblasts cannot grow. Fibroblasts do not contain the enzyme (D-amino acid oxidase) necessary to convert the D-amino acid to its essential L-form *(22)*.

11. Phenotypic changes may occur in cultured mesangial cells after about 10 passages with the loss of angiotensin II receptors. The morphology of the cells may change from stellate-shaped cells to large, flat cells with the development of stress fibers. In addition, cells at high passage number will no longer contract isotonically to vasoactive hormones *(28)*.

12. If mesangial or epithelial cell matrix proteins are required, the cells should first be dislodged in 1% EDTA/PBS (w/v). The matrix should then be removed in a volume of detergent (e.g., 0.5% SDS) with vigorous scraping using the barrel of a 1-mL syringe.

References

1. Sterzel, R. B. and Lovett, D. H. (1988) Interactions of inflammatory and glomerular cells in the response to glomerular injury, in *Contemporary Issues in Nephrology*, vol. 18 (Wilson, C. B., Brenner, B. M., and Stein, J. H., eds.), Churchill Livingstone, New York, p. 137.

2. Striker, L. J., Doi, T., Elliot, S., and Striker G. E. (1989) The contribution of glomerular mesangial cells to progressive glomerulosclerosis. *Semin. Nephrol.* **9,** 318–328.

3. Scheinman, J. I., Fish, A. J., Brown, D. M., and Michael A. J. (1976) Human glomerular smooth muscle (mesangial) cells in culture. *Lab. Invest.* **34,** 150–158.

4. Holdsworth, S. R., Glasgow, E. F., Atkins, R. C., and Thomson, N. M. (1978) Cell characteristics of cultured glomeruli from different animal species. *Nephron* **22,** 454–459.

5. Kreisberg, J. I. and Karnovsky, M. J. (1983) Glomerular cells in culture. *Kidney Int.* **23,** 439–447.

6. Andrews, P. M. and Coffey, A. K. (1980) In vitro studies of kidney glomerular epithelial cells. *Scanning Electron Microsc.* **2,** 179–184.

7. Burlington, H. and Cronkite, E. P. (1973) Characteristics of cell cultures derived from renal glomeruli. *Proc. Soc. Exp. Biol. Med.* **142,** 143–149.

8. Krakower, C. A. and Greenspon, S. A. (1954) Factors leading to variation in concentration of "nephrotoxic" antigens of glomerular basement membranes. *Arch. Pathol.* **58,** 401–432.

9. Striker, G. E. and Striker L. J. (1985) Biology of disease. Glomerular cells in culture. *Lab. Invest.* **53,** 122–131.

10. Norgaard, J. O. (1987) Rat glomerular epithelial cells in culture: parietal or visceral epithelial origin? *Lab. Invest.* **57,** 277–282.
11. Stewart, K. N., Roy-Chaudhury, P., Lumsden, L., et al. (1991) Monoclonal antibodies to cultured human glomerular mesangial cells. Reactivity with normal kidney. *J. Pathol.* **163,** 265–272.
12. Soukupova, M. and Holeckova, E. (1964) The latent period of explanted organs of newborn, adult, and senile rats. *Exp. Cell. Res.* **33,** 361–367.
13. Fish, A. J., Michael, A. F., Vernier, R. L., and Brown, D. M. (1975) Human glomerular cells in tissue culture. *Lab. Invest.* **33,** 330–341.
14. Abbott, F., Jones, S., Lockwood, C. M., and Rees, A. J. (1989) Autoantibodies to glomerular antigens in patients with Wegner's granulomatosis. *Nephrol. Dial. Transplant.* **4,** 1–8.
15. Rees, A. J. (1989) Proliferation of glomerular cells, in *New Clinical Applications in Nephrology:* Glomerulonephritis (Catto, G. R. D., ed.), Kluwer, London, pp. 163–193.
16. Müller, G. A. and Müller, C. (1983) Characterisation of renal antigens on distinct parts of the human nephron by monoclonal antibodies. *Klin. Wochenschr.* **61,** 893–902.
17. Stewart K. N., Hillis G., Roy-Chaudhury P., Brown P. A. J., Simpson, J. G., and MacLeod, A. M. (1995) Integrin distribution in normal kidney and cultured human glomerular cells. *Exp. Nephrol.* **3,** 140–141.
18. Green, D. F., Resnick, L., and Bourgoignie, J. J. (1992) HIV infects glomerular endothelial and mesangial but not epithelial cells in vitro. *Kidney Int.* **41,** 956–960.
19. Holthofer, H., Miettinen, A., Lehto, V. P., Lehtonen E., and Virtanen I. (1984) Expression of vimentin types of intermediate filament proteins in developing and adult human kidneys. *Lab. Invest.* **50,** 552–559.
20. Weinstein, T., Cameron, R., Katz, A., and Silverman, M. (1986) Rat glomerular epithelial cells in culture express characteristics of parietal not visceral epithelium. *J. Am. Soc. Nephrol.* **125,** 493.
21. Ringstead, S. and Robinson, G. B. (1994) Cell culture from rat renal glomeruli. *Vichows Arch.* **425,** 391–398.
22. Gilbert, S. F. and Migeon, B. R. (1987) D-valine as a selective agent for normal human and rodent epithelial cells in culture. *Cell* **5,** 11–17.
23. Sterzel, R. B., Loyett, D. H., Foelimer, H. G., Perfetto, M., Biemesderfer, D., and Kashgarian, M. (1986) Mesangial cell hillocks: nodular foci of exaggerated growth of cells and matrix. *Am. J. Pathol.* **125,** 130–140.
24. Cordell, J. L., Falini, B., Erber, W. N., et al. (1984) Immunoenzymatic labelling of monoclonal antibodies using immune complexes of alkaline phosphatase and monoclonal anti-alkaline phosphatase (APAAP complexes). *J. Histochem. Cytochem.* **32,** 219–229.
25. Harper, P. A., Robinson J. M., Hoover, R. L., Wright, T. C., and Karnovsky, M. J. (1984) Improved methods for culturing rat glomerular cells. *Kidney Int.* **26,** 875–880.

26. Mendrick, D. L. and Kelly, D. M. (1993) Temporal expression of VLA-2 and modulation of its ligand specificity by rat glomerular epithelial cells in vitro. *Lab. Invest.* **69,** 690–702.
27. Yaoita, E., Kurihara, H., Sakai, S., Ohshiro K., and Yamamoto, T. (2001) Phenotypic modulation of parietal epithelial cells of Bowman's capsule in culture. *Cell Tissue Res.* **304,** 339–349.
28. Kreisberg, J. L., Venkatachalam, M. A., and Troyer, D. A. (1985) Contractile properties of cultured glomerular mesangial cells. *Am. J. Physiol.* **249,** F229–F234.

19

Isolation and Culture of Human Renal Cortical Cells with Characteristics of Proximal Tubules

Gabrielle M. Hawksworth

1. Introduction

The kidney is an extremely heterogeneous organ from a morphological and a functional point of view. In spite of this heterogeneity, several methods have been described in an attempt to isolate and culture homogeneous populations of renal epithelial cells, seeking to maintain normal differentiated characteristics of the nephron segment of origin.

Several techniques have been used to isolate and culture human proximal tubular cells (PTC). Detrisac et al. first devised a simple method of culturing human renal epithelial cells from cortical tissue explants resulting in cells that retain the characteristics of PTC after several passages *(1)*. Other methods used involved the enzymatic digestion of the cortical tissue with collagenase *(2–5)*, removal of contaminating glomeruli and larger fragments by filtration, further purification of the cells by isopycnic centrifugation using Nycodenz *(6)* or Percoll *(7)*, or the more sophisticated technique of microdissection without collagenase *(8)* or after proteolytic digestion of the tissue *(9–11)*.

The method presented here is a modification of that of McLaren et al. *(7)*. It is based on enzymatic digestion using a two-step collagenase digestion followed by mechanical disruption of the tissue, further purification by filtration through a 75-µm sieve, and sedimentation of the cells and fragments obtained in a continuous density gradient formed with Percoll. This technique allows for the separation of a cellular fraction that is highly enriched in proximal tubular cells and fragments. This method produces a far higher yield than microdissection, and because the cells are purified, primary cultures can be grown in the

From: *Methods in Molecular Medicine, vol. 107: Human Cell Culture Protocols, Second Edition*
Edited by: J. Picot © Humana Press Inc., Totowa, NJ

presence of serum without fibroblast overgrowth being a significant problem. Isolating cells by this method typically produces a yield of 6–12 × 10^6 PTC/g of cortical tissue with relatively high viability (between 70 and 90%).

Cultures initiated from this fraction can be maintained for several passages, expressing functional characteristics of proximal tubules, such as the formation of polarized monolayers and domes, parathyroid hormone responsiveness, and expression of brush border enzymes (γ-glutamyl transpeptidase, alkaline phosphatase), and do not show characteristics of other renal cells, such as responsiveness toward calcitonin and vasopressin. The brush border is sparse compared to freshly isolated cells and expression of brush border enzymes is higher in primary cultures than in subsequent passages.

Cells cultured in this manner can be used for a wide range of purposes (biochemical, physiological, or toxicological studies) where human PTC are required, thus, eliminating the need for extrapolation from animal cell cultures or cell lines of uncertain origin *(12,13)*.

2. Materials (*see* Note 1)

1. Balanced salt solution (BSS): Prepare 1 L containing 5.37 m*M* KC1, 0.44 m*M* KH_2PO_4, 137 m*M* NaC1, 0.34 m*M* Na_2HPO_4, 1.35 m*M* $NaHCO_3$, 5.56 m*M* D-glucose, 25 m*M* (*N*-[2-hydroxyethyl]piperazine-*N*′-[2-ethanesulphonicacid]) (HEPES), 0.5 m*M* ethylene glycol-*bis*-(β-aminoethyl ether) *N,N,N′N′*-tetraacetic acid (EGTA), and 0.5% bovine serum albumin (BSA). Adjust the pH to 7.0–7.2, sterilize by filtration through a 22-μm filter, and store at 4°C.
2. Phosphate-buffered saline (PBS): Prepare 1 L containing 2.68 m*M* KC1, 1.47 m*M* KH_2PO_4, 137 m*M* NaCl and 8.12 m*M* Na_2HPO_4. Adjust the pH to 7.0–7.2, sterilize by filtration through a 22-μm filter, and store at 4°C.
3. Cell-culture medium: Dulbecco's modified Eagle's medium/nutrient mixture Ham's F-12 (DMEM/Ham's F-12) mixture (1:1) with 15 m*M* HEPES and 14.28 m*M* sodium bicarbonate obtained from Sigma (Poole, U.K.). Store in the dark at 4°C.
4. Fetal bovine serum (FBS) obtained from Biowest, UK. On receipt, aliquot the serum and store at –20°C.
5. Penicillin/streptomycin, 50 U/50 μg/mL. The stock solution is obtained from Gibco-BRL (5000 U/mL/5000 μg/mL). Store aliquots at –20°C to avoid repeat freeze-thawing (*see* **Note 2**).
6. Supplements for defined medium: 5 μg/mL insulin from bovine pancreas, 5 μg/mL human transferrin (iron-free), 5 ng/mL sodium selenite (Na_2SeO_3), and 18 ng/mL hydrocortisone are all obtained from Sigma. Insulin, transferrin, and sodium selenite are reconstituted in medium, whereas hydrocortisone is dissolved in a small volume of absolute ethanol and brought to the desired volume with medium. Store aliquots of the stock solution at –20°C for no longer than 3 mo.

7. Versene: 0.2% (w/v) EDTA in BPS. Sterilize by filtration and store at 4°C.
8. Trypsin/EDTA solution: Obtained from Gibco-BRL as a 10X stock solution containing 0.5% porcine trypsin and 0.2% EDTA. This solution is aliquoted and stored at –20°C. Prior to use, combine an aliquot of the trypsin/EDTA solution with an equal volume of Earle's BSS without calcium and without magnesium obtained from Gibco-BRL.
9. Percoll solution: Percoll is obtained from Sigma, autoclaved in 18-mL aliquots, and stored at 4°C. Prior to use, combine 18 mL of sterile Percoll with 42 mL of culture medium.
10. Collagenase A solution: Just before use, prepare a solution of collagenase (*Clostridium histolyticum*, Sigma, U.K.) in medium (0.2% [w/v]) and filter through a 22-μm pore filter. Typically, we use 100 mg of collagenase (activity 0.5 U/mg) in a total volume of 50 mL of medium/each 10 g of cortex to be digested.
11. Incubation vessel: The incubation vessel consists of a glass-jacketed flask (200 mL) with a magnetic bar in it. The temperature of the collagenase solution in the inner reservoir is maintained at 37°C by circulating warm water through the jacket of the flask. The flask and the magnetic bar are autoclaved prior to use.
12. Stainless-steel test sieves obtained from Endecotts (London, U.K.) with mesh sizes of 300, 150, and 75 μm. The sieves are autoclaved to ensure sterility.
13. 50-mL autoclaved centrifuge tubes obtained from Nalgene (Rochester, NY).
14. Trypan blue solution: Stock solution (0.4% sterile Trypan blue) is obtained from Sigma. Just before use, this is diluted to 0.2% in PBS.

3. Methods

3.1. Isolation of Human PTC

Pieces of human renal tissue are obtained from nephrectomy specimens removed because of renal cell carcinoma. Ethical approval was obtained from the Grampian Research Ethics Committee. The tissue is always obtained from a fragment as distant from the neoplastic lesion as possible and is confirmed as being normal by the Department of Pathology, University of Aberdeen. The method described here is ideally suited for the isolation of PTC from approx 10 g of cortex.

1. Transport the tissue under sterile conditions to the tissue-culture cabinet, and place it in a sterile Petri dish containing BSS. Using sterile forceps peel off the renal capsule, and dissect the cortical tissue from the adjacent medulla with a sterile scalpel (*see* **Note 3**).
2. Make incisions on the surface of the cortical tissue to facilitate penetration of the buffer into the tissue. Place the cortex in a preweighed disposable and sterile Universal container with BSS.
3. Weigh the Universal container.

4. Bring the cortex back to the cabinet, and place it in a plastic Petri dish containing fresh BSS. Chop the tissue coarsely with a scalpel.
5. Dissolve the required amount of collagenase in approx 50% of the final volume of medium, sterilize it by filtration, and place it in the incubation vessel.
6. Transfer the tissue fragments to a sterile Universal, and wash them thoroughly with several changes of BSS. Then cut the tissue to obtain pieces of approx 3 mm^3. Wash the cortex with several changes of BSS until the solution remains clear (*see* **Note 4**).
7. Wash the cortical fragments three times in tissue-culture medium. During each wash shake the Universal vigorously (*see* **Note 5**).
8. Resuspend the cortical fragments in prewarmed (37°C) medium, and combine it with the collagenase solution in the incubation vessel. Adjust the volume of the collagenase mixture by the addition of further prewarmed medium.
9. Incubate the cortical fragments in the collagenase solution for a variable period of time (between 20 and 30 min) depending on the consistency of the tissue.
10. Pour off the digested mixture onto the first sieve (300 µm), and force it through with the plunger of a 20-mL syringe. Then wash the sieve through with fresh medium. The same procedure is then applied to the second sieve (150 µm). Wash the material collected onto the third sieve with medium, and if necessary, stir the contents with a glass rod (*see* **Note 6**).
11. Pipet out the suspension thus collected into the Petri dish into a series of Universals, and rinse the Petri dish with medium.
12. Centrifuge the Universals at 400g in a bench centrifuge to pellet the digested material.
13. Resuspend the pellets obtained in cold medium, and combine them into one or two Universals, which are centrifuged again as before.
14. Pour off the medium, and resuspend the pellet in 10 mL of medium (5 mL/tube to be used during the isopycnic separation). Load each of the two 50-mL centrifuge tubes with 5 mL of this cell suspension and 30 mL of the Percoll mixture.
15. Centrifuge the tubes at 21,500$g$max for 30 min at 34° angle rotor at 4°C. This separates the material into four distinct bands, three bands (A, B, and C) composed of renal fragments and cells, and a fourth band (D) that contains the blood cells. Band A is very diffuse and contains mostly cell debris, nonviable cells, and cells of uncertain origin. Bands B and C contain tubular cells and tubular fragments. Band C contains a mixture of cells and tubules, composed mainly of proximal tubules and proximal tubular cells (*see* **Notes 7** and **8**). Band D contains blood cells. The density of the gradient formed was measured using density marker beads (Pharmacia). Band A sediments at 1.019 g/mL and band C sediments just above the 1.062 g/mL marker.
16. Pipet out the contents of band C, and wash it in medium. Centrifuge this suspension as in **step 12** to remove traces of Percoll and then resuspend the pellet in a given volume of medium (*see* **Note 9**). Add an aliquot to the trypan blue solution, and estimate cell viability and number with an improved Neubauer hemocytometer (*see* **Note 10**).

3.2. Culture of PTC

1. Initiate the primary cultures by seeding cells into plastic culture dishes or flasks or on type IV-coated polyethyleneterephthalate (PET) inserts at an approximate density of 5×10^4 cells/cm^2 in medium supplemented with 10% FCS to promote attachment.
2. Change the medium after approx 24 h and at 48-h intervals thereafter, changing, if desired, to defined medium once attachment of the cells has taken place (*see* **Notes 11** and **12**).
3. When seeded and cultured under these conditions, the cultures reach confluence in approx 5–6 d, displaying by then the typical cobblestone appearance characteristic of epithelial cells.

3.3. Subculture of PTC

1. PTC cultures can be subcultured at approx 7-d intervals (*see* **Note 13**).
2. Wash the cell monolayer three times with PBS. This step is of particular importance if the cells have been cultured in medium containing FCS in order to remove serum.
3. Incubate the cultures for 3–5 min with Versene solution (0.2% EDTA in PBS) to weaken cell adhesion through the action of the chelating agent.
4. After removal of the Versene, incubate the cultures at 37°C in a trypsin solution for approx 10 min or until the cells have detached from the plate.
5. Inactivate the trypsin by the addition of an equal volume of medium supplemented with 10% FBS.
6. Centrifuge the cell suspension at $400g$ for 5 min, and resuspend the pellet in a given amount of medium (*see* **Note 14**).
7. Determine viability and cell number as described using the Trypan blue solution.
8. Seed the cells onto tissue-culture plates or dishes at a density or 1×10^4 cells/cm^2.

4. Notes

1. Although some of the solutions are stable for a long time, we use them within 3 mo of preparation.
2. The medium supplements and the antibiotics should be added to the medium prior to use for tissue culture and used within 2 wk of preparation, otherwise poor cell growth may result.
3. It is important to maintain the tissue and all solutions (other than those used for collagenase digestion) in ice at all times. This increases cell viability and helps to prevent clumping of the cells, particularly at the later stages of the isolation procedure.
4. The enzymatic digestion of the tissue using collagenase is a two-step procedure. First, the tissue is washed thoroughly with BSS, a solution containing EGTA as a chelating agent. Then the tissue is digested in a solution of collagenase containing calcium, which is essential for optimal enzymatic activity of the collagenase. The advantages of this two-step collagenase digestion are thought to lie in the

    loosening of tight junctions between the cells via the removal of calcium by the chelating agent and therefore producing a better digestion by collagenase *(14)*.

5. It is essential to rinse the tissue thoroughly with medium before collagenase digestion. The EGTA present in the BSS is an inhibitor of collagenase activity, whereas the calcium present in the medium activates the enzyme.

6. Sieving the digested material through the 300- and 150-μm sieves further disrupts the tissue, whereas the function of the 75-μm sieve is primarily to remove contaminating glomeruli. This is the reason for not forcing the digested material through this latter sieve.

7. Band A contains mainly damaged cells and debris together with a number of viable cells of uncertain origin. Band B is composed of cells and tubular fragments, and although some of them are proximal tubules, the vast majority are not. Band C is mainly composed of proximal tubules and cells.

8. During the isopycnic separation of the tubular cells, the centrifuge tubes should not be overloaded with the digested material. Otherwise distinguishing between bands B and C can become extremely difficult.

9. The cells and fragments in band C are sometimes clumped together. Normally, these clumps can be dislodged by gently pipeting the pellet after resuspension in medium. However, on some occasions, this is not sufficient to produce a homogenous cell suspension. In these cases, we found that the suspension can be further homogenized by forcing it through a 19-gauge needle, without causing a significant reduction in the viability of the suspension. These clumps may also be prevented by treating the suspension with DNase (0.05 mg/mL).

10. The assessment of viability and cell number in the isolated suspensions can only be approximate, because the preparation contains a substantial amount of tubular fragments and the number of cells in each fragment can only be estimated under the microscope.

11. In some instances, the primary cultures may be slightly contaminated with fibroblast-like cells. It is, therefore, convenient to culture the cells in defined medium as soon as attachment to the substratum has taken place, thus greatly reducing the problem of fibroblast overgrowth *(15,16)*. Fibroblast overgrowth can be completely eliminated by substituting L-valine and arginine by D-valine and ornithine, respectively, in the culture medium. Epithelial cells, and therefore PTC, can convert D-valine into L-valine and ornithine into arginine, whereas fibroblasts cannot *(17,18)*.

12. With this technique, there is a risk that cell types other than proximal tubules may grow in culture. A better preparation may be obtained if the cultures are grown in glucose-deficient medium. Within the renal cortex, gluconeogenesis primarily occurs in the convoluted proximal tubules *(19,20)*. This method has been successfully used to culture rabbit PTC *(21)*.

13. When cultured on a plastic substratum, the cultures can be passaged three times before they differentiate and senesce. However, by culturing these cells on a collagen matrix, the number of passages may be increased *(1)*. Activities of PT enzymes such as γ glutamyl transpeptidase, renal dipeptidase, and glutamine transaminase K are significantly lower in cultured cells compared with freshly

isolated cells, although there is little difference between primary cultures and passage 1 cultures. Differences from freshly isolated cells are less marked for amino peptidase-N.

14. Both the isolated cells and/or fragments and the cell suspension obtained after trypsinisation of monolayers can be cryopreserved in cell-culture medium containing 10% FBS and 10% dimethyl sulfoxide (DMSO) using standard methods.

15. Following collagenase digestion and Percoll density centrifugation, tubular epithelial cells of the proximal and distal segments can be isolated with an immuno-magnetic method using MACS microbeads *(22)*.

Acknowledgment

Richard Wainford and Kevin Duffy, Departments of Medicine and Therapeutics and Biomedical Sciences, University of Aberdeen, are acknowledged for their contribution to and evaluation of these protocols.

References

1. Detrisac, C. J., Sens, M. A., Garvin, A. J., Spicer, S. S., and Sens, D. A. (1984) Tissue culture of human kidney epithelial cells of proximal tubule origin. *Kidney Int.* **25,** 383–390.

2. States, B., Foreman, J., Lee, J., and Segal, S. (1986) Characteristics of cultured human renal cortical epithelia. *Biochem. Med. Metab. Biol.* **36,** 151–161.

3. Kempson, S. A., McAteer, J. A., Al-Mahrouq, H. A., Dousa, T. P., Dougherty, G. S., and Evan, A. P. (1989) Proximal tubule characteristics of cultured human renal cortex epithelium. *J. Lab. Clin. Med.* **113,** 285–296.

4. Trifillis, A. L., Regec, A. L., and Trump, B. F. (1985) Isolation, culture and characterisation of human renal tubular cells. *J. Urol.* **133,** 324–329.

5. Yang, H. A., Gould-Kostka, J., and Oberley, T. D. (1987) In vitro growth and differentiation of human kidney tubular cells on a basement membrane substrate. *In Vitro Cell. Dev. Biol.* **23,** 34–46.

6. Boogard, P. J., Nagelkerke, J. F., and Mulder, G. J. (1990) Renal proximal tubular cells in suspension or in primary culture as in vitro models to study nephrotoxicity. *Chem. Biol. Interact.* **76,** 251–291.

7. McLaren, J., Whiting, P. H., and Hawksworth, G. M. (1990) Maintenance of glucose uptake in suspensions and cultures of human renal proximal tubular cells. *Toxicol. Lett.* **53,** 237–239.

8. Wilson, P. D., Dillingham, M. A., Breckon, R., and Anderson, R. J. (1985) Defined human renal tubular epithelia in culture: growth, characterization, and hormonal response. *Am. J. Physiol.* **248,** F436–F443.

9. Blaehr, H., Andersen, C. B., and Ladefoged, J. (1993) Acute effects of FK506 and Cyclosporine-A on cultured human proximal tubular cells. *Eur. J. Pharmacol. Environ. Toxic.* **228,** 283–288.

10. Blaehr, H. (1991) Human renal biopsies as source of cells for glomerular and tubular cell cultures. *Scand. J. Urol. Nephrol.* **25,** 287–295.

11. McAteer, J. A., Kempson, S. A., and Evan, A. P. (1991) Culture of human renal cortex epithelial cells. *J. Tissue Cult. Methods* **13,** 143–148.
12. Seglen, P. O. (1976) Preparation of isolated rat liver cells. *Methods Cell. Biol.* **13,** 29–83.
13. Carvalho, M., Hawksworth G., Milhazes N., et al. (2002) Role of metabolites in MDMA (ecstasy)-induced nephrotoxicity: and in vitro study using rat and human renal proximal tubular cells. *Arch. Toxicol.* **76,** 581–588
14. McGoldrick, T. A., Lock, E. A., Rodilla, V., and Hawksworth, G. M. (2003) Renal cysteine conjugate C-S lyase mediated toxicity of halogenated alkenes in primary cultures of human and rat proximal tubular cells. *Arch. Toxicol.* **77,** 365–370
15. Taub, M. and Sato, G. H. (1979) Growth of kidney epithelial cells in hormone-supplemented, serum free medium. *J. Supramol. Struct.* **11,** 207–216.
16. Taub, M. and Livingston, D. (1981) The development of serum-free hormone-supplemented media for primary kidney cultures and their use in examining renal functions. *Ann. NY Acad. Sci.* **372,** 406–421.
17. Leffert, H. and Paul, D. (1973) Serum dependent growth of primary cultured differentiated fetal rat hepatocytes in arginine-deficient medium. *J. Cell. Physiol.* **81,** 113–124.
18. Gilbert, S. F. and Migeon, B. R. (1975) D-valine as a selective agent for normal human and rodent epithelial cells in culture. *Cell* **5,** 11–17.
19. Guder, W. G. and Ross, B. D. (1984) Enzyme distribution along the nephron. *Kidney Int.* **26,** 101–111.
20. Castellino, P. and De Fronzo, R. A. (1990) Glucose metabolism and the kidney. *Semin. Nephrol.* **103,** 458–463.
21. Jung, J. C., Lee, S. M., Kadaia, M., and Taub, M. (1992) Growth and function of primary rabbit kidney proximal tubule cells in glucose-free serum-free medium. *J. Cell. Physiol.* **150,** 243–250.
22. Baer P. C., Nockher W. A., Haase W., and Scherberich, J. E. (1997) Isolation of proximal and distal tubule cells from human kidney by immunomagnetic separation. *Kid. Int.* **52,** 1321–1331.

20

Culture of Parathyroid Cells

Per Hellman

1. Introduction

The human parathyroid glands are comprised of several cell types, and the parathyroid chief cell is the most frequent, consisting of approx 40–70% in a normal parathyroid gland. Other cell types are fibroblasts, oxyphil cells, and fat-storing cells. The oxyphil cells are mitochondria-rich acidophilic cells interspersed between the chief cells and spread throughout the gland *(1)*. The fat-storing cells in some respects resemble the Ito cells in the liver *(2)*. The amount of fat correlates to the activity of the gland, meaning that an active hyperplastic gland contains less fat than a normal parathyroid cell. This phenomenon has been used as a tool for diagnosing pathological features in extirpated parathyroid glands by performing the rather easy oil red O fat staining *(1)*.

The parathyroid glands are small, with a fat-free dry weight of 40–50 mg each. Studies of parathyroid cell physiology have involved characterization of parathyroid hormone (PTH) release in response to various extracellular stimuli, although calcium is the most important regulator of PTH secretion *(3)*. The most common pathological disorder of parathyroid cells is a benign adenoma or hyperplasia due either to a primary genetic dysregulation or secondary to another disorder, most frequently, uremia. Only extremely rarely malignant parathyroid carcinomas ensue.

Physiological studies of parathyroid cells have mostly been performed in short-term cultures. Studies of PTH release during 30-min to 2-h incubations have been frequent *(4,5)*. Only a few groups have reported long-term cultures of parathyroid cells in order to study, e.g., proliferation *(6–8)*. An obvious reason is the lack of success when trying to establish proliferating parathyroid

From: *Methods in Molecular Medicine, vol. 107: Human Cell Culture Protocols, Second Edition*
Edited by: J. Picot © Humana Press Inc., Totowa, NJ

cells in culture, and many different protocols have been tried. Normal parathyroid cells seem to be especially difficult to proliferate in a culture flask, including normal bovine, as well as human, cells. Hyperplastic or adenomatous human cells have offered more success in this matter. On the other hand, parathyroid cells—normal or pathological—even though they fail to proliferate, may attach to a plastic surface and function in terms of secreting PTH in a calcium-dependent way. Attachment to glass is more difficult, although not impossible (*see* **Subheading 3.4.**). In Uppsala, we have used glass cover slips in a fluorescent microscope to study intracellular calcium concentrations after loading cells with the intracellular indicator fura-2. In this setting, the cells attach to the glass cover slips during the time of the experiment, but longer than 24 h culture on this surface has been difficult. Attachment chemicals such as serum have an intrinsic fluorescence interfering with the wavelengths (340 and 380 nm) used for fura-2 determinations. In this chapter, the method for dispersion of glands for short-term cultures will be described. In addition, the presently used method offering reliable and easily repeated cultures of proliferating human parathyroid cells will also be described.

2. Materials

1. Phosphate-buffered saline (PBS): 1 X PBS: 8 g NaCl (137 mM), 0.2 g KCl (2.7 mM), 2.9 g Na_2PO_4 X 12 H_2O (3.1 mM), and 0.24 g KH_2PO_4 (1.5 mM) in 1000 mL distilled H_2O, adjusted to pH 7.4.
2. Transport buffer: HEPES-buffered Ham's F10 with 10% fetal calf serum (FCS).
3. Sharp scissors.
4. Conical-shaped container.
5. Collagenase (Sigma Chemical Co., St. Louis, MO): Collagenase may be stored in –20°C at a ready-to-use solution in vials usable for the shaking incubator (in our laboratory, scintillation vials); 10 mg collagenase in 10 mL Ham's F10 HEPES-buffered medium or a digestion buffer [digestion buffer: 8 g NaCl (137 mM), 0.35 g KCl (4.7 mM), 0.16 g $MgSO_4$ • $7H_2O$ (0.65 mM), 0.18 g $CaCl_2$ • $2H_2O$ (1.22 mM), and 6 g HEPES (25 mM) in 1000 mL sterile water, adjusted to pH 7.45]. Both have been used for digestion of human cells with success. Bovine serum albumin (BSA) is preferably added if the digestion will be performed within short time.
6. DNase (Sigma): stored at –20°C as stock solution at 50 mg/mL in Ham's F10 or buffer as above (add 10 µL to 10 mL).
7. BSA (Sigma): preferably added to freshly prepared collagenase solution (*see* **item 5** above).
8. Calcium chloride: Stock solution (200 mM): 2.94 g in 200 mL sterile water stored at –20°C.
9. Shaking incubator allowing 300 rpm.
10. EGTA dispersion buffer: 1 mM EGTA in 25 mM HEPES buffer (pH 7.4) containing 142 mM NaCl and 6.7 mM KCl.

11. Percoll (Pharmacia, Stockholm, Sweden).
12. Secretion buffer: 125 mM NaCl, 5.9 mM KCl, 5 mM MgCl$_2$ 0.625 mM HEPES, pH 7.4.
13. 24-well plates (Nunclon, Lincoln Park, NJ).
14. Keratinocyte culturing medium (Gibco).
15. Immunoradiometric assay/radioimmunoassay for measurements of PTH (e.g., Nichols Institute, San Clemente, CA).

Optional (*see* **Note 7**)

16. Epidermal growth factor (EGF; Gibco, Uxbridge, U.K.).
17. Bovine pituitary extract (Gibco).
18. Dulbecco's minimal Essential medium (DMEM; Gibco).
19. RPMI (Gibco).

3. Methods

3.1. Parathyroid Cell Dispersion

Human parathyroid glands are obtained during surgery for primary or secondary hyperparathyroidism. Biopsies of the excised human glands are placed in an ice-cold buffer and transported to the laboratory. Informed consent of the patients is obtained, and approval of the Local Ethics Committee to store human tissue in a biobank and use for physiological studies are ensured.

1. Place biopsies of the excised human glands in an ice-cold transport buffer and transfer to the laboratory (*see* **Note 1**).
2. On arriving in the laboratory, human glands are minced with small and sharp scissors after removal of visible fat and connective tissue surrounding the glands. It is essential to remove all periparathyroid tissue and to leave the naked capsule before mincing of the actual parathyroid gland (*see* **Note 2**).
3. Digest the minced preparation in 10 mL (per approx 100 mg minced tissue) of 1 mg/mL collagenase, 0.05 mg/mL DNase, 1.5% BSA, and 1.25 mM Ca^{2+} at 37°C (*9*). Transfer to a shaking incubator and digest at 300 rpm for 20–30 min (*see* **Note 3**).
4. After 20–30 min, wash the cell suspension two times with transport buffer to inactivate and dilute the collagenase (*see* **Note 4**).
5. Eventually, one may expose the cell suspension to EGTA dispersion buffer for 10–20 s, which allows further dispersion to single cell suspensions, if needed (*see* **Note 5**).
6. Remove dead cells and debris by centrifugation (5 min; 300g) through 25% standard isotonic Percoll diluted in PBS. Wash the suspensions thoroughly with PBS (two or three times) (*see* **Note 6**).
7. Determine cell viability by trypan blue exclusion. Add a small amount, e.g., 5 µL trypan blue and 5 µL cell suspension, and count the cells in which the dye is not excluded (=dead cells). In our hands, this routinely exceeded 95%.

8. The entire procedure generally yields small clusters of 2–20 cells, which may be used for short-term cultures for measurements of, e.g., PTH release, or for long-term cultures. Needless to mention are the demands of performing the whole procedure in a sterile fashion.

3.2. Short-Term Cultures for PTH Measurements

1. Suspend the cells in a balanced buffer solution ("secretion buffer") or in culture medium (preferably RPMI with low calcium concentration allowing additions of calcium to be made). Cells are incubated at 37°C in duplicate or triplicate, preferably using a 24-well plate (*see* **Note 8**). Always include incubations at different calcium concentrations (0.5, 1.25, and 3.0 m*M*) to characterize the actual cell batch and its responsiveness to calcium (*see* **Note 9**). For physiological studies of other compounds, additions should be made at an external calcium concentration of 1.25 m*M*, resembling the physiological level.
2. After the incubation period, aspirate the contents of the wells and centrifuge quickly to separate nonattached cells and debris from the supernatant.

3.2.1. Measurements of PTH

1. Determine the PTH concentrations of the supernatants from above. Always use at least duplicates, preferably triplicates (*see* **Note 11**).
2. Analyses of PTH may be correlated to the amount of cells either by using at least duplicates and a thorough addition of similar number of cells in each well (500,000), or by determining the total protein amount in each well after separation of the cell pellet.
3. Interpretation and presentation of the assay results usually is done by setting the PTH release at 0.5 mmol/L to 100%, and maximal inhibition of PTH release set to the amount at an external Ca^{2+} concentration of 3.0 mmol/L. Typically, the PTH release at external Ca^{2+} is about 60% of that at 0.5 m*M* Ca^{2+} (*see* **Note 12**).

3.3. Long-Term Cultures

1. Suspend the cells in the abovementioned keratinocyte culturing medium in a 25-cm² cell-culture flask. An alternative is overnight culture in DMEM with 10% fetal calf serum (FCS) to allow the cells to attach to the culture vial plastics somewhat more efficiently than in the keratinocyte medium, despite the negative effect on the growth of fibroblasts (*see* **Note 13**).
2. After the eventual overnight culture in DMEM/10% FCS, cells are fed with the keratinocytes medium as above. In this environment, parathyroid cells may be cultured up to 60 d at our laboratory. The most success has been when using pathological parathyroid cells, in terms of achieving dividing proliferating parathyroid cells (*see* **Note 14**). Normal human parathyroid cells attach to the plastic surfaces but do not divide in this environment. However, they still secrete PTH in a calcium-dependent manner.
3. When the cells fill the bottom layer of the culture flask, contact inhibition ensues. At this stage, detach the cells with or without trypsin and culture them further in

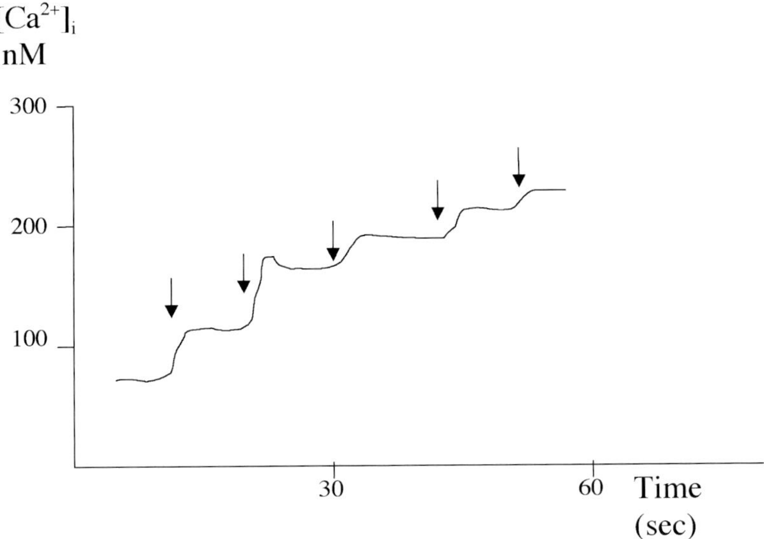

Fig. 1. Intracellular calcium concentration measured using fura-2 loaded human parathyroid cells. The external calcium concentration is stepwise added in steps of 0.5 mM. Baseline concentration is 0.5 mM, and the baseline [Ca^{2+}]$_i$ is approx 80 nM. In the literature, parathyroid cells are generally characterized by using the interval between 0.5 and 3.0 mM, allowing calculation of the set-point, i.e., the external calcium concentration at which half-maximal increase of [Ca^{2+}]$_i$ occurs within this interval. Reproduced from **ref. 8** by permission of the Society for Endocrinology.

new flasks. Regardless of the procedure at this stage, the cells usually cease to proliferate in the next passage, although they still function in terms of secreting PTH and responding to different external calcium concentrations *(8)*. Descriptions of the responsiveness to external calcium at different stages of culture are seen in **Figs. 1–5**.

3.3.1. Long-Term Cultures in the Literature

Reports of long-term cultures in the literature are scarce. An impressive 140 doublings were achieved with cells isolated from bovine parathyroid glands cultured in Coon's modified Ham's F-12 medium containing low (0.3 mM) concentrations of calcium and supplements of bovine hypothalamic extract, bovine pituitary extract, epidermal growth factor, insulin, transferrin, selenous acid, hydrocortisone, triiodothyronine, retinoic acid, and galactose *(6)*. We have made our own attempts in trying to repeat such cultures without obvious success. Attempts of coculture with irradiated 3T3 fibroblast cell lines and even irradiated isolated parathyroid fibroblasts have also been fruitless in order to

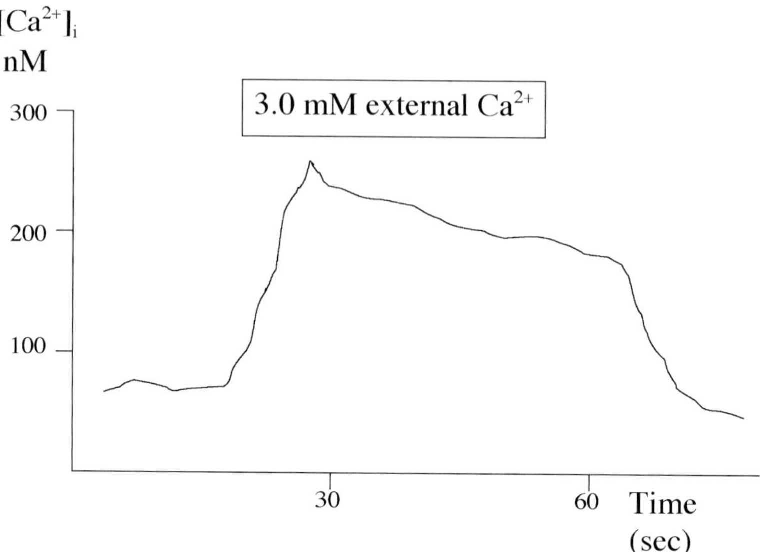

Fig. 2. A typical response of $[Ca^{2+}]_i$ to an increase of external calcium concentration from 0.5 to 3.0 mM in a fura-2 loaded human parathyroid cell recently dispersed. Reproduced from **ref. 8**.

achieve long-term cultures of proliferating and functioning parathyroid cells. Another successful report of long-term cultures of parathyroid cells used clusters of cells derived from patients with secondary hyperparathyroidism due to uremia *(7)*. These cells may be easier to culture because of their inborn proliferating tendency, in parallel to our experience with hyperplastic cells derived from MEN-1 patients.

3.4. Measurements of Physiological Function

The main task for the parathyroid cells are to secrete PTH, which may be measured as described above. However, determination of the intracellular calcium concentration ($[Ca^{2+}]_i$) using the intracellular fluorescent indicator fura-2 is an excellent method for characterization of physiological function in parathyroid cells. These cells are equipped with cell-surface bound calcium receptors *(10)* signaling via $[Ca^{2+}]_i$.

1. Plate cells on glass cover slips onto which the cells generally attach.
2. Load the fura-2 esher into the cells by incubating the cells for 30 min at 37°C (*see* **Note 15**).
3. Use excitations wavelengths of 340 and 380 nm and emission at 510 nm in a dual-wavelength fluorescence microscope. The excitation wavelengths are in our setting automatically shifted about once every second, and the software calculates

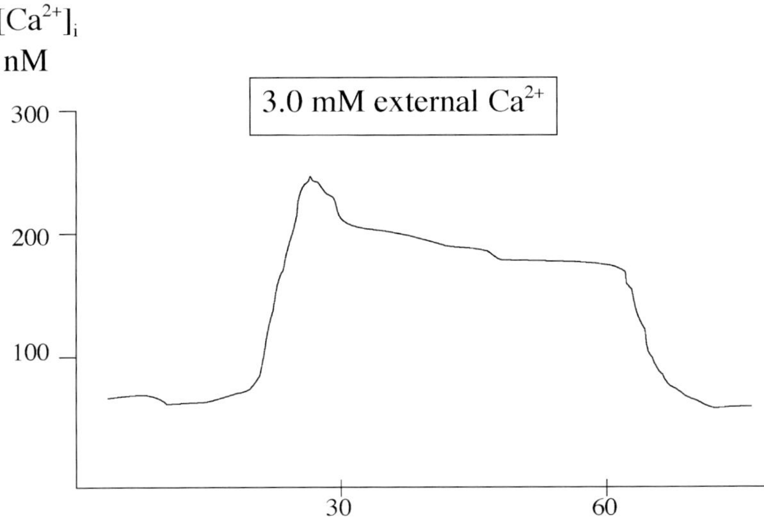

Fig. 3. A typical $[Ca^{2+}]_i$ response in a fura-2 loaded human parathyroid cell cultured for 2 d in low-calcium containing medium. Reproduced from **ref. 8**.

a ratio of the respective emitted signals achieved after excitation at 340 and 380 nm. This ratio correlates to $[Ca^{2+}]_i$, which may then be determined. The detailed description of this procedure is beyond the scope of this chapter, but is described elsewhere *(11)*. Typical results are seen in **Figs. 1–5**.

4. Notes

1. Because the time of warm ischemia is important, it is crucial to remove the parathyroid glands as soon as possible and place them in an ice-cold buffer after an initial, and at this stage, rather rough, mincing with a sharp pair of small scissors. Transport of the glands from the operating theater to the laboratory should be done in cold HEPES-buffered Ham's F10 with 10% FCS on ice, which keeps pH stable and metabolism low.
2. We have used a conical-shaped small container for this procedure, enabling the glands to be kept together in the narrow bottom of the container and facilitating the mincing.
3. Our incubator uses a circulating shaking movement, which we found to be superior to transverse shaking. The digestion takes place in 10 mL collagenase solution in regular scintillation vials. The goal for us has been to achieve clusters of approx 2–8 cells and not a fully dispersed suspension. Single-cell suspensions seem to have been more harmed during the dispersion, expressing fewer numbers of cell-surface receptors and so on, compared to small clusters of cells in which the presumably important cell contact is still present.
4. The parathyroid gland dispersion is the most critical step in obtaining cells for further culture. The collagenase treatment is harmful to the cells, and should be kept

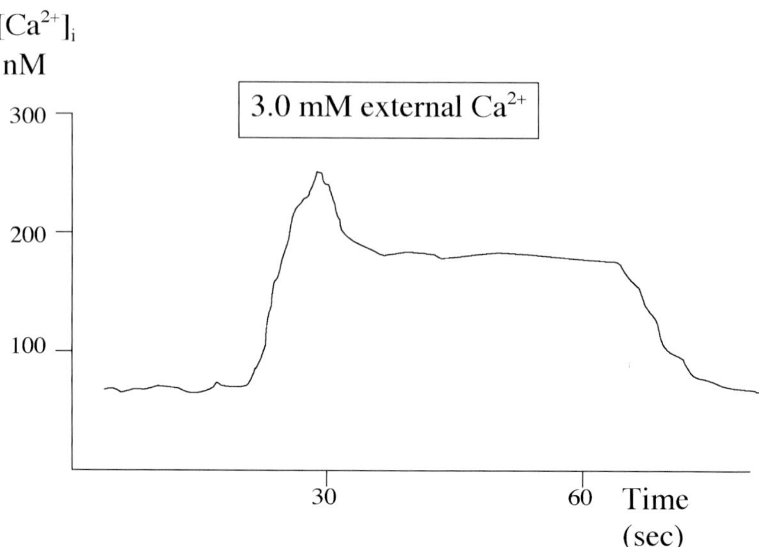

Fig. 4. [Ca^{2+}]$_i$ response in human parathyroid cell cultured for 10 d. Reproduced from **ref. 8**.

at as short time as possible. To reduce the exposure time to collagenase we also use manually pipetting every 5–15 min.

5. This procedure further disrupts bounds between the cells, but should be limited in time because the somewhat sensitive cells may otherwise be disrupted and a high rate of dead cells follows. The EGTA treatment is been performed if, for instance, measurements of intracellular calcium concentration is planned.

6. Because the collagenase treatment is harmful to the cells, a significant number of dead cells are present after the initial steps. Debris has a lower density than viable cells, and therefore, is collected on top of the Percoll gradient, whereas the viable cells remain in the bottom pellet. In case of a cell suspension rich in erythrocytes, a possible additional step is to include a 50% standard isotonic Percoll gradient as well. The comparably high-density erythrocytes will be collected in the bottom pellet with debris on the top, while the viable cells are collected in the interphase between the 25 and 50% Percoll cushions. It is essential to add these Percoll cushions thoroughly, starting either with the 50% in the bottom of a tube, followed by the 25% on top, and last, the cell suspension to be separated—or to first add the cell suspension followed by 25% Percoll using a long syringe to add the Percoll underneath the cell suspension. The 50% Percoll may be added similarly.

7. For studies of PTH release, incubation times of between 30 min and h are generally sufficient. For studies of immediate responses, we have used 30-min incubations (occasionally shorter; 5–15 min), whereas the impact on PTH release after genomic interactions (e.g., vitamin D$_3$) takes longer *(4,8,12)*. Similar short-term cultures have been made after incubation of cells in a serum-free medium con-

(%)

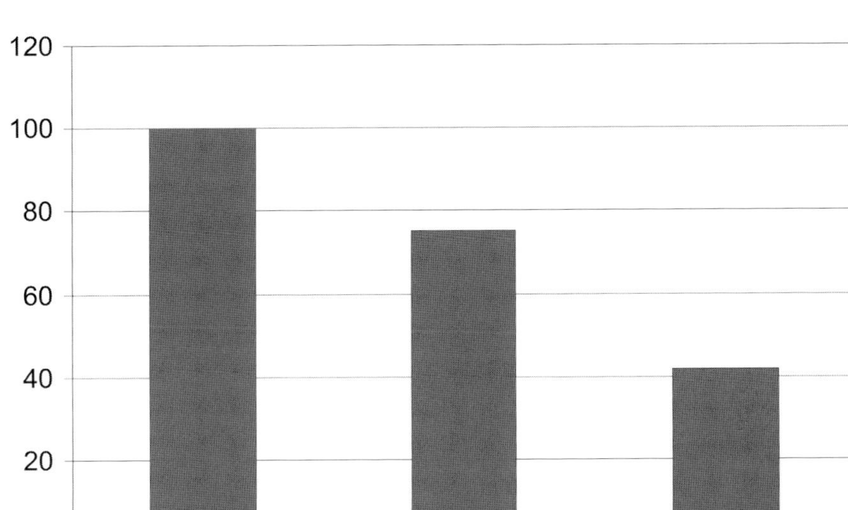

Fig. 5. Number of human parathyroid cells after culture at different calcium concentrations for keratinocytes in a specially defined medium (Gibco) in which additions of calcium have been made.

taining 5 ng/mL EGF and 50 µg/mL bovine pituitary extract, as well as in regular DMEM or RPMI containing 5–10% FCS *(8)*.

8. A minimum of 250,000 cells in each well in a 24-well plate are needed, although 500,000 cells give many more reliable results, incubated in 500 µL buffer.

9. The ability of high external calcium concentrations to reduce PTH release has been used as an indication of the viability of the cells. Thus, healthy cells express functioning cell-surface calcium receptors and a functioning internal pathway, whereas damaged cells seem merely to leak PTH in an uncontrolled way. Incubation of cells with 3.0 m*M* calcium should reduce PTH release with approx 20–40% after 30 min compared to 0.5 m*M* external calcium concentration *(4)*. It is wise to check the concentration of ionized calcium in the buffer by, e.g., an ion-selective electrode, in order to make sure that the BSA, serum, and phosphate in the buffer solution or medium do not dramatically interfere with the free-calcium concentration. The used serum-free medium originally intended for keratinocytes has an extremely low calcium concentration of 0.09 m*M*, although DMEM has 1.8 m*M* and RPMI approx 0.4 m*M*.

10. For proliferation studies, a low-calcium concentration should be ideal for parathyroid cells, because this signal causes a diffuse parathyroid hyperplasia and presumably would initiate proliferation in the culture flask. Furthermore, fibroblasts

do not grow in extremely low calcium concentration, which is an advantage with the "keratinocyte" medium.

11. PTH may be analyzed in different ways. The active aminoterminal portion—PTH(1–34)—may be analyzed by a radioimmunoassay using anti-serum directed against this portion of the PTH molecule *(4)*, or—perhaps more accurate and if human PTH is to be measured—determination of intact PTH(1–84) using two anti-sera and an immunoradiometric method.

12. The set-point—the external calcium concentration at which half-maximal inhibition of PTH release is achieved—is shifted to the right in pathological cells. The right-shifted set-point is also seen after measuring the intracellular calcium concentration ($[Ca^{2+}]_i$) changes to external calcium variations. Thus, the relation between external and internal calcium concentrations is inverse and sigmoidal (**Fig. 1**), with a corresponding set-point as the calcium/PTH relation.

13. Cells may be cultured in different flasks, but owing to the usual limited number of cells, cell-culture flasks not larger than 25 cm^2 are recommended. The low-calcium concentration of this medium triggers parathyroid chief cell division and proliferation and also reduces the proliferation and survival of occasional fibroblasts. The usual problem by performing cultures in higher calcium concentration has been a rapid overgrowth of fibroblasts. Culture of parathyroid cell dispersions in DMEM/10% FCS leads to overgrowth of fibroblasts within 2–3 d. Nevertheless, this medium may be used to allow cells to attach to the culture flask plastics if used over night. The use of surface coating with polylysine or serum and so on has not been proven superior to culturing without a coating.

14. Parathyroid cells derived from hyperplastic glands seem to be the easiest from which to obtain cell cultures. In particular, cells from patients suffering from multiple endocrine neoplasia type 1 (MEN-1) are especially prone to proliferate in this environment. The reason may be their inborn tendency for proliferation, which is further stimulated in the low-calcium concentration *(8)*. The parathyroid pathophysiology of MEN-1 is characterized by a proliferative disorder rather than a set-point shift.

15. The esther diffuses through the lipid cell membrane and is cleaved by intracellular estherases to fura-2 acid which is trapped intracellularly.

Acknowledgment

The development of our culture procedures would not have been possible without a number of collaborators throughout the years: Birgitta Bondeson, Göran Åkerström, Peter Ridefelt, Jonas Rastad, Claes Rudberg, Wei Liu, and others, who are greatly acknowledged.

References

1. Grimelius, L., Åkerström, G., Johansson, H., Juhlin, C., and Rastad, J. (1991) The parathyroid glands, in *Functional Endocrine Pathology* (Kovacs, K., and Asa, L., eds.), Blackwell Scientific, Boston, pp. 375–395.

2. Liu, W., Hellman, P., Li, Q., et al. (1996) Biosynthesis and function of all-trans- and 9-cis-retinoic acid in parathyroid cells. *Biochem. Biophys. Res. Commun.* **229,** 922–929.

3. Hellman, P., Carling, T., Rask, L., and Akerstrom, G. (2000) Pathophysiology of primary hyperparathyroidism. *Histol. Histopathol.* **15,** 619–627.

4. Ridefelt, P., Nygren, P., Hellman, P., et al. (1992) Regulation of parathyroid hormone release in normal and pathological parathyroid cells exposed to modulators of protein kinase C. *Acta Endocrinol.* **126,** 505–509.

5. Brown, E. (1991) Extracellular Ca^{2+} sensing, regulation of parathyroid cell function, and role of Ca^{2+} and other ions as extracellular (first) messengers. *Physiol. Rev.* **71,** 371–411.

6. Brandi, M. L., Fitzpatrick, L. A., Coon, H. G., and Aurbach, G. D. (1986) Bovine parathyroid cells: Cultures maintained for more than 140 populations doublings. *Proc. Natl. Acad. Sci. USA* **83,** 1709–1713.

7. Roussanne, M., Gogusev, J., Hory, B., et al. (1998) Persistance of Ca2+-sensing receptor expression in functionally active, long-term human parathyroid cultures. *J. Bone Min. Res.* **13,** 354–362.

8. Liu, W., Ridefelt, P., Akerstrom, G., and Hellman, P. (2001) Differentiation of human parathyroid cells in culture. *J. Endocrinol.* **168,** 417–425.

9. Rudberg, C., Grimelius, L., Johansson, H., et al. (1986) Alterations in density, morphology and parathyroid hormone release of dispersed parathyroid cells from patients with hyperparathyroidism. *Acta Path. Microbiol. Immunol. Scand. Sect. A* **94,** 253–261.

10. Brown, E. M., Gamba, G., Riccardi, D., et al. (1993) Cloning and characterization of an extracellular Ca(2+)-sensing receptor from bovine parathyroid. *Nature* **366,** 575–580.

11. Ridefelt, P., Rastad, J., Åkerström, G., Hellman, P., and Gylfe, E. (1996) Imaging of Ca2+-induced cytoplasmic Ca^{2+} responses in normal and pathological parathyroid cells. *Eur. J. Clin. Invest.* **26,** 1166–1170.

12. Ridefelt, P., Hellman, P., Wallfelt, C., Åkerström, G., Rastad, J., and Gylfe, E. (1992) Neomycin interacts with Ca2+ sensing of normal and adenomatous parathyroid cells. *Mol. Cell Endocrinol.* **83,** 211–218.

21

Long-Term Culture and Maintenance of Human Islets of Langerhans in Memphis Serum-Free Media

Daniel W. Fraga, A. Osama Gaber, and Malak Kotb

1. Introduction

Tissue culture refers to the in vitro growth or maintenance of organs, tissues, or cells. Methods of tissue culture in most cases have the dual goals of preservation of physical integrity and viability *(1,2)*. As a nonreplicating tissue responsible for significant endocrine hormone production, islets of Langerhans represent a unique challenge for the tissue-culture laboratory. Physical integrity must be maintained not only on a cellular level, but also in terms of the three-dimensional (3D) matrix and multicellular composition of the intact islets. The maintenance of viability with respect to islet culture requires not only the preservation of cellular integrity, but also the retention of an ability to respond to external stimuli in a physiologically appropriate manner. Specifically, with respect to the suitability of tissue for islet transplantation, this refers to preserving the ability to produce and release insulin in response to secretagogues.

Presented here is a simple but effective technique to maintain human islet tissue for prolonged periods of in vitro culture *(3,4)*.

2. Materials

1. CMRL 1066 with 100 mg/dL glucose (Gibco BRL cat. no. 320-1530AJ).
2. Sodium hydroxide, 10 N (NaOH) (Fisher cat. no. SS255-1).
3. Hydrochloric acid, concentrated (HCl) (Fisher cat. no. A144-500).
4. HEPES buffer (Sigma cat. no. H0891).
5. Gelman culture capsule, 0.22 μm (CN 12170).
6. $ZnSO_4 \cdot 7H_2O$ (Sigma cat. no. Z0251).
7. Antibiotic/antimycotic solution (Cellgro cat. no. MT 30-004-CI).
8. L-Glutamine (Cellgro cat. no. MT 25-005CI).

From: *Methods in Molecular Medicine, vol. 107: Human Cell Culture Protocols, Second Edition*
Edited by: J. Picot © Humana Press Inc., Totowa, NJ

9. ITS+ Premix (Collaborative Biomedical cat. no. 40352).
10. T-175 tissue culture (suspension) flask (Sarstedt cat. no. 83.1812.502).
11. Sterile media bottles (Nalge/Nunc cat. no. 2019-1000).
12. Printed labels.
13. Sterile 50-mL conical tubes (Sarstedt cat. no. 62.547.004).
14. Sterile Tygon tubing, 8 ft (Cole Parmer cat. no. 96410-15).
15. Peristaltic pump (Cole Parmer cat. no. 7520-25).
16. Digital scale.
17. 70% Ethanol.

3. Methods

The methods described below cover the areas of: (1) stock reagent preparation; (2) working reagent final preparation; (3) islet culture; (4) media change for prolonged islet culture maintenance; and (5) reagent quality assurance testing.

3.1. Stock Reagent Preparation

Stock reagent preparation presented here is for a batch size of 20 L. Preparation volumes, however, may be scaled to the needs of each laboratory.

1. Open 40 (500 mL) bottles of CMRL 1066 and transfer to a sterile 20-L carboy.
2. Add a larger sterile stir bar to the carboy and place on magnetic stir plate.
3. Add 107.1 g of HEPES buffer to the reagent carboy.
4. Add 200 mg $ZnSO_4 \cdot 7H_2O$ to the reagent carboy.
5. Add 23.4 mL of NaOH to the reagent carboy.
   CAUTION: Sodium hydroxide is a potent caustic reagent and must be handled according to good laboratory practices.
6. Allow solution to mix thoroughly and pH to stabilize (*see* **Note 1**).
7. Verify media pH is between 7.35 and 7.45. Adjust pH as needed with NaOH or HCl. Once the pH is stabilized, it is ready for packaging into stock reagent bottles.
   CAUTION: Hydrochloric acid is a potent acid and must be handled according to good laboratory practices. Handle reagent in certified fume hood.
8. While solution is mixing, disinfect laminar flow hood according to standard operating procedures.
9. Disinfect and transfer 20 sterile 1-L bottles into the hood.
10. Aseptically set up filter sterilization apparatus in the hood (*see* **Fig. 1**).
11. Remove cap and place one of the sterile bottles on a digital scale. Zero the scale.
12. Using the peristaltic pump, pump reagent from the 20-L carboy through the Gelman Micro Capsule sterilization filter (*see* **Figs. 2** and **3**).
13. Fill each bottle to a weight of 990–1010 g.
14. Recap the bottles as they are filled. Place identification labels on all bottles (*see* **Note 2**).

Fig. 1. Media filter sterilization setup inside laminar flowhood. Batch reagent reservoir on magnetic stirrer. Culture filter capsule, sterile media bottle, and digital scale.

15. After every five bottles of reagent have been filtered, transfer a 10-mL Quality Control aliquot into the 50-mL sterile conical (*see* **Note 3**).
16. Move the tubing in the pump rollers 2–3 in. to prevent tubing leakage. Resume filtration for the next five bottles.
17. Continue process until all 20 L have been filtered.
18. Place media in quarantined refrigerated storage until Quality Control testing has been completed and the reagent clears Quality Assurance (QA).
19. Perform QA testing as outlined in **Subheading 3.5.**
20. Stock media is stable for 12 mo when stored at 4–8°C.

3.2. Working Reagent Final Preparation

1. Obtain the current lot number of stock media that has passed QA testing from the refrigerator and transfer into a disinfected laminar flow hood.
2. Aseptically add 10 mL of ITS+ premix to each liter of stock media being prepared to achieve the following working concentrations of supplement ingredients:

Insulin	6.25 µg/mL
Selenious acid	6.25 ng/mL
Transferrin	6.25 µg/mL
Linoleic acid	5.35 µg/mL
Bovine serum albumin (BSA)	1.25 mg/mL

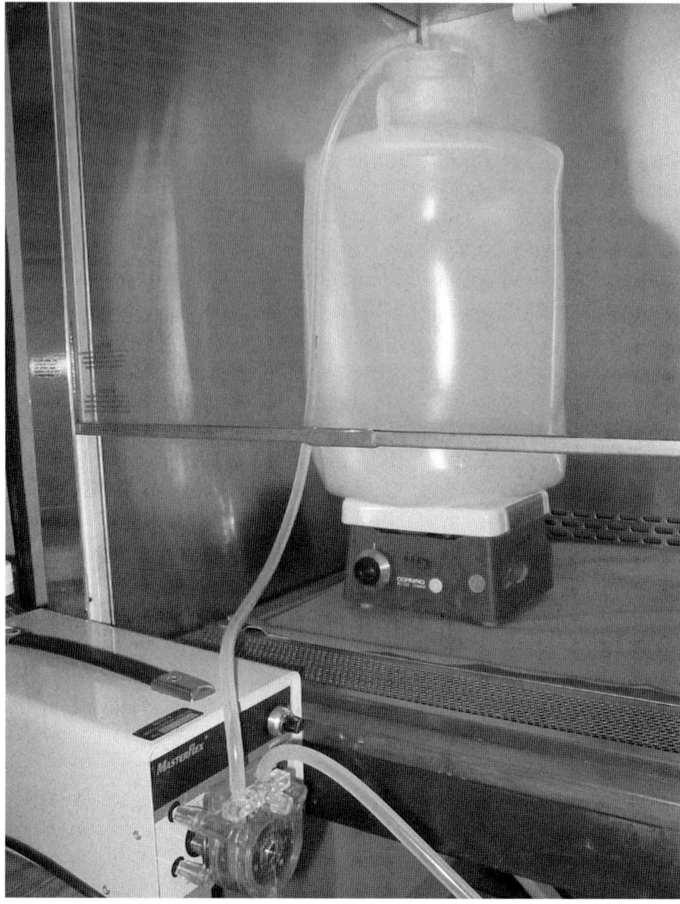

Fig. 2. Batch reagent reservoir and peristaltic drive pump for media filtration.

3. Aseptically add 10 mL of L-glutamine to each liter of stock media being prepared to achieve a working concentration of 0.292 mg/mL.
4. Aseptically add 10 mL of antibiotic/antimycotic solution to each liter of stock media being prepared to achieve the following working concentrations of supplement ingredients (*see* **Note 4**):
   Penicillin   100 U/mL
   Streptomycin   100 µg/mL
   Amphotericin B   0.25 µg/mL
5. It is recommended that a 30-mL QA sample be taken for critical applications and for all transplant related media preparations.
6. Perform QA testing as outlined in **Subheading 3.5.**

Fig. 3. Media capsule filtration into sterile 1-L bottles. Bottles filled by weight.

3.3. Islet Tissue Culture

1. Isolate and purify islet tissue according to standard operating procedures *(5,6)*.
2. While waiting for the final count of the islet preparation, disinfect laminar flowhood according to standard operating procedures.
3. Disinfect and transfer bottles of the current lot number of working CMRL working reagents into the hood.
4. Disinfect and transfer T-175 suspension culture flask bags into the hood. Open the bags aseptically and stage the flasks in the hood.
5. After obtaining the final count, disinfect and transfer the final preparation conical into the hood.

Table 1
Media Change Flowchart

| | Isolation Purity | | | |
Day	<50%	50–70%	70–80%	80–100%
1	X	X	X	X
2	X			
3	X	X		
4	X		X	
5	X	X		
6				
7	X	X	X	X
Every 7 d	X	X	X	X

6. Determine the aliquot of final islet preparation required to obtain a tissue density of approx 400 islet equivalents per milliliter of culture media (50,000 islet equivalents per T-175 flask).
7. Transfer aliquots into T-175 suspension tissue-culture flasks that have been labeled with appropriate isolation information (*see* **Note 5**).
8. Transfer additional media to achieve a total volume of 130 mL per flask.
9. After setting up culture flasks, allow the tissue to settle for 15 min and take a 5-mL sample from each flask. Transfer these samples into a single 250-mL conical for QA testing of the final preparation.
10. Replace vented caps on culture flasks and transfer them to a CO_2 incubator.
11. Incubate at 37°C, 5% CO_2 for 24–48 h. Transfer to low temperature (24–30°C, 5% CO_2) incubator for the remainder of time on culture.
12. Perform media changes as outlined in **Subheading 3.4.**

3.4. Media Change for Prolonged Tissue-Culture Maintenance

Perform media changes at regular intervals using aseptic technique according to the Media Change Flowchart (**Table 1**). The frequency of media changes is higher during the first week after islet isolation for islet preparations with high residual aciner content. This increased frequency is designed to remove potentially harmful exocrine enzymes which can be released in the culture as aciner tissue dies off on culture.

1. Disinfect laminar flow hood according to standard operating procedures.
2. Disinfect and transfer bottles of the current lot number of CMRL working reagent into the hood by wiping down flasks for media change with 70% ethanol.
3. Disinfect and transfer culture flasks into the hood by wiping down flasks for media change with 70% ethanol.

4. Stand the flasks on end and loosen caps.
5. Allow the flasks to stand on end for 15 min.
6. Aseptically remove 75 mL of supernatant from each flask and transfer to a sterile bottle (*see* **Note 6**).
7. Pour fresh media into a sterile 250-mL conical and aseptically transfer 75 mL of fresh media to each flask.
8. Replace caps on the flasks and return them to the 24–30°C incubator.
9. Perform QA testing on the supernatant as outlined in **Subheading 3.5.**

3.5. Reagent QA Testing

Successful tissue culture is predicated on the use of adequate aseptic or sterile technique. This applies to all aspects of the tissue-culture process, from stock reagent preparation through media changes and final tissue disposition *(2,7)*. Documentation of successful implementation of such techniques is accomplished through routine testing of media for the absence of bacterial, fungal, and mycoplasm contamination. In addition, media should be routinely assayed for endotoxin contamination.

There are a large variety of methods available for sterility testing. Each laboratory must select and validate those methods which meet their needs for testing. Recommendations given here are therefore general in nature.

3.5.1. Stock and Media QA Testing

1. Following preparation of a batch of stock media, set up cultures for aerobic, anaerobic, and fungal contaminants.
2. Following preparation of a batch of stock media, set up cultures for mycoplasm contamination.
3. Following preparation of a batch of stock media, test a representative sample of the batch for endotoxin levels.
4. Hold the media in quarantine until testing results are complete.

3.5.2. Working Media QA Testing

Testing of working media is not required prior to use because it will be performed *de facto* when islet final preparations are tested. Hold the working reagent QA sample in storage for follow-up testing if needed (*see* **Note 7**).

3.5.3. Islet Culture Final Preparation QA Testing

1. Following placement of islet tissue on tissue culture, take the final preparation QA sample and set up cultures for aerobic, anaerobic, and fungal contaminants.
2. Following placement of islet tissue on tissue culture, take the final preparation QA sample and set up cultures for mycoplasm contamination.
3. Following placement of islet tissue on tissue culture, take the final preparation QA sample and test for endotoxin levels.

3.5.4. Media Change Islet Culture QA Testing

1. Following media change, use the removed supernatant media to set up cultures for aerobic, anaerobic, and fungal contaminants.
2. Following media change, use the removed supernatant media to set up cultures for mycoplasm contamination.
3. Following media change, use the removed supernatant media and test for endotoxin levels.

4. Notes

1. Allow adequate time for pH stabilization. In practice, this takes at least 30–45 min. Make sure that the solution is being well mixed by the magnetic stirrer.
2. Reagent labels should include as a minimum, "Reagent Identification," "Lot Number," "Date Prepared," "Expiration Date," and a place to note additives that might be used. An example is shown here.

```
CMRL        PREP: 04/30/03        Lot: CM03-01
ITS: ___________ AB/AM: ___________ L-GLU: ___________
Other: _____________________________________________
Date: ____________________ Tech: _________________
Expires: _________________ 4/30/04 ________________
```

3. Take additional QA samples if the filter is changed during the filling sequence. Samples should reflect the last media processed through a given filter.
4. AB/AM and L-glutamine will remain active for 14 d. If the reagent is being used after 2 wk, an additional 10 mL of each must be added. This may be repeated for up to three additions *(1)*.
5. It is critically important to use a consistent technique when aliquoting islet preparations. Because islet tissue will settle rapidly, continuous gentle, but thorough, mixing of the final preparation during the process will ensure a homogeneous distribution of tissue throughout all culture flasks.
6. Use sterile 25- or 50-mL disposable plastic pipets. Do not use glass pipets owing to the possible shedding of painted volume markings.

Acknowledgments

This work was supported by the Assisi Foundation Grant 99-067, JDF Grant 1-2000-416, and NIH Grants RO1-DK57700 and UU@-RR16602.

References

1. Barnes, D. (1987) Serum-free animal cell culture. *BioTechniques* **5**, 534–541.
2. Unchern, S. (1999) Basic techniques in animal cell culture. *Drug Deliv. Syst. Workshop*, August 19–20, 1999, Bangkok, Thailand.
3. Fraga, D. W., Sabek, O., Hathaway, D. K., and Gaber, A. O. (1998) A comparison of media supplement methods for the extended culture of human islet tissue. *Transplantation* **65(8)**, 1060–1066.

4. Gaber, A. O., Fraga, D. W., Callicut, C. S., Gerling, I. C., Sabek, O. M., and Kotb, M. Y. Improved in vivo pancreatic islet function after prolonged in vitro islet culture. *Transplantation* **72(11),** 1730–1736.
5. Ricordi, C., Lacy, P. E., Finke, E. H., Olack, B. J., and Scharp, D. W. (1988) Automated method for isolation of human pancreatic islets. *Diabetes* **37,** 413–420.
6. Lake, S. P., Bassett, P. D., Larkins, A., et al. (1989) Large-scale purification of human islets utilizing discontinuous albumin gradient on IBM 2991 cell separator. *Diabetes* **38(Suppl. 1),** 143.
7. Class II special controls guidance document: tissue culture media for human ex vivo tissue and cell culture processing applications; Final guidance for industry and FDA reviewers. U.S. Dept. of Health and Human Services, Center for Devices and Radiological Health, May 16, 2001.

22

Primary Culture of Human Antral Endocrine and Epithelial Cells

Susan B. Curtis and Alison M. J. Buchan

1. Introduction

The mucosal endocrine cells in the antrum are found as individual elements interspersed among the surrounding epithelial cells, the majority of these being the gastric mucous cells. To establish the factors regulating either endocrine or mucous cell function, the cells have to be separated not only from nonepithelial cells, but also from circulating and neuronal elements within the stomach.

A major problem in obtaining cultures of gastric endocrine cells is their diffuse distribution in the stomach and the nonsterile nature of the gastric lumen. To overcome these problems, we have used a combination of collagenase digestion of the mucosal layer with centrifugal elutriation to remove small particles, such as bacteria and fungi, and provide an enriched preparation of endocrine cells. The technique represents a modification of the methodology originally developed to isolate endocrine cells from the canine stomach *(1)*.

Unfortunately, none of the techniques developed so far produce a 100% pure culture of an individual endocrine cell type, with the antral gastrin cell cultures ranging from 20 to 45% purity. However, by modifying the single-cell fractionation protocol cultures containing 95% gastric mucous cells can be obtained.

These epithelial cell cultures have been used for a number of different experimental techniques: release studies examining regulation of hormonal secretion, intracellular ion flux in response to stimulation (e.g., Fura-2 measurement of intracellular calcium levels) *(2)*, molecular studies of gene expression patterns using array technology, or the interaction between antral epithelial cells and *Helicobacter pylori (3–8)*.

From: *Methods in Molecular Medicine, vol. 107: Human Cell Culture Protocols, Second Edition*
Edited by: J. Picot © Humana Press Inc., Totowa, NJ

2. Materials

2.1. Tissue Collection

A 100-mL screw-topped container with 50 mL of chilled buffer (*see* **Subheading 2.5., item 6**), 1 pair scissors (8–12 in.), 1 pair forceps (6–8 in.).

2.2. Dissection of Stomach

1. Plastic container full of ice.
2. 50 mL of chilled Hank's balanced salt solution (HBSS) containing 10 mM N-2-hydroxyethyl piperazine-N-2-ethane sulfonic acid (HEPES), with 0.1% bovine serum albumin (BSA), referred to as HBSS/BSA from now on.
3. Scissors (6–8 in.).
4. Sharp, fine scissors (4–6 in.).
5. Forceps (6–8 in.).
6. Fine forceps (6 in.).
7. 140-mm glass Petri dish.
8. Latex gloves (mandatory for handling human tissue).
9. Face mask, if uncertain of the status of the patient (i.e., hepatitis, virus infection, and so on).

2.3. Digestion of Mucosa

1. Shaking water bath at 37°C.
2. 250-mL nonsterile conical flask with lid (1–12 g of stripped mucosa).
3. Sigma (St. Louis, MO) type I collagenase (300 U/mg, stored as 100-mg aliquots dry powder at –20°C), and type XI collagenase (1750 U/mg, stored as 50-mg aliquots dry powder). Dissolve each collagenase type in 1 mL of basal medium eagle (BME), and store on ice until ready to use.
4. 300 mL BME containing 10 mM HEPES, 24 mM $NaHCO_3$, and 0.1% BSA.
5. 1–2 L HBSS/BSA supplemented with 100 mg/L dithiothreitol (DTT) and 10 mg/L DNase.
6. 5 mL 0.5 M ethylenediaminetetraacetic acid (EDTA).
7. O_2 gas cylinder.
8. Sharp scissors (4–6 in.).
9. Scalpels (2-No.4) and blades (2-No. 21).
10. 100-mL beaker.
11. 250-mL glass/plastic beaker (1).
12. 250-mL plastic beaker (1).
13. Squares of 400 µm of Nytex™ mesh cut to fit over the 250-mL beaker.
14. Plastic ring cut from a 250-mL beaker to fit into the rim of a 250-mL beaker.
15. 50-mL nonsterile centrifuge tubes (6–8).
16. 1-mL Gilson pipet (1).
17. 0.4% Trypan blue.
18. Hemocytometer.

2.4. Centrifugal Elutriation

1. Beckman centrifuge with elutriator rotor.
2. 5-mL standard separation chamber.
3. Pump (Cole Parmer [Niles, IL], Masterflex Model 7520-20 or similar).
4. Laminar flow hood adjacent to centrifuge/rotor assembly.
5. 70% ethanol.
6. 100-mL graduated cylinder.
7. 1 L sterile water.
8. 2 L sterile HBSS/BSA.
9. 50-mL sterile centrifuge tubes (Falcon) (four per elutriation load).

2.5. Tissue Culture

1. 6-, 12-, or 24-well Costar plates.
2. 24-well plates with APES (3-aminopropyltriethoxy-silane, Sigma) coated cover slips (12 mm) are optimal for fluorescence ICC and confocal microscopy. Prior to coating the cover slips with APES, they are first cleaned with 1% HCl/70% ethanol for 30 min, rinsed well with dH_2O, soaked in 70% ethanol for 30 min, rinsed with H_2O, then dried in an oven. Two percent APES in dry acetone is poured over the cover slips, and cover slips are swirled in a Petri dish for 5 s. The APES solution is poured into a waste bottle and the cover slips are washed thoroughly in dH_2O, and dried in an oven. Place the coated cover slips into each well of a 24-well culture plate and expose the plates to UV light in the hood for 30 min. Store plates at 4°C until required.
3. 35-mm Falcon or Costar dishes with APES coated sterile cover slips are optimal for imaging of intracellular ion concentration (Fura-2 ratio measurement).
4. 90-mm Falcon plates or 6-well plates are used to collect cells for ultrastructural studies and to isolate mRNA for molecular studies, or protein for Western blotting.
5. Growth medium: Dulbecco's modified Eagle's medium (DMEM) containing 5.5 mM glucose: Ham's F10 medium (1:1), 10 mM HEPES, 2 mM glutamine, 8 µg/mL insulin, 1 µg/mL hydrocortisone, 5% heat-inactivated fetal calf serum (FCS) (Invitrogen), 1 mL/100 mL penicillin/streptomycin, 100X stock (Invitrogen).
6. University of Wisconsin (UW) Collection Buffer: 10.0 mM NaCl, 57.7 mM Na_2PO_4, 115.0 mM KCl, 19.0 mM glucose, and 10.0 mM $NaHCO_3$. Osmolarity = 330 mosm/kg, pH 7.0.

3. Methods

3.1. Collection of Antral Material

The distal portion of the stomach, 3–4 in. proximal to the pyloric sphincter, is required. The optimal situation would be to obtain material from the retrieval program of the local Transplant Society. (Note: ethical permission forms must be in place for the University/Research Institute, and the permission of the next-of-kin for use of material for research should be obtained where required.)

The critical factor in the tissue collection is to minimize warm ischemic time. In this case, because the tissue was collected during an organ harvest for transplantation, the venous supply was replaced by ice-cold UW buffer prior to clamping the aorta. Thus, there was little, if any, warm ischemia. If the material has to be collected from surgery, care must be taken to ensure that as soon as the antral material is removed from the abdominal cavity, it is immediately immersed in ice-cold buffer.

3.2. Transport to the Laboratory

The transport time should be kept to a minimum. If material has been collected from surgery, once the tissue is out of the operating room, open up the antrum and flush out the lumen with ice-cold buffer. This is vitally important if there is bile present. The bile acts as a detergent and digests the tissue even at 4°C. Once cleaned, it should be transferred to an intracellular buffer at 4°C (such as UW buffer), which slows enzyme activity and minimizes damage from free radicals. Antral tissue once cleaned and immersed in the ice-cold buffer can be stored at 4°C for up to 10h prior to digestion (*see* **Note 1**).

3.3. Initial Separation of Mucosa (see Fig. 1)

1. The antral mucosa is blunt dissected from the underlying submucosa and muscle layers by taking the whole piece of tissue and removing any adherent fat from the serosal surface. With the mucosa face downward in a Petri dish with approx 5 mL of HBSS/BSA, the muscle layer is gripped using forceps and sharp scissors are used to cut through the submucosal layer. This should be done as close to the mucosal layer as possible without cutting into the mucosa. Once all of the muscle layer is removed, any remaining adherent submucosa appears as a whitish layer over the beige mucosa. As far as possible, this white layer should be removed using sharp fine scissors and fine forceps. It is imperative to remove the submucosa with muscularis mucosae to ensure that no nerve or muscle cells are retained in the preparation. The effectiveness of the dissection can be checked by taking a small sample of the tissue, freezing for cryostat sectioning and staining 20-μm sections with hematoxylin and eosin. It takes several practice sessions to perfect the removal process.
2. The muscle and submucosal tissue are placed in an autoclave bag (all waste from the isolation procedure contaminated with human cells, including Nytex filters, must be autoclaved prior to disposal; check the regulations governing disposal of human tissue at your Institution) and the mucosa turned face upward. At this point, the antral region can be discriminated from the corpus by color. The corpus mucosa is thicker and is darker beige compared to the thinner antral region. The darker regions are cut away and discarded, leaving the stripped antral mucosa. Again, this can be checked by taking a small piece of mucosa, freezing, and cryostat sectioning to ensure no corpus tissue is included.
3. The antral tissue is then weighed; the average weight is usually around 7–8 g.

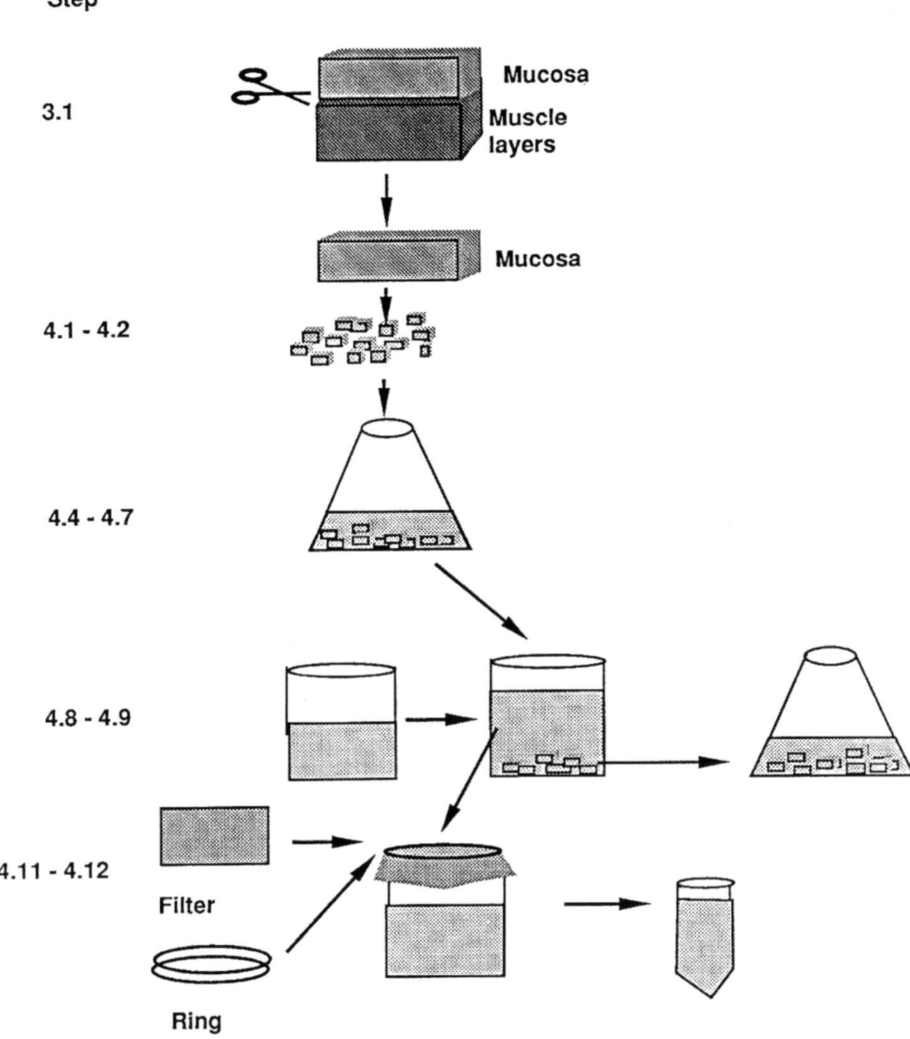

Fig. 1. A schematic drawing of the initial steps in the isolation procedure. The numbers refer to the individual steps in the protocol section.

3.4. Isolation of Single Cells

1. Place the mucosa in a small beaker or 50-mL Falcon tube with no buffer, and mince with sharp scissors.
2. Transfer the minced tissue into a Petri dish, and continue to chop using scalpels (No. 4) until the average size of the pieces is 5 mm^3.

3. If the tissue is <12 g in weight, use a single 250-mL flask for the digestion, if >12 g, use two flasks.
4. For the first digest, add 50 mL BME containing 100 mg (300 U/mg) of Sigma type I collagenase, and 10 mg (1750 U/mg) of Sigma collagenase type XI to each flask.
5. Gas with 5% CO_2/95% O_2 for 30 s prior to capping.
6. Incubate in the 37°C shaking water bath at 200 rpm for 45 min.
7. Add 500 μL 0.5 *M* EDTA and return to the water bath shaker for a further 15 min.
8. Remove flask(s), empty contents into the 250-mL beaker, and add 2 or 3 volumes of HBSS/BSA/DDT/DNase (*see* **Note 2**). Allow nondigested material to settle to the bottom of beaker and discard supernatant from first digest (*see* **Note 3**). Leave a maximum of 5 mL buffer in the beaker.
9. Replace the undigested material in the flask(s), and add 50 mL of BME/BSA containing 50 mg collagenase type I and 10 mg collagenase type XI. Gas as before and return to shaking water bath for 45 min.
10. Add 500-μL 0.5 *M* EDTA for 15 min.
11. Remove flask(s), empty contents into the 250-mL beaker, add 2 or 3 volumes of HBSS/BSA/DDT/DNase. Allow nondigested material to settle, and pour supernatant into a second beaker leaving a maximum of 5 mL of buffer with the undigested material in the original beaker. Filter the supernatant through 400-μm Nytex mesh suspended over a plastic 250-mL beaker. The Nytex mesh can be anchored over the beaker using a ring made from part of another plastic beaker. The undigested material should be returned to the flask for a third digest, as in **step 9**.
12. Separate supernatant from second digest into 50-mL centrifuge tubes (nonsterile) and centrifuge at 200*g* for 5 min. Discard supernatant, and resuspend pellet in 50 mL HBSS/BSA. Cells should always be initially resuspended in a small volume (e.g., 5 mL) to aid dispersion of the clumped cells and then brought to final volume.
13. Repeat **steps 11** and **12** to ensure collagenase is removed from the cell suspension.
14. Repeat **steps 8–13** until no undigested material remains (usually four to five times) (*see* **Note 4**).
15. Take the 50-mL resuspensions from digests 2–5, centrifuge at 200*g* for 5 min, resuspend each in 5 mL, and combine. Add HBSS/BSA and take the total volume through the fine Nytex mesh. A hole can be made in the side of the 250-mL plastic beaker at the 225-mL level and a 20-gauge needle inserted attached to a 50-mL syringe. Air drawn into the syringe creates suction and speeds up the process. If the digest was not effective at producing a single-cell preparation, the 40-μm mesh will clog rapidly. Replace with a fresh square.
16. Check viability of cells, and count cells using trypan blue. A 100-μL sample of the cell suspension is added to 100 μL of 0.4% trypan blue, placed in a hemocytometer, and examined under the microscope. Viable cells are unstained. Dead cells have a blue nucleus. At this point, count all cells alive or dead.
17. Dilute the cell suspension to 1.5×10^8 cells/20–30 mL in HBSS/BSA/DDT/DNase. The number of elutriation loads is calculated by dividing the total number of cells by 1.5×10^8.

3.5. Elutriation

1. Centrifugal elutriation: rotor assembly and calibration of flow rates.
   a. Assemble rotor; ensure separation chamber is free of debris.
   b. Sterilize rotor, chamber, and input and output lines with 70% alcohol. Leave in 70% alcohol until 30 min prior to separation.
   c. Replace 70% alcohol with sterile water, flush out air bubbles, and set flow rates. Flow rates are calibrated at 2500 rpm for 25 mL/min, 2100 rpm for 40 mL/min, and 1800 rpm for 55 mL/min. Care was taken to ensure that at low flow rates (25–35 mL/min), pressure in the line was <3 psi. If higher, this indicated that bubbles were trapped either in the separation chamber or in the spindle assembly of the rotor. To remove bubbles, the elutriator rotor was run at 150g with a flow rate at 75 mL/min, the rotor was stopped while the flow rate was maintained, and all bubbles were forced out of the chamber and spindle.
   d. Replace sterile water with sterile HBSS/BSA.
2. Elutriation of antral cells.
   a. 1.5×10^8 cells are loaded into the chamber at a flow rate of 25 mL/min and a speed of 2500 rpm. Loading volume should be between 20 and 30 mL.
   b. Once cells are in the chamber (checked using the stroboscope), wash for 3 min at 25 mL/min.
   c. Increase the flow rate to 40 mL/min while decreasing the centrifuge speed to 2100 rpm, and collect 2×50 mL fractions in sterile tubes. This fraction (F1) contains the majority of the gastrin containing G-cells.
   d. Increase the flow rate to 55mL/min while decreasing the centrifuge speed to 1500 rpm, and collect 2×50 mL fractions in sterile tubes. This fraction (F2) contains the gastric mucous cells.
   Note: From this point on, care should be taken to ensure sterile techniques are followed.
   e. Repeat a–d until all cells have been run through the elutriator (*see* **Note 5**).
   f. Centrifuge all collected fractions at 200g for 5 min.
   g. Resuspend in 10 mL growth medium (for composition, *see* **Subheading 2.6.**) and complete a cell count using trypan blue as before. This time count only live cells (*see* **Note 6**).

3.6. Tissue Culture

1. The F1 fraction containing the G-cells is resuspended at a final concentration of 1×10^6 cells/mL in growth medium.
2. The cell suspension is plated on 24-well plates with or without cover slips at 1 mL/well for release experiments or immunocytochemistry, on 35-mm dishes (or 6-well plates) with or without cover slips at 2 mL/dish (or well) for immuno-cytochemical and intracellular ion measurement (e.g., Fura-2) experiments, and 90-mm dishes at 10 mL/dish. The average yield of cells in the F1 fraction is 200×10^6 cells.

3. The F2 fraction containing the mucous cells is resuspended at a final concentration of 1×10^6 cells /mL in growth medium.

4. The F2 cell suspension is plated on 24-well plates at 1 mL/well for infection studies with *H. pylori*, and 35-mm dishes (or 6-well plates) with or without cover slips at 2 mL/dish for immunocytochemical and collection of mRNA or protein for array experiments. The average yield of cells in the F2 fraction is 100×10^6 cells (*see* **Note 7**).

3.7. Characteristics of Resultant Cultures

Approximately 50% of the F1 and F2 cells will adhere to the culture dishes overnight and are phase bright. The cells will remain >95% viable for approx 2 mo, as shown by trypan blue exclusion experiments conducted throughout this period. Note that when the cells are cultured on a rat tail collagen substrate (as described in the first edition of *Human Cell Culture Protocols*), the cell viability is significantly reduced, to approx 1 wk. Cells aggregate into small clusters containing a mixture of endocrine and mucous cells. In the authors' experiments, 48 h was chosen for the majority of the experimental protocols designed to examine endocrine cell function, this time period has been demonstrated to allow reformation of cell-surface receptors and cell polarity. In the experiments designed to examine the interaction with *H. pylori*, the gastric cells are allowed to attach to the culture plates and recover from the isolation procedure overnight prior to infection with the bacterium (*see* **Fig. 2**).

In studies of receptor and ion-channel expression by the cultured cells, it has proved useful to collect RNA samples from the cells immediately after elutriation and prior to the culture period. In many cases, the addition of specific growth factors (insulin or epidermal growth factor) at the high concentrations used in these cultures can induce expression of receptors or channels not normally present on the cells in vivo. To control for this possibility, the presence of the receptor or ion channel in the cells prior to culture can be determined using mRNA collected from the post-elutriation samples for the polymerase chain reaction (PCR).

4. Notes

1. One advantage of the studies carried out in the AMJB's laboratory was the availability of tissue from the B.C. Transplant Society in the province of British Columbia. Patients were screened before acquiring tissue to ensure their suitability as organ donors and did not have any known pathophysiological conditions. However, as is often the case with human studies, and unlike studies of laboratory animals, there was variability with respect to age, size, and sex of the donors. This information should be noted and reported in any subsequent publications.

   Antral tissue is not as heavily vascularized as the small intestine, and once the tissue is placed in ice-cold buffer, the cells can withstand a delay of up to 6 h

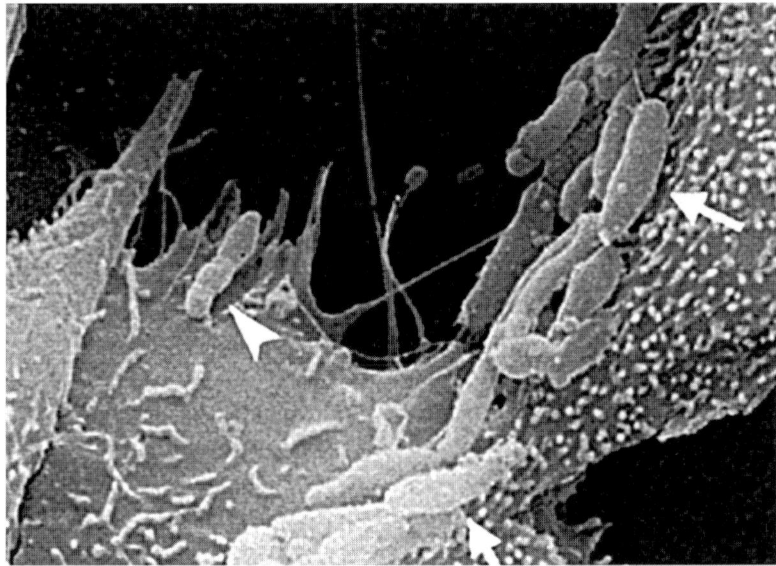

Fig. 2. A scanning electron micrograph showing the attachment of spiral *H. pylori* to antral mucous cells. The arrows indicate the attachment sites.

before starting the isolation procedure. Longer storage times (>8 h) result in a lower yield of viable cells and should be avoided if possible.

2. The volume of the digestion solution is at least doubled with cold HBSS/BSA/ DDT/DNase at the end of each 1-h incubation period prior to the centrifugation step to wash the tissue properly. This quickly reduces the activity of the collagenase. The DDT and DNase are added to minimize clumping of the single cells. DDT counteracts the aggregation of the mucin released from the gastric mucous cells, and DNase stops DNA released from lysed cells sticking the live cells together.

3. The reason for discarding the cells collected from the first digest is that these represent the surface epithelial cells, which have a reduced viability. You can check the viability by trypan blue exclusion. In addition, there are few if any endocrine cells in the upper layers of the human antral mucosa. Therefore, this cell population can be discarded without affecting the total yield of endocrine cells.

4. The isolation protocol described produced the highest yield of viable cells. The choice of collagenase can be difficult. The combination of Sigma type I and type XI collagenase gave the best yield of antral G cells. However, each batch of collagenase should be tested prior to its use, because the activity and level of different additional enzymes in the preparation vary between batches. It is well worth testing several batches from Sigma and then putting a large amount of a good batch on hold. It is important to remember that if you change batches of collagenase, problems may be encountered either with a loss of viable cells in the initial digest or the isolated cells may not survive in culture.

5. The Beckman elutriator centrifuge is equipped with a stroboscope, which allows the separation chamber to be monitored during each run. To speed up the process, it is usual not to stop the rotor completely between runs, but to wash the cells remaining in a pellet out of the chamber by decreasing the rotor speed to 700 rpm and increasing the flow rate to give a pressure of 5 psi on the gauge. However, this does not always work, and the pellet remaining in the chamber can be seen when the rotor is returned to 2500 rpm before the next load. If the pellet is retained, **do not attempt to load the next batch of cells**. Stop the rotor with a high flow rate (>100 mL/min), open the centrifuge, and check that the pellet has been washed clear. If stopping the rotor still does not remove the pellet, then the entire rotor assembly will have to be taken out of the centrifuge, the chamber removed, and the pellet dislodged with a fine needle. Once the pellet is removed, the elutriation process can be restarted.

6. This procedure increases the number of viable cells, and removes the majority of the other cell types and the usual bacterial and fungal contaminants introduced through the nonsterile environment of the antral lumen.

    Receptor damage due to hyperosmolality *(9)* is avoided because cell separation in these experiments was carried out using elutriation, which permits the use of isotonic solutions *(10)*. Centrifugal elutriation utilizes centrifugal force and flow, which act in opposing directions, to separate cells on the basis of their volume. Different fractions of cells can be removed by altering the flow rate of fluid passing through the elutriation chamber by altering the pump rate or by changing the speed of the centrifuge. Appropriate flow rates, centrifugation speeds, and washing times can be determined empirically, and can easily be altered to choose different populations of cells for study.

    Peptide-containing cells can be used as indicators for the enrichment by taking samples before and after elutriation centrifugation and carrying out radioimmunoassay. For example, gastrin content of 10 mg of undigested antral mucosa can be compared to the content of the cell suspension prior to elutriation and the two cell fractions collected during elutriation (usually determined for 1×10^6 cells). The elutriation procedure can then be altered to maximize enrichment of G-cells.

7. An initial plating density of 1×10^6 cells/mL/plate was chosen for the F1 fraction. This gives optimal attachment of the cells to the substrate, and the content of gastrin is well within the range required for detection by the available radioimmunoassays. This plating density has to be doubled for the F2 fraction if it is intended to complete release experiments examining the regulation of somatostatin release. The number of somatostatin containing cells and their content are lower than that of gastrin. Therefore, the plating density has to be increased to bring the peptide content up to the detection level of the available radioimmunoassays. This is vital to detect basal somatostatin levels to provide the control level for comparison with subsequent stimulation or inhibition of the D-cells.

References

1. Soll, A., Amirian, D., Park, J., Elashoff, J., and Yamada, T. (1985) Cholecystokinin potently releases somatostatin from canine fundic mucosal cells in short-term culture. *Am. J. Physiol.* **248,** G659–G573.
2. Squires, P. E., Meloche, R. M., and Buchan, A. M. J. (1999) Bombesin-evoked gastrin release and calcium-signaling in human antral G-cells in culture. *Am. J. Physiol.* **276,** G227–G237.
3. Buchan, A. M. J., Curtis, S. B., and Meloche, R. M. (1990) Release of somato-statin-immunoreactivity from human antral D cells in culture. *Gastroenterology* **99,** 690–696.
4. Richter-Dahlfors, A., Meloche, R. M., Finlay, B. B., and Buchan, A. M. J. (1998) *Helicobacter pylori* infected human antral primary cell cultures: effect on gastrin cell function. *Am. J. Physiol.* **275,** G393–G401.
5. Buchan, A. M. J. (1991) Effect of sympathomimetics on gastrin secretion from antral cells in culture. *J. Clin. Invest.* **87,** 1382–1386.
6. Buchan, A. M. J., MacLeod, M. D., Meloche, R. M., and Kwok, Y. N. (1992) Mus-carinic regulation of somatostatin release from primary cultures of human antral epithelial cells. *Pharmacology* **44,** 33–40.
7. Buchan, A. M. J., Meloche, R. M., Kwok, Y. N., and Kofod, H. (1993) Effect of CCK and secretin on somatostatin release from cultured antral cells. *Gastroen-terology* **104,** 1414–1419.
8. Backhed, F., Rokbi, B., Tortensson, E., et al. (2003) Gastric mucosal recognition of Helicobacter pylori is independent of Toll-like receptor 4. *J. Inf. Dis.* **203,** 829–836.
9. Guarnieri, M., Krell, L. S., McKhann, G. M., Pasternak, G. W., and Yamamura, H.I. (1975) The effects of cell isolation techniques on neuronal membrane receptors. *Brain Res.* **93,** 337–342.
10. Meinstrich, M. L. (1983) Experimental factors involved in separation by cen-trifugal elutriation, in *Cell Separation Methods and Selected Application,* vol. 2 (Pretlow, T. G. and Pretlow, T. P., eds.), Academic, New York, pp. 33–61.

23

Conjunctiva Organ and Cell Culture

Monica Berry and Marcus Radburn-Smith

1. Introduction

Conjunctiva, the mucous membrane of the eye, covers its surface from the limbus, the junction with the cornea, to the edges of the eyelids where it meets the skin, thus forming a blind sac that permits free movement of the eye. Topologically, the conjunctiva is divided into bulbar (covering the sclera, adherent at the limbus), tarsal (lining the lids), and fornical (from the medial "corner" of the eye). At the limbus, where putative corneal stem cells reside, a barrier whose nature has yet to be determined stops conjunctival epithelium from migrating over the cornea. When this barrier is damaged, the migration of conjunctiva over the cornea is accompanied by invasion of blood vessels and, consequently, loss of sight. There is a change in morphology from the fornix, where cylindrical cells give rise to a columnar stratified conjunctival epithelium, to the limbus and lid margins where flattened cells result in a squamous stratified tissue *(1)*. In rodents and rabbits, goblet cells appear in clusters. Their apical openings are decorated with actin collars *(2)*. These are not seen in human tissue, where goblet cells may appear either singly (*see* **Fig. 1**) or in clusters. Although these cells are larger than the surrounding squames, they do not necessarily span the entire epithelial thickness.

The equal efficiency of expanding primary epithelial cultures from every region of the human conjunctiva, i.e., limbal, bulbar, tarsal, and fornical, suggested that this epithelium does not have stem cells *(3,4)*. On the other hand, phorbol ester treatment and pulse ^{3}H-thymidine labeling of conjunctival epithelium in vivo indicated the presence of slow cycling cells, consistent with a stem cell population, in the fornix *(5)*. Clonal expansion of cells from every part of the conjunctiva yielded cultures in which goblet cells differentiated after a

From: *Methods in Molecular Medicine, vol. 107: Human Cell Culture Protocols, Second Edition*
Edited by: J. Picot © Humana Press Inc., Totowa, NJ

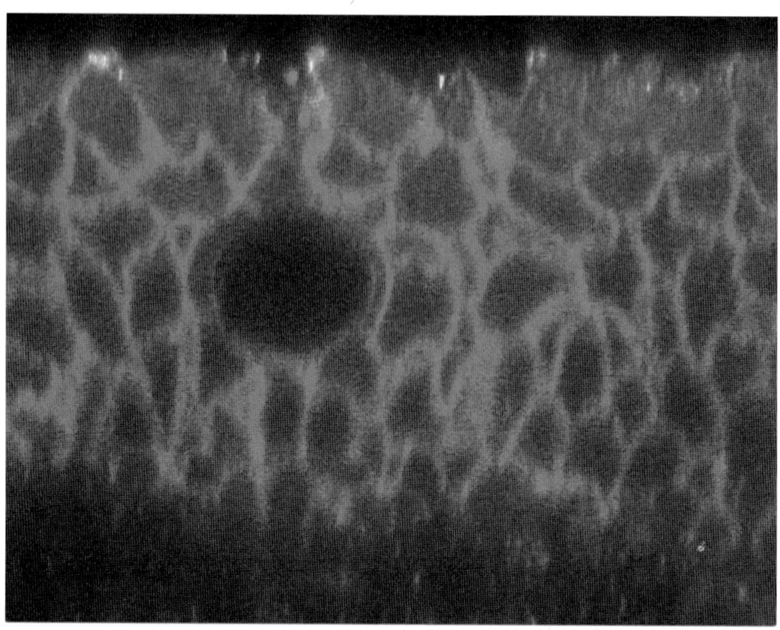

Fig. 1. Confocal *x-z* (transversal) scan of normal human conjunctival epithelium. Actin in the cell cytokeleon has been visualized with Phalloidin. The goblet cell can be recognised by its shape and cross-reaction with antibody 45M1 to MUC5AC. The latter is difficult to distinguish on this gray-scale image. The stippling at the apical surface of the superficial epithelial cells are intercellular junctions.

specific number of cell divisions *(3)*. Expansion from explants also resulted in epithelial cultures where goblet cells were observed again after a period of absence *(4)* (*see* **Fig. 2**). Electron microscope images have been interpreted as refilling of goblet cells *(6)*. Goblet cells fulfill the following criteria:

1. Positive staining of secretory granules with alcian blue or periodic acid Schiff.
2. Reactivity with: (a) cytokeratin (CK)-7; (b) lectins *Ulex europea* I (UEA-I) and *Helix pomatia agglutinin* (HPA); (c) mucin MUC5AC (or detection of MUC5AC mRNA using RT-PCR); (d) the M(3) muscarinic receptor.
3. Nonreacting with: (a) cytokeratin CK-4; (b) *Banderia simplicifolia* lectin; (c) M(1) muscarinic receptor *(7)*.

They have been isolated from explant-derived cultures and passaged without loss of goblet cell characteristics. However, a spontaneous mutation in a normal human conjunctival epithelial culture resulted in an immortal cell line that is totally epithelial *(8)*. There is some evidence that conjunctival fibroblasts are different from the keratocytes of the cornea. In vitro they can be included in a

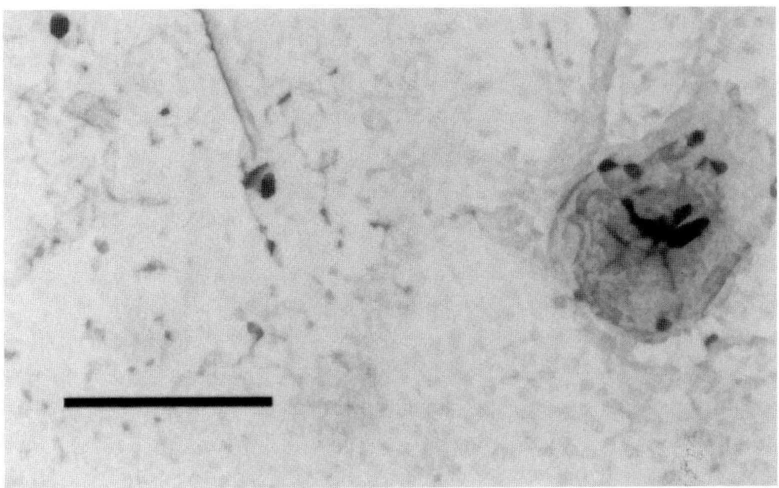

Fig. 2. Conjunctival epithelium migrated out from the explant onto the plastic cell culture vessel. Goblet cells, reacting with antibody 45M1 to MUC5AC, can be seen on the surface of the explant and among the squamous epithelial monolayer. Scale bar 20 μm.

collagenous matrix to function as a stromal equivalent which improves epithelial stratification *(9)*.

Cadaveric tissue or conjunctival biopsies can be used to set up primary epithelial or organ cultures. Cultured epithelia can be stratified and included in organotypic cultures on an artificial stroma, a collagen gel populated with stromal fibroblasts. Innervation and vascularisation of these cultures are outside the scope of this chapter.

2. Materials

2.1. Legal Framework

Before any human material can be used for research it is necessary to satisfy all the legal requirements governing this application, e.g., obtain the approval of the relevant committee for ethics in research. If biopsy material is used, a signed informed consent form should be obtained from each tissue donor. The readers should inquire about requirements and procedures that apply to their institution.

2.2. Media

2.2.1. Disinfection

1. Dulbecco's phosphate-buffered saline (DPBS, Invitrogen Corporation, www. invitrogen.com; Gibco cat. no. 1040-091) sterile.

2. 3% (w/v) polyvinylpyrrolidone-iodine (Sigma, www.sigmaaldrich.com; cat. no. PVP-I) in distilled sterile water; keep in the dark and discard solution after 14 d.
3. 0.3% (w/v) sodium thiosulfate 5-hydrate (BDH, www.bdh.com; cat. no. 102684G) in distilled sterile water. Keep protected from light.

2.2.2. Tissue Dissociation

1. Dispase Invitrogen Corporation (Gibco, cat. no. 17105).
2. DPBS (Invitrogen Corporation, Gibco cat. no. 1040-091) sterile.

2.2.3. Growth Media

2.2.3.1. PRIMARY CELL CULTURE

It is common practice to establish a primary cell culture in a medium supplemented with animal serum, usually fetal bovine serum (FBS). Established cultures can be transferred to defined media after a number of passages.

1. RPMI 1640 (Invitrogen Corporation, Gibco 21875-034); keep refrigerated at 4°C. This medium contains glutamine and has a limited shelf life.
2. Nonessential amino acids [RPMI-1640 Amino Acids Solution (50X), Sigma]; keep refrigerated at 4°C. Filter through 0.2-μm filter before adding to medium.
3. Epidermal growth factor (human; recombinant E9644 Sigma). Aliquot on arrival and keep refrigerated at less than –20°C. Do not filter EGF! All animal-derived products should be treated as potential biohazards and handled with appropriate care and protection. Dispose of used containers in incineration bins.
4. FBS (Invitrogen Corporation, Gibco 26140-079). Keep at less than –20°C. Aliquot after defrosting overnight at 4°C, thus avoiding multiple freeze-thaw cycles. The volume of the aliquots is determined by its use. Filter through 0.2 μm filter before adding to medium.
5. Antibiotic antimycotic solution 100X (Sigma, cat. no. A9909). Aliquot and keep at –20°C. Avoid repeated freezing and thawing. Filter through 0.2-μm filter before adding to medium.

2.2.3.2. DEFINED MEDIA

A large number of defined media are available. We have been most successful using Keratinocyte Growth Medium (KGM) BulletKit (Cambrex Bio Science, www.cambrex.com; cat. no. CC-3107). Basal medium should be kept at 4°C. Additives included with the BulletKit should be stored at –20°C and added prior to use. These are:

1. Recombinant human epithelial growth factor hEGF (cat. no. CC-4015).
2. Bovine pituitary extract (cat. no. CC-4002).
3. Insulin (cat. no. CC-4021).
4. Gentamicin GA1000 (cat. no. CC-4081). If possible, avoid the use of gentamicin, because it reduces cell growth.

2.2.4. Cell Dissociation and Passage

1. Ca- and Mg-free PBS (Dulbecco's PBS, Sigma cat. no. D5527).
2. Trypsin-EDTA solution: 0.25% trypsin; 0.02% EDTA (Sigma cat. no. T4049)—aliquot aseptically and store at –20°C. Filter sterilize (0.2-μm filter) before use.

2.2.5. Stratification Media

2.2.5.1. HIGH-CALCIUM MEDIUM

KGM (Cambrex Bio Science, cat. no. CC-3107) supplemented with 1 mM CaCl$_2$ to give a final concentration of 1.15 mM CaCl$_2$.

2.2.5.2. COLLAGEN GEL MATRIX

1. 10X tissue-culture medium without supplements (*see* **Note 1**).
2. Collagen type I: acid soluble rat tail collagen (Sigma cat. no. C7661). The collagen should be solubilized overnight and stored at 4°C (*see* **Note 2**).
3. 1 N sodium hydroxide.
4. 50% glutaraldehyde solution (Sigma, cat. no. 7651); this should be handled in laminar flow hood.
5. Chondroitin-6-sulfate, chondroitin sulfate C (Sigma, cat. no. C4384).
6. Glycine (Sigma, cat. no. G8790). Glycine "mops up" any unreacted glutaraldehyde. Make up 25% glycine in tissue-culture medium and store at room temperature in small aliquots.

2.2.6. Cryopreservation Media

For primary cultures or cultures grown in media with animal sera, the freezing medium consists of:

1. RPMI-1640 (Invitrogen Corporation, Gibco cat. no. 21875-034).
2. 20% FBS (Invitrogen Corporation, Gibco cat. no. 26140-079).
3. Dimethyl sulfoxide 10% (v/v) (DMSO, Sigma cat. no. D8779); this should be kept at room temperature. Ensure it is liquid before dissociating the cells.
4. Antibiotic antimycotic mixture (Sigma, cat. no. A9909). Final dilutions of antibiotics are 50 U/mL penicillin, 50 μg/mL streptomycin and 125 ng/mL amphotericin.

For cultures grown in defined media, use the complete KGM with 10% FBS (16000-036, Gibco) and 7.5% DMSO (*see* **Note 3**).

2.2.7. Organ Culture Medium

For organ culture, we have used tissue-culture medium:

1. RPMI-1640 (21875-034 Gibco).
2. 10% FBS (Invitrogen Corporation, Gibco cat. no. 26140-079).
3. Antibiotic antimycotic mixture (Sigma cat. no. A9909).

2.3. Histochemistry

2.3.1. Cell Fixing and Staining

1. Paraformaldehyde (Sigma P6148) 1% solution pH 7.5. The paraformaldehyde solution has to be prepared in a fume hood while wearing protective clothing and gloves:
   a. Dissolve 1 g paraformaldehyde in 50 mL distilled water.
   b. Heat to 65°C.
   c. Add 1 N NaOH until solution clears.
   d. Adjust pH to 7.5 with 1 N NaOH or 1 N HCl.
   e. Bring final volume up to 100 mL.
   f. Filter through a 0.2-µm filter.
2. 1 N NaOH.
3. 1 N HCl.
4. PBS.
5. Triton X-100 (Sigma, cat. no. T2199).

2.3.2. Assessment of Epithelial Nature and Stratification

1. Bovine serum albumin (BSA, Sigma, cat. no. A3912) should be stored at 4°C. Prepare solutions close to the time of use, store them at 4°C, and discard them after 1 wk. Inspect for bacterial growth and discard if any is detected.
2. Polyoxyethylene-sorbitan monolaurate (Tween-20, P1379 Sigma).
3. Anticytokeratin antibodies: The mixture of anit-human cytokeratin antibodies AE1/AE3 stains keratinized epithelial and corneal epithelium (Dako, www.dako.co.uk; cat. no. M3515). Monoclonal anticytokeratin peptide 19 antibody (Sigma, cat. no. C6930) should detect conjunctival epithelial cells only.
4. Phalloidin–tetramethylrhodamine C isothiocyanate conjugate (TRITC- Phalloidin, Sigma, cat. no. P1951) stains F-actin found in the cytoskeleton. **This material is very poisonous and should used be with great care.** All procedures should be carried out in a laminar flow hood. Protective clothing should be worn and all slides disposed of according to local safety procedures.

2.3.3. Assessment of Goblet Cell Differentiation

1. Cytokeratin 7 is believed to be specific to goblet cells. The monoclonal antibody to cytokeratin peptide 7 (Sigma, cat. no. C6417) can thus be used to assess goblet cell differentiation.
2. An antibody against the main goblet cell mucin, MUC5AC, antibody 45M1 (Novocastra Laboratories Ltd. www.novocastra.co.uk; cat. no. NCL-HGM-45M1), can also be used to detect goblet cells.
3. A histochemical stain, periodic acid-Schiff (PAS, Sigma, cat. no. 395B) will also detect glycoproteins and mucins.

2.4. Plastics

1. 1-mL plugged pipets (www.bibby-sterilin.com; Sterilin, cat. no. 47105) (*see* **Note 4**).
2. 10-mL plugged pipets (Sterilin, cat. no. 47110).
3. 25-mL plugged pipets (Sterilin, cat. no. 47125).
4. Sterile tips suitable for the pipettor used.
5. Sterile wide orifice tips (Rainin, cat. no. HR-250WS).
6. BD Falcon sterile tissue-culture dishes 35 × 10 mm (BD Biosciences, cat. no. 353001).
7. BD Falcon sterile organ culture dishes 60 × 10 mm (BD Biosciences, cat. no. 353037).
8. BD Falcon sterile 25 cm² vented tissue-culture flasks (BD Biosciences, cat. no. 353107).
9. BD Falcon sterile 75 cm² vented tissue-culture flasks (BD Biosciences, cat. no. 355001).
10. BD Falcon sterile 24-well tissue-culture insert companion plates (BD Biosciences, cat. no. 3535404).
11. BD Falcon tissue-culture inserts 0.4 μm pore, transparent membrane (BD Biosciences, cat. no. 353495).
12. Stratification supports: A large variety of tissue culture inserts is available. We have successfully used Cellagen™ Discs CD-24 (www.icnbiomed.com; cat. no. 152316). The Cellagen™ Discs are also referred to as inserts.
13. Organ culture supports: Triangular stainless-steel mesh supports, which fit the central well of an organ culture dish, side length to fit the organ culture central well should be sterilized before use. Squares of lens paper are cut with equal sides to the steel mesh and sterilized before use.

3. Methods

3.1. Tissue Retrieval

3.1.1. Cadaver Tissue

1. Ensure all appropriate donation and consent forms have been obtained.
2. To collect conjunctiva from a cadaver donor, first insert an eyelid speculum to part the eyelids. Use iris scissors to cut the conjunctiva all around the anatomical limbus and detach it from adherent Tenon's fascia. Make a further incision into the conjunctiva away from the limbus and cut around to obtain an annulus of tissue.
3. Place conjunctiva in sterile saline or tissue-culture medium for transport.
4. Tissue should be obtained using aseptic techniques (*see* **Note 5**). However, the ocular surface is exposed to the environment and, therefore, not necessarily free of microorganisms. Postmortem there is an increase in the ocular surface flora, even under closed lids and with refrigeration. It is therefore necessary to disinfect the fragments of tissue before culture.

3.1.2. Disinfection

1. Thorough rinsing in sterile PBS is necessary to dislodge any particles or loosely adhering organisms. Rinse in sterile PBS for 5 min three times.
2. Incubate in 3% PVPI for 1 min.
3. Dip quickly in 0.3% sodium isothiosulfate (for 15 s).
4. Final rinse in fresh portions of PBS 3 times for 5 min each.

3.2. Dissociation

3.2.1. Epithelial Cells

There are a number of enzymes that can be used to detach epithelial cells from the underlying tissue. The use of Dispase ensures a good yield of viable epithelial cells (*see* **Note 6**).

1. Fragment the tissue with a sterile scalpel or scissors.
2. Wash the tissue fragments in sterile PBS.
3. Incubate the fragments in the Dispase solution (2.4 to 0.6 U/mL) at 37°C. Make sure that the tissue fragments are well covered by the solution.
4. Stir slowly at 37°C until the tissue is sufficiently dissolved. When using Dispase for the first time, determine the total reaction time by counting the cells after 20–25 min. If necessary, separate the dispersed cells from residual tissue by lightly scraping the surface.
5. Check that cell clumps are coming off. If they are not, replace cornea in Dispase and wait another 5–10 min.
6. Repeat scraping.
7. Dilute cells with a large volume of medium (this inhibits Dispase activity) in a centrifuge tube.
8. Spin at 800*g* for 10 min at room temperature.
9. Discard supernatant; resuspend the pellet in 1mL medium and gently pipet up and down to obtain a single-cell suspension.
10. Plate in one 25-cm^2 flask in 1 mL medium. Leave for a few hours or overnight.
11. Add 1–3 mL medium.
12. Check every 2 d for cell growth.

3.2.2. Stromal Fibroblasts

Explants, approx 1–2 mm^3 should be cut with a sterile blade and spread on the bottom of the culture vessel. Adhesion to the vessel is improved if only a small amount of medium, <0.5 mL, sufficient to maintain a humid atmosphere, is added for the first 15–30 min. Nonadhering explants can be picked the next day and re-laid on a new flask. The remaining stroma after dissociation of epithelial cells is a good source for fibroblasts. Explants should be left on until sufficient growth is obtained, sometimes for up to 14 d, only refeeding when fibroblasts can be seen migrating out of the explants. Thereafter, medium

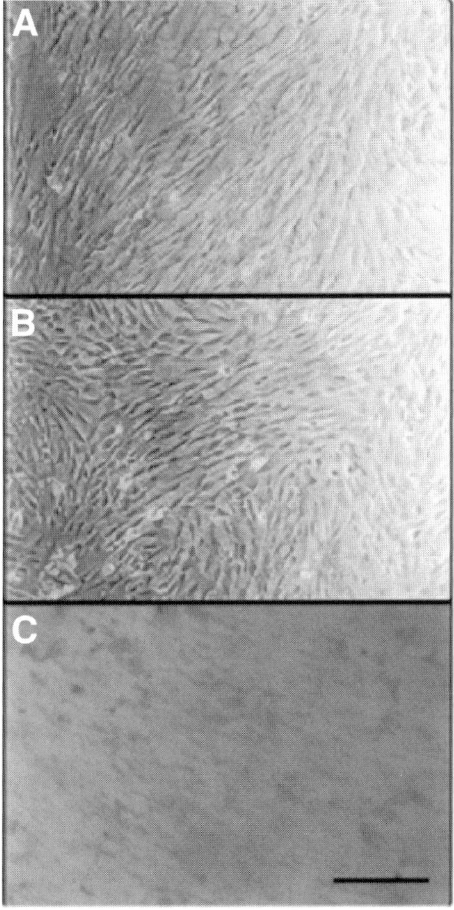

Fig. 3. Human conjunctival fibroblasts after **(A)** 7. **(B)** 10. **(C)** 24 d in culture. Scale bar 20 μm.

should be changed two or three times a week. Examples of conjunctival fibroblast cultures are shown in **Fig. 3**.

3.3. Cell Expansion

3.3.1. Epithelial Cultures in Media Containing Animal Sera

To prepare the tissue-culture medium, mix the following ingredients in a laminar flow hood ensuring that all ingredients are sterile. It is advisable to filter sterilize all previously opened or aliquoted ingredients, except EGF, through a 0.2-μm filter.

To 500 mL RPMI-1640, add:

1. 50 mL FBS (10% final concentration).
2. 5 mL antibiotic/antimycotic mixture, final concentration: 50 U/mL penicillin, 50 µg/mL streptomycin, and 125 ng/mL amphotericin.
3. 5 mL nonessential amino acids (1% final concentration).
4. 5 µg EGF. Do not filter EGF!

Aliquot into smaller sterile bottles to avoid repeated warming and improve sterility. Medium should be warmed to 35°C before use. The medium should be changed every other day. Confluence is reached within 7–14 d. At this stage, it is advisable to maintain the antibiotics and antifungal additions to the medium.

Examples of primary cultures of normal human conjunctival epithelial cells grown in this medium are shown in **Fig. 4**. Most primary cell cultures have a limited life-span. Inspection of cell morphology should be undertaken to detect changes indicative of senescence, e.g., increase in cell size and varying shape (*see* **Fig. 5**). Passaging when the very first signs of morphology change are detected may, sometimes, result in nonsenescent cultures.

3.3.2. Epithelial Cultures in Defined Media

Basal KGM medium should be kept at 4°C. Additives shipped with the medium should be stored at –20°C, thawed at room temperature, and added prior to use. It is always advisable to aliquot the complete medium into smaller volumes in order to avoid repeated warming and cooling. To reduce the probability of infection it might be desirable to use a fresh aliquot with each feed.

3.3.3. Passage

At confluence, around d 14 for epithelial cultures, or when the cells cover 70–80% of the bottom of the flask, the cultures can be expanded or passaged by enzymatic dissociation from the culture vessel and replated over a larger area, usually three times the original. The passage number of the culture increases by one unit every time this procedure is carried out.

1. Discard the entire volume of spent medium.
2. Rinse with sterile-filtered Ca^{2+}/Mg^{2+}-free PBS, 3 mL of PBS for a 25-cm^2 flask. Repeat three times, each time exchanging the entire volume of buffer used. Discard supernatant.
3. Add 1 mL sterile-filtered Trypsin-EDTA to a 25 cm^2 flask, or sufficient trypsin to just cover the cell layer.
4. Place at 37°C for 4 min.
5. Check rounding (*see* **Note 7**). If cells are not detached, return to incubator and check every 1 min thereafter.
6. When cells are rounded and detached, add 3 mL medium or sterile PBS with 10% FBS, and gently mix the detached cells with it. Excess protein and Ca inhibit trypsin activity.

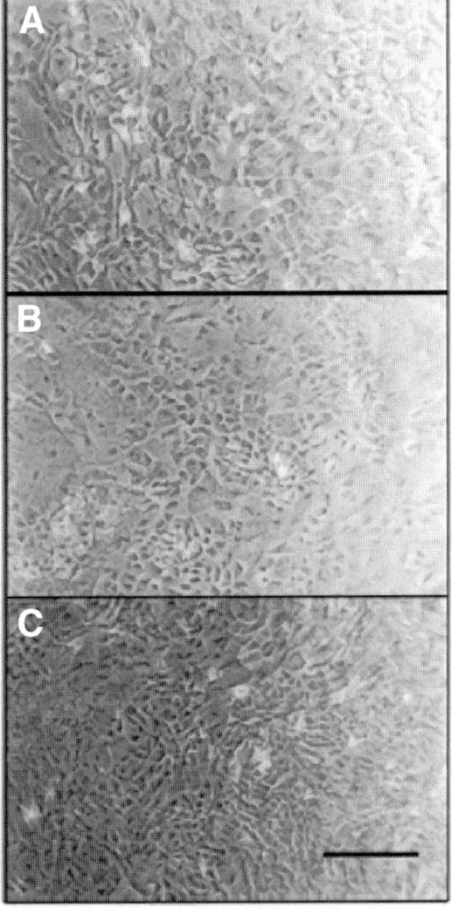

Fig. 4. Human conjunctival epithelium in primary culture in RPMI-1640 with 10% FBS. (**A**) 4. (**B**) 7. (**C**) 10 d in culture. Scale bar 20 μm.

7. Gently transfer the medium containing cells into an appropriate sterile centrifuge tube.
8. Spin for 10 min at 500*g*, to obtain a cell pellet (*see* **Note 8**).
9. Discard supernatant avoiding the cell pellet at the bottom of the tube, which might not be very well formed.
10. Add 1 mL tissue-culture medium, gently pipet up and down to obtain single-cell suspension.
11. Take a 20-μL aliquot, to assess cell concentration, dilute as required with PBS ± Trypan blue.
12. Count cells with hemocytometer: count five 1 mm^2 squares; the number of cells is *mean count × 10^4 × dilution factor* cells/mL (*see* **Note 9**).
13. Plate as required, or passage 1:3, i.e., from a 25-cm^2 flask to a 75-cm^2 flask.

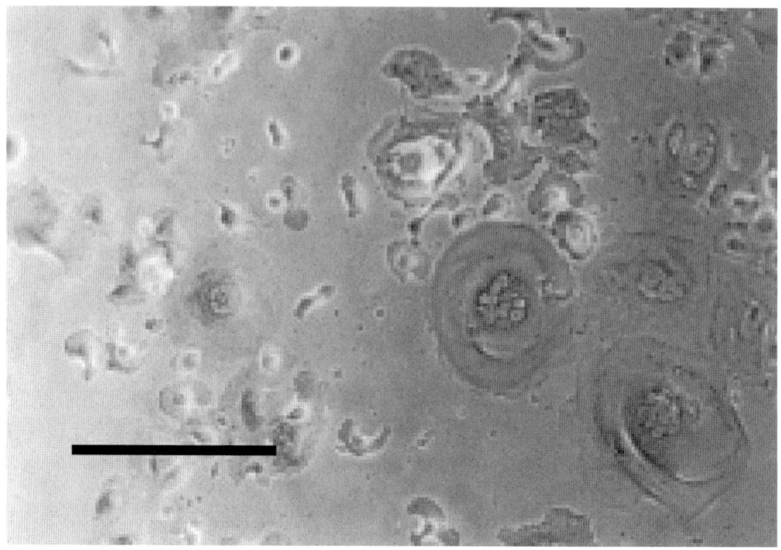

Fig. 5. Senescent conjunctival epithelium. Note the very large cells and the dismorphic squamous epithelium. Scale bar = 20 μm.

3.4. "Weaning" to a Defined Medium

A weaning regime is not necessary when changing to a fully defined medium; cultures can be fed with the defined medium without any transitional period.

3.5. Stratification

Stratification can be achieved either on tissue-culture inserts, artificial stroma (collagen gel), or stromal equivalent (collagen gel populated with stromal fibroblasts).

3.5.1. Epithelial Stratification

To obtain stratified growth, epithelial cells will be seeded on collagenized tissue culture inserts, allowed to establish and grow, and then air-lifted.

1. Resuspend dissociated cells to 7.5×10^5 cells/mL in tissue-culture medium. Gently pipet up and down to obtain a homogeneous suspension.
2. Immerse Cellagen™ inserts in tissue-culture medium, then discard the medium.
3. Seed each insert with 1.5×10^5 cells per insert, by adding 200 μL of the 7.5×10^5 cells/mL stock to each insert. Cells should be seeded using a 200-μL sterile wide-orifice tip (*see* **Note 10**).
4. Incubate cultures at 37°C, 5% CO_2, 90% humidity. This is day 0.

5. 12–24 h after seeding, remove medium from the inserts. Aspirate first around and then carefully inside the insert. Tilting the plate helps remove the medium without touching the cells.
6. Add 1 mL tissue-culture medium to the well and 200 μL of KGM to the inserts. It is important to avoid air bubbles. Tilting the plate and pipetting slowly diminishes the risk of air bubbles. Air bubbles must be removed before replacing the plate in the incubator.

3.5.1.1. AIRLIFTING

Cells are brought to liquid–air interface and fed with high-calcium medium, which improves the formation of intercellular junctions. This can be done 3 d after seeding.

1. Remove medium from wells and inserts. Add 1 mL high-calcium medium to the well, avoiding air bubble formation. Do not add medium to the inside of the inserts.
2. Feed twice a week by replenishing the medium in the well. Add 50 μL of medium to the cellular surface to minimize keratinization.

The cultures should be used within 7–14 d of air-lifting.

3.5.2. Epithelial Stratification on Stromal Equivalents

This organotypic culture is useful when interactions between epithelium and stroma are being investigated. In vivo, there are multiple interactions between stroma and epithelium maintaining tissue homeostasis. This method has been adapted from **ref. *10***.

To make a collagen gel as a matrix for stromal fibroblasts:

1. Dissolve the collagen in 0.01% acetic acid, overnight at 4°C at a concentration of 2–4 mg/mL (3 mg/mL works well).
2. To make collagen gels, prepare cold pipets, universals, plates, inserts if used, and an ice pack by placing all in the freezer for a few hours or overnight.
3. To delay gel setting, use glass containers and mix keeping cold:
   a. 8 mL collagen (3–5 mg/mL), ice cold.
   b. 1.8 mL of 10X tissue-culture medium, cold.
   c. 0.2 mL 1 *N* NaOH.
   d. Adjust pH to 7.2–7.4 with 0.1 *N* HCl.
4. Add while gently and thoroughly mixing, 520 μL of chondroitin sulphate C 20% (w/v) in Ca^{2+}/Mg^{2+}-free PBS.
5. In a separate Eppendorf, make up a 1.5% glutaraldehyde solution by mixing
   a. 6 μL of 25% aqueous glutaraldehyde.
   b. 94 μL of 20% dextran in DMEM.
   Prepare the glutaraldehyde solution just before use and ensure that the aliquots are not exposed to air (seal tight!). Oxidized glutaraldehide does not cross-link.

6. Add the 1.5% glutaraldehyde mixture to the collagen matrix to give a final concentration of 0.02% glutaraldehyde. Mix thoroughly, but gently.
7. Cross-link on ice for 2 h—stir a few times.
8. To remove excess glutaraldehyde, add 500 μL of 25% glycine in tissue culture medium. Mix well and incubate for 2 h on ice.
9. Mix 10^5 fibroblasts/mL in 3 mL medium with the cold pH-ed collagen mixture.
10. Dispense 0.5 mL/well in cold 24-well plate.
11. Incubate at 37°C overnight.
12. Seed epithelial cells as described above.

3.6. Cryopreservation

Dissociated cells can be preserved in liquid nitrogen until use. It is prudent to keep samples of early passages as comparators for morphological and physiological changes that might occur with time during cell culture.

1. Pellet dissociated cells by centrifugation for 10 min at 800g at room temperature.
2. Resuspend in 1–1.5 mL cold freezing medium at a density of 10^6 cells/mL in a sterile cryovial.
3. Freeze at a rate of −1°C/min.
4. Place in liquid nitrogen (*see* **Note 11**).

To defrost an aliquot of cells, it is beneficial to raise the temperature fast and remove the freezing medium efficiently.

1. Heat the cryovial quickly in a 37°C water bath.
2. As soon as the liquid has thawed, add 5 mL of appropriate tissue culture.
3. Plate cells.

3.7. Organ Culture

Organ culture is used to follow the metabolic fate of radioactive precursors in tissue fragments (*see* **Note 12**).

1. Place 2 mL tissue-culture medium in the central well of an organ-culture dish.
2. Place lens tissue squares on triangular stainless-steel grids and saturate the lens squares with medium by gently touching the meniscus of the medium. Place the grid over the well ensuring that the paper is in contact with the medium.
3. Cut the conjunctiva in pieces 4–6 mm^2.
4. Place conjunctival fragments on the grid.
5. Add radioactive precursors to the medium in the central well. Add 1 mL sterile distilled water to the outside well, to maintain a moist atmosphere for the excised tissue (*see* **Note 13**).
6. Incubate for at least 24 h and up to 96 h in a 5% CO_2 incubator at 35°C (*see* **Note 14**).

3.8. Identification of the Phenotype of Cultured Cells

3.8.1. Cell Fixing and Staining Protocols for Epithelial Differentiation

The method described below was optimized for confocal microscopy, but can equally be used for other modalitlies of visualising immunocytochemistry.

Cytokeratins are part of the cytoskeleton differentially expressed in epithelial cells from different sources. Monoclonal anti-epithelial keratin AE1 and AE3 recognize the intermediate filaments of the cytoskeleton of epithelial cells and help differentiate cultured epithelium from fibroblasts (*see* **Note 15**).

To determine the epithelial type of the cultures using anti-cytokeratin antibodies:

1. Wash cells with PBS to remove adhering medium.
2. Add 750 µL of 1% paraformaldehyde to each culture well or insert of a 6-well plate and fix for 20–30 min at room temperature.
3. Rinse twice with PBS, for 5 min each.
4. Permeabilize with 500 µL 0.5% Triton X-100 for 30 min.
5. Rinse with PBS for 5 min.
6. Nonspecific binding can be decreased by a 10-min incubation with PBS containing 10% serum appropriate to the antibody used (or FBS) and 0.1% Triton X-100.
7. Rinse for 5 min with PBS with 0.1% Triton X-100 (PBT).
8. Dilute primary antibody (e.g., antibody to epithelial cytokeratins AE1/AE3) 1:100 in PBT with 5% serum (or FBS). Add 150 µL of primary antibody to each insert and incubate at 37°C in dark for 1 h (*see* **Note 16**).
9. Rinse three times with PBT for 5 min each.
10. Dilute secondary antibody (e.g., tagged antibody to the Ig type of the primary antibody) 1:200 in PBT with 5% serum (or FBS). Add 250 µL of secondary antibody to each insert, leave at 37°C (in dark, if fluorescently tagged) for 1 h.
11. Rinse with PBT for 5 min.
12. Rinse PBS for 5 min.
13. Add mounting medium and cover.

To visualize cell boundaries and the organization of F-actin, phalloidin gives beautiful results, especially if fluorescently tagged. Precautions should be taken when handling this very poisonous chemical:

1. Follow **steps 1–5** above.
2. Dilute phalloidin 1:100 in PBT.
3. Add 300 µL phalloidin to each insert and incubate for 45–60 min at room temperature in the dark.
4. Rinse with PBS + 0.1% Triton X-100 for 5 min, twice.
5. Rinse with PBS for 5 min.
6. Mount, cover, and image.

3.8.2. Goblet Cell Identification with Periodic Acid-Schiff (PAS)

PAS is a useful and quick staining method for identifying glycoproteins and mucins, and in particular goblet cell secretory granules.

1. Incubate cell cultures in 1% periodic acid in 3% acetic acid for 30 min.
2. Rinse twice in 0.1% sodium metabilsufite in 1 mM hydrochloric acid for 2 min.
3. Dip in Schiff reagent for 15 min.
4. Rinse twice with 0.1% sodium metabisulfite for at least 5 min.

Counterstaining with hematoxylin-eosin can be used if desired.

4. Notes

1. If the medium used for tissue culture is not available in 10X form, use 10X RPMI.
2. Bovine skin type I collagen (Sigma, cat. no. C9791) can also be used to form the gel. Its advantages are that it cross-links better with glutaraldehyde and produces gels that are transparent. This method originates with May Griffith *(10)*.
3. This FBS has been chosen because of its low cytotoxicity.
4. It is useful to have a supply of various pipets in close proximity to the laminar flow hood in which cell cultures are manipulated. All plastics used in tissue or organ culture have to be incinerated. Institutions usually provide special rigid containers for the incineration of pipets.
5. The time postmortem should be kept to a minimum, although conjunctival cells grow well from tissue collected within 24 h of death, if maintained at 4°C, in a moist chamber.
6. The time of incubation and concentration of Dispase should be optimized for each application. Too concentrated a solution will cause excess cell death and may release fibroblasts from the underlying stroma which would then overtake the epithelial cells.
7. Detaching cells appear rounded and floating in the trypsin. A tap to the side of the flask may help dislodge cells that are detached but still in place. Prolonged exposure to trypsin is damaging to cells and should be kept to a minimum, unless there is a good reason not to do so. Some authors consider that stem cells are the most insensitive to trypsin, and use trypsinization as a cell-selection technique.
8. Centrifugation should be gentle to ensure that cells are not mechanically harmed. If necessary, increase spin length rather than speed.
9. The optimal cell number is 20–50 per square; densities above and below this range lead to errors in the cell count.
10. Standard 200 µL pipet tips should not be used because the opening is too small and might damage the cells.
11. Good results can also be obtained by placing the cryotubes in an insulated box at –70°C overnight and then quickly transferring to liquid nitrogen. There are a number of desk-top devices that enable a degree of controlled-rate freezing at low expense.
12. This is a modification of the method described by Corfield and Paraskeva *(11)*.

13. ^{14}C-Glucosamine and ^{3}H-ethanolamine were incorporated in both secreted and membrane-bound human conjunctival glycoproteins. A small fraction of the ^{35}S-sodium sulfate radioactivity was recovered from mucin-like material *(12)*.
14. We have observed goblet cells migrating within a sheet of epithelial cells for at least 4 d. Maintenance of organ cultures is limited by media depletion and diffusion of nutrients within the tissue.
15. Second layer antibody only controls should also be prepared to assess the level of nonspecific binding of these antibodies.
16. Or incubate overnight at 4°C.

References

1. Setzer, P. Y., Nichols, B. A., and Davson, C. R. (1987) Unusual structure of rat conjunctival epithelium: Light and electron microscopy. *Invest. Ophthalmol. Vis. Sci.* **27,** 531–537.
2. Gipson, I. K. and Tisdale, A. S. (1997) Visualization of conjunctival goblet cell actin cytoskeleton and mucin content in tissue whole mounts. *Exper. Eye Res.* **65,** 407–415.
3. Pellegrini, G., Golisano, O., Paterna, P., et al. (1999) Location and clonal analysis of stem cells and their differentiated progeny in the human ocular surface. *J. Cell Biol.* **145,** 769–782.
4. Ellingham, R. B. (1999) Human copnjunctival mucins. Ph.D Thesis, Univ. Bristol.
5. Lavker, R. M., Wei, Z.-G., and Sun, T.-T. (1998) Phorbol ester preferentially stimulates mouse fornical conjunctival and limbal epithelial cells to proliferate in vivo. *Invest. Ophthalmol. Vis. Sci.* **39,** 301–307.
6. Aitken, D., Friend, J., Thoft, R. A., and Lee, W. R. (1988) An ultrastructural study of rabbit ocular surface transdifferentiation. *Invest. Ophthalmol. Vis. Sci.* **29,** 224–231.
7. Shatos, M. A., Rios, J. D., Tepavcevic, V., Kano, H., Hodges, R., and Dartt, D. A. (2001) Isolation, characterization, and propagation of rat conjunctival goblet cells in vitro. *Invest. Ophthalmol. Vis. Sci.* **42,** 1455–1464.
8. Diebold, Y., Calonge, M., Fernandez, N., et al. (1997) Characterization of epithelial primary cultures from human conjunctiva. *Graefe's Arch. Clin. Exper. Ophthalmol.* **235,** 268–276.
9. Taliana, L., Evans, M. D., Dimitrijevich, S. D., and Steele, J. G. (2001) The influence of stromal contraction in a wound model system on corneal epithelial stratification. *Invest. Ophthalmol. Vis. Sci.* **42,** 81–89.
10. Griffith, M., Trinkaus-Randall, V., Watsky, M. A., Liu, C.-H., and Sheardown, H. (2002) Cornea, in *Methods of Tissue Engineering* (Atala, A., ed.), Academic, New York, pp. 927–941.
11. Corfield, A. P. and Paraskeva, C. (1993) Secreted mucus glycoproteins in cell and organ culture, in *Methods in Molecular Biology: Glycoprotein Analysis in Biomedicine* (Hounsell, E. F., ed.), vol. 14, Humana, Totowa, NJ.
12. Corfield, A. P., Myerscough, N., Berry, M., Clamp, J. R., and Easty, D. L. (1991) Mucins synthesized in organ culture of human conjunctival tissue. *Biochem. Soc. Trans.* **19,** 352S.

24

Establishment, Maintenance, and Transfection of In Vitro Cultures of Human Retinal Pigment Epithelium

Martin J. Stevens, Dennis D. Larkin, Eva L. Feldman, Monte A. DelMonte, and Douglas A. Greene

1. Introduction

The retinal pigment epithelium (RPE) is a layer of multipotential cells of neural ectoderm origin lying between Bruch's membrane and the neural retina. The RPE subserves several essential ocular functions, including phagocytosis of shed photoreceptor outer segments, maintenance of the blood–retinal barrier, absorption of stray light, regulation of the biochemical, metabolic, and ionic composition of the subretinal space, and induction of embryonic differentiation of adjacent neural retina and choroids *(1)*. Experimental evidence indicates that early in embryonic life the neural retina can regenerate from the pigment epithelium *(2)*. In vitro cultures of pure RPE provide a vehicle for studying RPE function in both normal and diseased states, and may also serve as a model for other neural cells *(3,4)*. Multiple techniques have been described for culturing human RPE *(5–12)*. The authors describe here a modification of the technique of DelMonte and Maumenee *(10)*, which is simple and effective in establishing primary cultures and extended cell lines of human RPE for research.

2. Materials

1. Hank's balanced salt solution (HBSS) without Ca^{2+} and Mg^{2+}, and Earle's balanced salt solution (EBSS) are supplied by Gibco (Grand Island, NY), and are stored at 4°C or room temperature.
2. Ham's F12 Nutrient Mixture and Dulbecco's minimum essential medium (DMEM) are supplied by Gibco and stored at 4°C.

From: *Methods in Molecular Medicine, vol. 107: Human Cell Culture Protocols, Second Edition*
Edited by: J. Picot © Humana Press Inc., Totowa, NJ

3. Bovine calf serum (defined and iron-supplemented) and fetal calf serum (FCS) (defined) are supplied by Hyclone Laboratories (Logan, UT), sterilely divided into 50-mL aliquots in 50-mL tubes, and stored at –20°C.

4. L-Glutamine, penicillin/streptomycin, and sterile sodium bicarbonate solutions are supplied by Gibco and stored at –20°C.

5. Establishment of primary cultures of human RPE is accomplished with the following recipe: Ham's F12 Nutrient Mixture, 16% fetal bovine serum (FBS), 0.02 mM L-glutamine, 100 U/mL penicillin/100 µg/mL streptomycin, 0.075% (w/v) sodium bicarbonate.

6. Trypsin-etheylenediaminetetraacetic acid (EDTA) is supplied by Gibco, sterilely divided into 10-mL aliquots in 15-mL tubes, and stored at –20°C.

7. Papain (cat. no. P3125) and L-cysteine-HCl (cat. no. C7880) are supplied by Sigma (St. Louis, MO). Papain stock solution is prepared in advance in Ca^{2+}- and Mg^{2+}-free HBSS by adding 3 mM L-cysteine HCl, 1 mM EDTA, and 10 µL/mL papain. The stock solution is stable for 2 wk when stored at 4°C.

8. Acid-soluble collagen (Sigma type III) is supplied by Sigma and stored at –20°C.

9. Silane (cat. no. M6514) for treating pipets is supplied by Sigma. Pipets are coated by rinsing in 0.2% silane, then chloroform, ethanol, and several water rinses, or by coating with silane vapor in a desiccator and then autoclaved to sterilize.

3. Methods

3.1. Preparation of Collagen-Coated Culture Vessels (see Note 1)

1. Coating of dishes with acid-soluble type 1 collagen is easily accomplished as follows: 0.5% solution of acid-soluble collagen in 0.1 M acetic acid is painted on the surface of the culture vessels with a fine brush.

2. The thin collagen coating is gelled by exposure to NH_3 fumes from ammonium hydroxide for 12 h.

3. The excess ammonia is neutralized with buffered saline.

4. Dishes are sterilized by overnight exposure to UV light.

3.2 Initial Dissection, Cell Isolation, and Primary Culture (see Note 2)

1. Postmortem eyes are obtained from an eye bank within 96 h of death, and the anterior segment of the eye is removed by circumferential incision 5 mm posterior to the limbus.

2. The vitreous and neurosensory retina are separated from the pigment epithelium, cut free at the optic nerve, and removed.

3. The eyecup is rinsed three times with HBSS without Ca^{2+} or Mg^{2+}. It is important to keep the eyecup moist throughout the isolation procedure.

4. The eyecup is filled with stock HBSS/papain solution and incubated for 40 min at 37°C or until the RPE cell layer appears marbled.

5. The papain reaction is stopped by the addition of 50 µL of FBS to the eyecup.

6. The HBSS/papain solution is removed from the eyecup and replaced with Ham's F12-medium without serum.

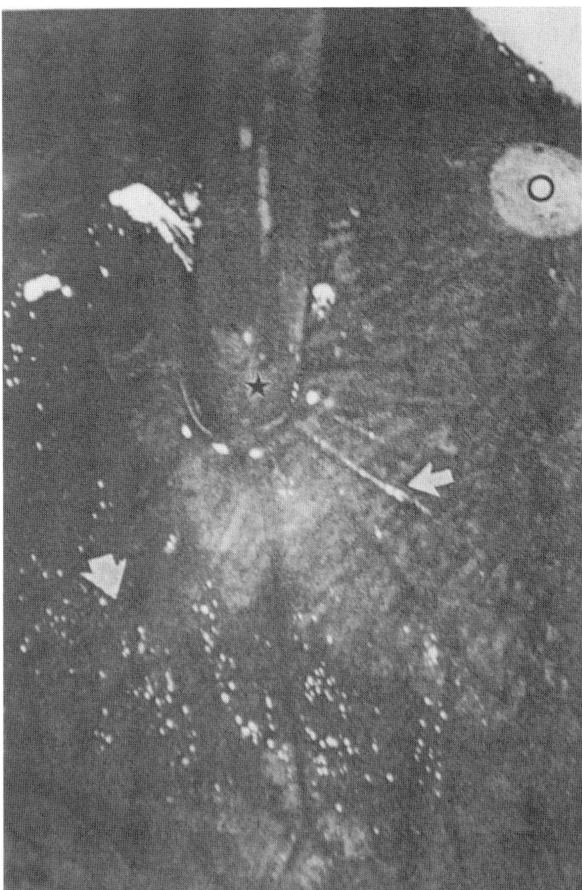

Fig. 1. Dissecting microscope—view of harvesting human RPE from the posterior of the eye using a fire-polished Pasteur pipet. Velvety appearing RPE (large arrow) is being gently vacuumed into the pipet (star) leaving behind the shiny Bruch's membrane (small arrow). Optic disk is marked with O for orientation.

7. The RPE cells are dislodged and suspended with a stream of medium from a fire-polished Pasteur pipet treated with 0.2% silane to prevent sticking. Remaining adherent RPE is gently vacuumed free from Bruch's membrane using the fire-polished Pasteur pipet under direct observation with a dissecting microscope as shown in **Fig. 1**.
8. The cell/papain solution is removed from the eye cup, placed in a sterile tube, and centrifuged at 50g for 3 min to pellet the cells.
9. The cells are resuspended and washed in HBSS.
10. The washed pellet is resuspended in complete media consisting of Ham's F12-M

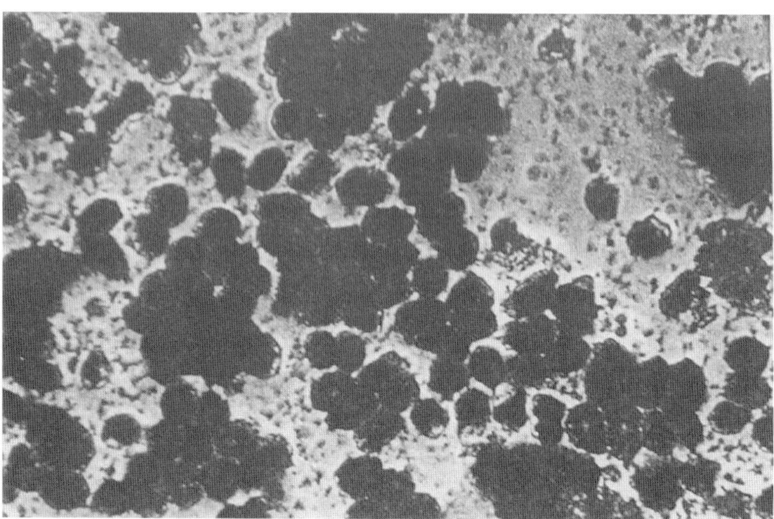

Fig. 2. Primary explants of human RPE as examined by phase-contrast microscopy immediately after harvesting (×140).

media supplemented with 16% FCS, 0.075% sodium bicarbonate, 0.02 m*M* glutamine, 100 U/mL penicillin, and 100 µg/mL streptomycin.

11. Cells are distributed evenly into sterile uncoated or extracellular matrix-coated 35-mm plastic culture dishes or the wells of 6- or 24-well culture plates. Twenty-four-well PRIMARIA culture plates (Falcon Plastics) provide excellent cell adhesion and colony formation characteristics, and are recommended for most primary culture applications. Other dishes or extracellular matrix coatings are fine for established cultures. Fresh RPE explants consist of single or small clusters of polygonal, deeply pigmented epithelial cells, which appear much as they do in vivo as seen in **Fig. 2**.

12. Cultures are incubated at 37°C in a humidified 95% air, 5% CO_2 atmosphere. After 2 d, the media are aspirated and replaced. Cell attachment and initiation of colony formation are monitored by phase contrast microscopy.

13. Alternatively, at **step 6**, medium with serum is added to the eyecup, and small patches of cells are dislodged and plated directly into culture dishes or plates without the centrifugation step. This technique is especially useful for obtaining cultures from specific locations of the globe, such as the macular area, midperiphery, or periphery, to study regional differences.

3.3. Establishment and Maintenance of RPE Cell Lines (see Notes 3–6)

1. The medium is aspirated, and the cells are rinsed with 5 mL HBSS without Ca^{2+} or Mg^{2+} for 5 min. Cells can then be removed from the dishes by addition of either 2 mL 0.05% trypsin in 0.53 m*M* EDTA or 2 mL of stock HBSS/ papain

solution. The cells are then incubated at 37°C for 2–15 min. until they are rounded and begin to detach from the plastic.

2. Trypsinization is terminated by addition of 1 vol of medium supplemented with 10–20% calf serum.

3. The cell suspension is centrifuged at 50g for 5 min and resuspended in medium. Established cell lines can be maintained in simpler media, such as DMEM with 16% FBS replaced by as low as 2–5% calf serum or even, for short defined periods of time, with defined serum-free medium.

4. Cells are subcultured at a density of 10,000–20.000 cells/cm^2 in T-25 or T-75 tissue-culture flasks.

5. Cells are maintained at 37°C and humidified 95% air, 5% CO_2 atmosphere with fresh medium changes three times weekly.

3.4. Transient Transfection of RPE Cells (see Notes 7–10)

1. When transfecting RPE cells, it is necessary to transfect cell cultures which are in exponential growth phase to maximize DNA uptake. Seeding density should be adjusted to allow cultures to reach approx 80–90% confluency at time of harvest. Therefore, our laboratory seeds cells at a density of 40,000 cells per well in 24-well cluster plates (Costar cat. no. 3524) containing 0.5 mL MEM, 5% bovine calf serum (BCS). It is important to exclude the use of antibiotics, to avoid increased toxicity or cell death caused by excess intracellular uptake of antibiotics during the DNA uptake period.

2. Cultures are transfected 16 h post-seeding with the minimum amount of DNA necessary to produce reliably measurable luciferase activity. In our laboratory, we have obtained good results using lipofectamine 2000 (Invitrogen cat. no. 52758) and transfecting 180 ng test DNA and 5 ng control DNA to normalize for variations in transfection efficiency. On the day of the transfection, combine test DNA and control DNA in 50 µL Opti-MEM (Gibco, cat. no. 31985-062) per well to be transfected. Control DNA should be 1/50 of test DNA or less (1–20 ng) to minimize potential interference by competing promoters.

3. Per well: Dilute 3 µL of lipofectamine 2000 (LF2000) into 50 µL opti-MEM. Incubate 5 min at room temperature.

4. Combine diluted LF2000 with the diluted DNA and allow to complex 20 min at room temperature.

5. Seed 100 µL of LF2000/DNA complex per well.

6. Cells should be re-fed after 6–8 h with standard medium with antibiotics. Alternatively, cells can be re-fed 16 h post-transfection, but increased toxicity may result.

7. Allow 24–48 h expression prior to assay.

4. Notes

1. Cell attachment and colony formation are enhanced, especially in eyes from older donors and those obtained later after death, by coating the culture vessels with collagen or other artificial extracellular matrix *(10,14)*. However, with the develop-

Table 1
Donor Factors Influencing Human RPE Culture Success

Donor characteristics	Success/attempts	%
Age		
<40	3/3	100
40–60	6/7	85
60–80	6/8	75
>80	1/3	33
Sex		
Male	6/8	75
Female	10/13	77
Time from death, h		
<24	5/5	100
24–48	9/10	90
48–72	2/4	50
>72	0/2	0

Adapted from **ref.** *13* with permission.

ment of specially treated tissue-culture plastics, such as PRIMARIA (Falcon Plastics), cell adhesion and colony formation are excellent. Therefore, the use of artificial extracellular matrix is usually unnecessary today unless required for specific experimental conditions.

2. Donor characteristics influence the establishment of successful cultures: The highest rates of viable cultures are found if patients are <60 yr of age and deceased <48 h. **Table 1** summarizes donor characteristics that correlated with viable cultures in 21 consecutive pairs of eyes.

3. After initiation of primary culture, cell adhesion occurs during the first 24–48 h. Viable cells begin to divide and establish heavily pigmented primary colonies within 5–14 d as seen in **Fig. 3**. Dividing cells become large, flattened, and polygonal with pigment granules clumped around a prominent nucleus with one or two nucleoli (*see* **Fig. 4**). The peripheral cytoplasm is thin and relatively transparent with prominent and often retractile cell/cell boundaries.

4. In successful cultures, dense colonies expand to fill 25–50% of the culture surface within 14–28 d. Cell lines can be established and maintained for over 20 passages.

5. Improved cell adhesion and colony formation in established cultures and cell lines are encouraged by using dishes coated with artificial extracellular matrix, such as collagen Type I (Sigma), as described above *(10)*, or collagen Type IV, laminin, and Matrigel (all available from Collaborative Research, Bedford, MD) *(14)*.

6. Established human RPE cell lines from adult donors produced in this way maintain polygonal epithelioid morphological characteristics and many in vivo functional characteristics, making them excellent models for study of the cell biology of these important neural-derived cells under normal and pathological conditions.

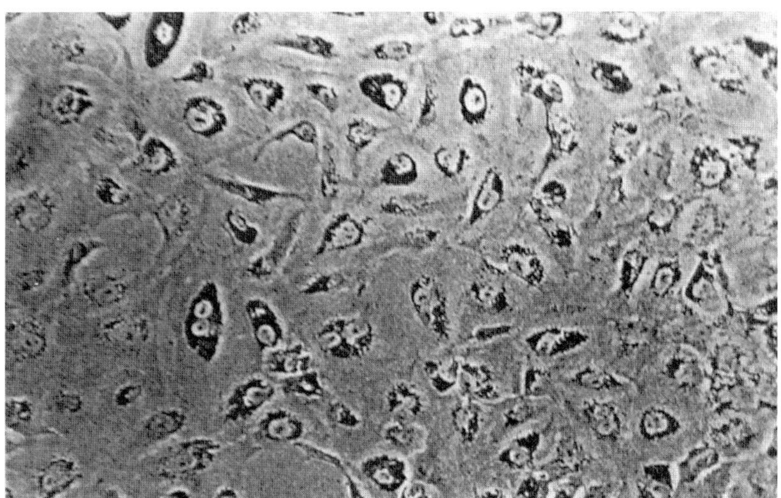

Fig. 3. Phase-contrast micrograph of primary colony formation after 14 d in culture showing pigmented, epithelioid cells, some with dividing nuclei (×100).

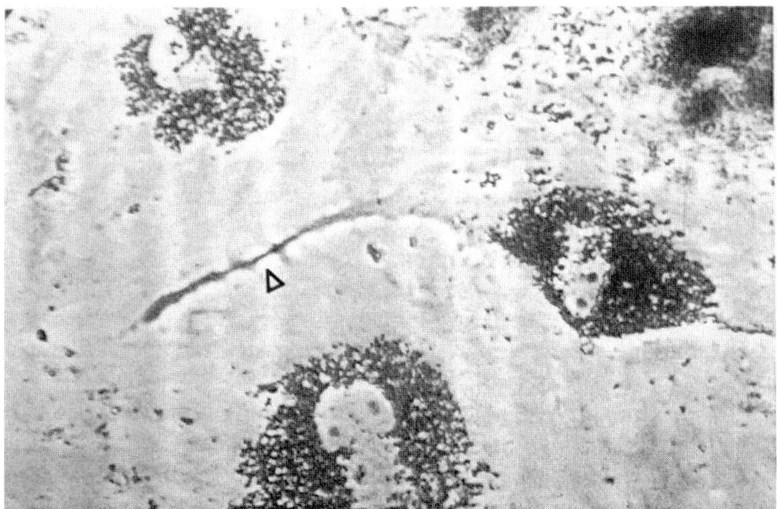

Fig. 4. Higher power view of primary culture of human RPE after 21 d in culture showing dividing epithelioid cells with dense pigment surrounding the nucleus, attenuated transparent peripheral cytoplasm with retractile cell/cell boundaries (phase contrast ×220).

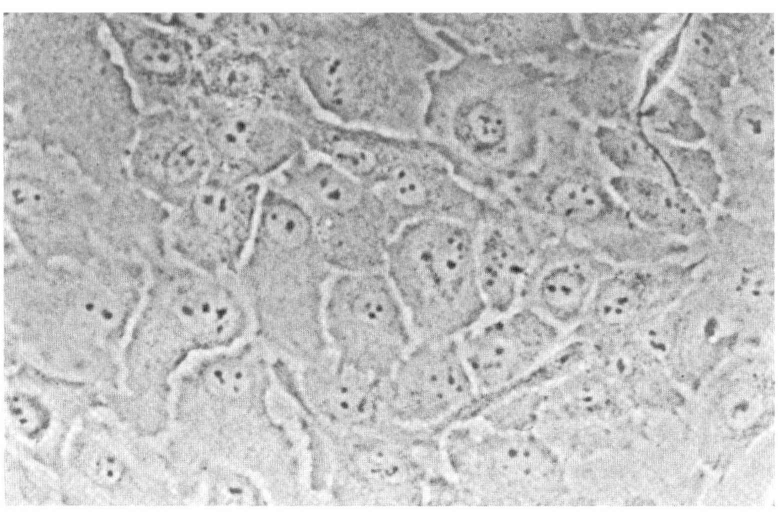

Fig. 5. Established cell line from human RPE after 13th passage. Epithelioid features are maintained in spite of total loss of pigment by dilution during cell division (phase contrast ×160).

In addition, future refinements and improvements in culture conditions should augment the utility of these cells in vision and neurobiology research. However, melanin synthesis generally does not occur after the fetal period and cells in established cultures become slowly depigmented by pigment granule dilution during successive cell divisions, such that third or fourth passage cells are almost completely devoid of pigment as shown in **Fig. 5**.

7. Most established RPE cell lines are relatively robust and are not easily damaged when handling in preparation for transfection experiments. Some RPE lines require higher concentrations of trypsin and/or longer trypsinization periods to properly trypsinize. Routinely, 0.05% trypsin-EDTA is highly effective in trypsinizing RPE cells, however, some RPE cell lines require 0.25% trypsin-EDTA and extended trypsinization times (10–30 min).

8. Some RPE lines demonstrate heterogeneous growth rate and cell volume. This should be taken into account when seeding RPE cells for transfection experiments in order to ensure cells are at desired confluencies at harvest time.

9. The optimum amount of test and control DNA needs to be determined empirically for a given set of plasmid DNA constructs.

10. RPE cells demonstrate some resistance to complete dissolution by mild detergents used in transfection experiments (Promega, passive lysis buffer cat. no. E194A). Therefore, at harvest time, cells should be thoroughly scraped to maximize lysis and dissolution. Lysed cell preparations can be stored at –80°C for extended periods (several months) until ready to assay. However, some luciferase activity will be lost due to the freeze thaw cycle.

Acknowledgment

These studies were supported by the USPHS Research Grants R29 NS32843 (E.L. Feldman), and the Michigan Diabetes Research and Training Center P60-DK20572 (E.L. Feldman and M.J. Stevens), and a grant from the Skillman Foundation (M.A. DelMonte).

References

1. Zinn, K. M. and Benjamin- Henkind, J. (1991) Retinal pigment epithelium, in *Biomedical Foundations of Ophthalmology*, vol. I (Duane, T. D., ed.), Harper and Row, Philadelphia, PA, ch. 21, pp. 1–20.
2. Coulombre, J. L. and Coulombre, H. A. (1965) Regeneration of neural retina from the pigmented epithelium in the chick embryo. *Dev. Biol.* **17,** 79–83.
3. Newsome, D. A. (1983) Retinal pigmented epithelium culture: current applications. *Trans. Ophthalmol. Soc. UK* **103,** 458–466.
4. Boulton, M. E., Marshall. J., and Mellerio, J. (1982) Human retinal pigment epithelial cells in tissue culture: a means of studying inherited retinal diseases. *Birth Defects* **19,** 101–118.
5. Albert, D. M. and Buyukmihci, N. (1979) Tissue culture of the retinal pigment epithelium, in *The Retinal Pigment Epithelium* (Zinn, K. M. and Marmour, M. F., eds.), Harvard Univ. Press, Cambridge, MA, ch. 16, pp. 277–292.
6. Albert, D. M., Tso, M. O., and Rabson, A. S. (1972) *In vitro* growth of pure cultures of retinal pigment epithelium. *Arch. Ophthalmol.* **88,** 63–69.
7. Mannaugh, J., Dhaarmendra, V. A., and Irvine, A. R. (1973) Tissue culture of human retinal pigment epithelium. *Invest. Opthalmol. Vis. Sci.* **12,** 52–64.
8. Edwards, R. B. (1982) Culture of mammalian retinal pigment epithelium and neural retina. *Methods Enzymol.* **81,** 39–43.
9. Flood, M. T., Gouras, P., and Kjeldbye, H. (1980) Growth characteristics and ultrastructure of human retinal pigment epithelium *in vitro. Invest. Opthalmol. Vis. Sci.* **19,** 1309–1320.
10. DelMonte, M. A. and Maumenee, I. H. (1981) *In vitro* culture of human retinal pigment epithelium for biochemical and metabolic study. *Vis. Res.* **21,** 137–142.
11. Pfeffer, B. A., Clark, V. M. Flanery, J. G., and Bok, D. (1986) Membrane receptors for retinal-binding protein in cultured human retinal pigment epithelium. *Invest. Ophthalmol. Vis. Sci.* **27,** 1031–1040.
12. Mircheff, A. K., Miller, S. S., Farber, D. B., Bradley M. E., O'Day. W. T., and Bok, D. (1990) Isolation and provisional identification of plasma membrane populations from cultured human retinal pigment epithelium. *Invest. Opthalmol. Vis. Sci.* **31,** 863–878.
13. DelMonte, M. A. and Maumenee, I. H. (1980) New techniques for *in vitro* culture of human retinal pigment epithelium, in *Birth Defects: Original Article Series*, vol. XVI, no. 2, pp. 327–338.
14. Dutt, K., Scott, M. M., Del Monte, M., et al. (1991) Extracellular matrix mediated growth and differentiation in human pigment epithelial cell line 0041. *Curr. Eye Res.* **10,** 1089–1100.

Index

Transferrin, *see* Media supplements
Trypan blue, *see* Cell counting
Trypsin, 4, 174, 192, 210, 285, 329
    inhibition by serum, 181, 245, 287, 334
    soybean trypsin inhibitor, 192
Trypsin-versene, *also see* Versene, 15

U

Ussing chambers, 200

V

Versene, 15, 285, 287
Viability, *see* Cell viability